25th International Conference of the Center for Chemical Process Safety 2010

Topical Conference at the 2010 AIChE Spring Meeting and 6th Global Congress on Process Safety

San Antonio, Texas, USA
21 – 25 March 2010

ISBN: 978-1-61738-435-6

Printed from e-media with permission by:

Curran Associates, Inc.
57 Morehouse Lane
Red Hook, NY 12571

Printed by Curran Associates, Inc. (2010)

For permission requests, please contact AIChE
at the address below.

AIChE
3 Park Avenue
New York, NY 10016-5991

Phone: (203) 702-7660
Fax: (203) 775-5177

www.aiche.org

Additional copies of this publication are available from:

Curran Associates, Inc.
57 Morehouse Lane
Red Hook, NY 12571 USA
Phone: 845-758-0400
Fax: 845-758-2634
Email: curran@proceedings.com
Web: www.proceedings.com

25th International Conference of the Center for Chemical Process Safety 2010

Topical Conference at the 2010 AIChE Spring Meeting
and 6th Global Conference on Process Safety

San Antonio, Texas, USA
21 - 25 March 2010

ISBN: 978-1-61738-435-6

AIChE

TABLE OF CONTENTS

Overview of Inherently Safer Technology

Dennis C. Hendershot
Process Safety Consultant
Bethlehem, PA
d.c.hendershot@att.net

Prepared for Presentation at
American Institute of Chemical Engineers
2010 Spring Meeting
6th Global Congress on Process Safety
San Antonio, Texas
March 22-24, 2010

Paper 21a

UNPUBLISHED

Overview of Inherently Safer Technology

Dennis C. Hendershot
Process Safety Consultant
Bethlehem, PA
d.c.hendershot@att.net

Keywords: Inherently Safer Design, Inherently Safer Technology, Process Safety, Process Design

Abstract

Inherently Safer Design (ISD) is a holistic approach to making the development, manufacturing, and use of chemicals safer. Over time, there have been many developments on the concept of inherent safety, however, currently there is a growing fixation on only one element of ISD: substitution. This paper will present an overview of ISD and its elements of minimize, substitute, moderate, and simplify. In addition, the life cycle of a process will be explained in context of ISD to further explain the most effective use of ISD as well as other risk mitigation methods and strategies.

1. What is Inherently Safer Design?

Inherently safer design (ISD) is a philosophy for addressing safety issues in the design and operation of chemical processes and manufacturing plants. When considering ISD, the designer tries to manage process risk by eliminating or significantly reducing hazards. Often, the traditional approach to managing chemical process safety has accepted the existence and magnitude of hazards in a process, and incorporated hardware, procedures, and management systems to reduce process risk. Where feasible, ISD provides more robust and reliable risk management, and has the potential to make the chemical processing technology simpler and more economical by eliminating the need for expensive safety systems and procedures. However, when one considers all of the multiple risks associated with any technology, including chemical processing, it is unlikely that any process or plant design can eliminate ***all*** hazards and risk. A combination of inherent, hardware, procedural, and management systems will always be required to adequately manage all process risks.

ISD addresses the immediate impact of single events (chemical accidents) on people, the environment, property, and business. In a chemical processing plant, this generally means the immediate impacts of fires, explosions, and the release of toxic materials. In many cases, an inherently safer design will also be beneficial for other types of process risk – for example, environmental risk, chronic health risk, or risk to consumers or users of a product. But this is not necessarily true – for example, a non-flammable solvent may be inherently safer from a fire and explosion risk viewpoint, but it may be a serious environmental contamination hazard, or it may be a chronic health hazard. While engineers recognize the potential benefits of ISD in managing other types of process risk, the main intent of ISD is to reduce the frequency and potential impact of chemical plant accidents – fires, explosions, and acute toxic exposures. Therefore, the application of ISD is one consideration in the selection of process and product technology, but the decision about what technology option is best overall must consider all risks.

Figure 1: Nypro caprolactam plant in Flixborough, UK after the deadly 1974 explosion.

2. History of ISD

The concept of ISD is not really new, and it is not really unique to the process industries. Technologists have always recognized the value of eliminating or reducing hazards. Applying ISD without calling it by that name, they simply considered it to be good design. For example, when a family of stone age cave dwellers decided to move to a cave higher above the river following a flood, they were practicing ISD by eliminating the risk of having their home flooded. They could have stayed in their old cave and managed the risk in other ways – for example, by building a dike around the cave mouth (hardware), or by assigning a family member to monitor the river level and warn everybody to move themselves and their possessions to higher ground when a flood was imminent (procedural).

The specific terminology "Inherently Safer Design" came into use in the process industries in the 1970s. Following a 1974 hydrocarbon vapor cloud explosion at Flixborough, England (Figure 1), Trevor Kletz, a senior safety advisor for ICI, questioned the need for such large quantities of flammable or toxic materials in a manufacturing plant, and the need for processing at elevated temperature and pressure. Kletz suggested that industry should re-direct its risk management efforts toward elimination of hazards where feasible. Instead of devoting extensive resources to safety systems and procedures to manage the resulting risks, industry could try to identify process modifications which reduce or eliminate hazards. This could be accomplished, for example, by reducing the quantity of hazardous material, using less hazardous materials, or developing technology which operates at less severe conditions. Kletz and others in the chemical industry established a set of principles for ISD, and provided many examples of its implementation. In 1996, the Center for Chemical Process Safety (CCPS) published a landmark book, *Inherently Safer Chemical Processes: A Life Cycle Approach*[1], compiling up to date information on industry thinking on ISD. In 2009, CCPS published a second edition of this book, incorporating the latest developments on ISD based on more than a decade of additional industrial experience[2].

3. Inherently Safer Design Basics

> **in·her·ent:** Adjective. Existing as an essential constituent or characteristic; intrinsic. From the Latin *inharens, inhaerent-*, present participle of *inhaerere,* to inhere.[3]

What do we mean by inherently safer design? The dictionary definition above indicates that something is "inherent" if it exists "as an essential constituent or characteristic." This means that safety is built in to the process or product, not added on. Hazards are eliminated, not controlled, and the means by which the hazards are eliminated are so fundamental to the design of the process that they cannot be changed or defeated without changing the process. In many cases this will result in simpler and cheaper plants. If extensive safety systems are required to control major hazards, they will introduce complexity to a plant, along with cost – both in the initial investment for the safety equipment and also for the ongoing operating cost for maintenance and operation of the safety systems.

Because the philosophy of ISD is to eliminate or reduce the hazard of a process, it is important to understand what we mean by the word "hazard". The Center for Chemical Process Safety (CCPS) has defined hazard as "an inherent physical or chemical characteristic that has the potential for causing harm to people, the environment, or property." [4] Hazards are intrinsic to a material or its conditions of use. Some examples of hazards are:

- Chlorine is toxic by inhalation
- Gasoline is flammable
- High pressure steam contains a large amount of potential energy, both from its elevated temperature and also from the high pressure

These hazards cannot be changed, except by changing the material or the conditions of use.

4. Chemical Process Safety Strategies

Chemical process safety strategies can be grouped in four categories: Inherent, Passive, Active, and Procedural (Figure 2). In general, inherent and passive strategies are the most robust and reliable, but elements of all strategies will be required for a comprehensive process safety management program when all hazards of a process and plant are considered.

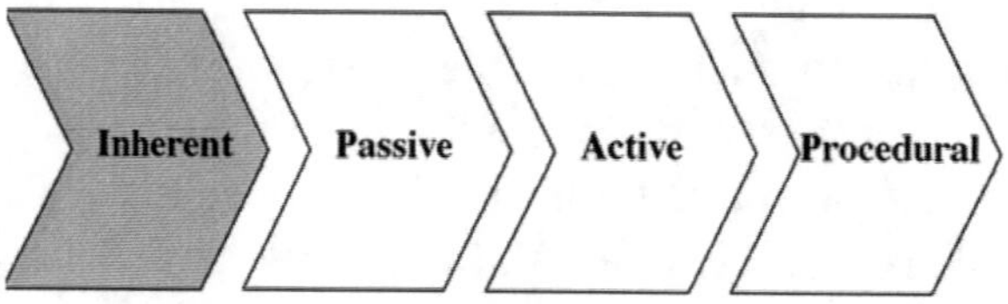

Figure 2: The four process safety strategies.

4.1 Inherent

The inherent approach to safety is, where feasible, to eliminate or greatly reduce the hazard by changing the process to use materials and conditions which are non-hazardous or much less hazardous. These changes must be integral to the process or product, and not easily defeated or changed without fundamentally changing the process or plant design. Substituting water for a flammable, and perhaps toxic, solvent as a carrier for a paint or coating – using water based latex paints instead of oil based paints – is an example. The elimination of the flammable or toxic solvent is an inherent characteristic of the product and its manufacturing process. The hazard of fire or exposure to toxic solvent vapors is eliminated, both in the manufacturing process and also throughout the manufacturing supply chain all the way to the product user.

4.2 Passive

Passive safety features minimize hazards using process or equipment design features which reduce either the frequency or consequence of an incident without the active functioning of any device. For example, a batch process uses a chemical reaction which has a maximum possible pressure of 5 bar in case of a runaway reaction. If this reaction is done in a reactor designed to contain a pressure up to 10 bar, the maximum runaway reaction pressure will be contained within the reactor vessel. The reactor contains the pressure because of its design and construction – the thickness and strength of the metal from which it is fabricated, the strength of the gaskets and bolts which hold it together, and it's other physical components. This containment is very robust and reliable – the reactor does not need to sense high pressure and take any action to contain the pressure, and no moving parts are required to contain the pressure. But the hazard (5 bar pressure) still exists, so there is some risk that an incident will occur. The reactor may be damaged, corroded, improperly constructed, or contain a faulty gasket, for example. It is possible that the reactor would fail to contain the pressure from the runaway reaction, even though it is designed to do so. The passive strategy would be considered less robust than an inherent strategy, which would change the process to eliminate or reduce the pressure from a runaway reaction.

4.3 Active

Active safety systems include process control systems, safety instrumented systems (SIS), and automatic incident mitigation systems such as sprinkler systems to extinguish a fire. These active systems are designed to sense a hazardous condition and take an appropriate action. Active systems may be designed to prevent an incident, or to minimize the consequences of an incident. For example, a tank might have a high level interlock which shuts off a pump feeding the tank and closes all feed valves – this system is designed to prevent a tank overflow. A fire sprinkler system is an active system designed to minimize the consequences of a fire – it does not prevent the fire, and may not even be activated unless a fire is detected.

4.4 Procedural

Procedural safety features include standard operating procedures, safety rules and procedures, operator training, emergency response procedures, and management systems. For example, an operator may be trained to observe the temperature in the reactor and apply emergency cooling if it exceeds a specified critical value. In general, for a high hazard system, procedural risk management systems do not, by themselves, provide adequate risk management. Human reliability is not high enough, and people often cannot diagnose a problem, determine the appropriate action, and take that action quickly enough. However, procedural safety systems will always be a part of a comprehensive risk management program – at a minimum they will be required to ensure ongoing maintenance and management of active and passive safety systems.

5. Designing Inherently Safer Processes

CCPS has categorized strategies for designing inherently safer processes into four groups:

- Substitute – use less hazardous materials, chemistry, and processes

  - Examples: an alternate synthesis chemistry for a molecule uses less toxic raw materials; water based latex paints eliminate fire, toxicity, and environmental hazards associated with solvent based paints
- Minimize – use small quantities of hazardous materials, reduce the size of equipment operating under hazardous condition such as high temperature or pressure
  - Example: nitroglycerine can be made in a continuous pipe reactor with a few kilograms of inventory instead of a large batch reactor with several thousand kilograms of inventory
- Moderate – reduce hazards by dilution, refrigeration, process alternatives which operate at less hazardous conditions
  - Example: a combustible solid was handled as a pellet instead of a fine powder, reducing the dust explosion hazard
- Simplify – eliminate unnecessary complexity, design "user friendly" plants
  - Example: Old piping which was no longer needed because of process modifications was removed from a plant, making it impossible to accidently transfer material into a reactor through that piping by operating error or leaking valves

6. ISD and the Process Design Life Cycle

Process design starts with the selection of a basic technology for a process operation. As the technology progresses through process development, conceptual plant design, scale up, engineering and detailed plant design, plant construction, startup, and ongoing operation and future modification, different kinds of choices and decisions are required by chemists, engineers, and other technologists (Figure 3). The philosophy of ISD applies at all stages, but the available options change. The best opportunities for implementation of inherently safer design are early in product or process research and development. At this point, there has not been any commitment to a particular technology, resources have not been expended on research and development which would have to be done over again, potential customers have not committed to using products produced by a certain technology, and developed their processes to fit this product, and capital has not been committed to build a plant to implement a particular technology. As the process moves through the life cycle, it becomes more difficult to change the basic technology. However, it is never too late to consider ISD – but options for implementation may be more limited in an existing plant. To illustrate how ISD can be applied at various levels of process development and design, disinfection of drinking water will be used as an example process.

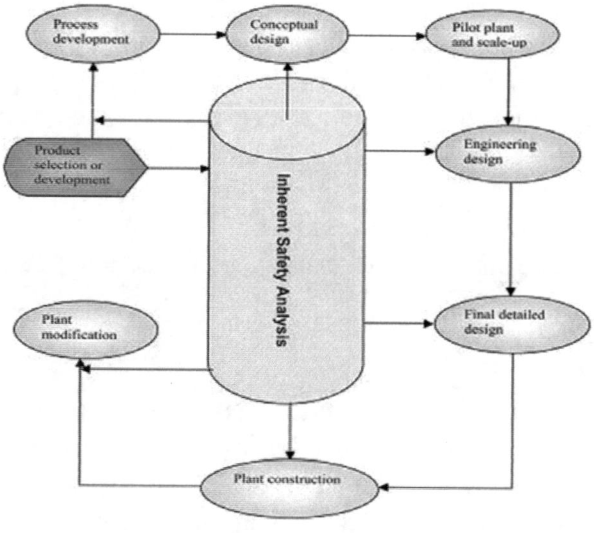

Figure 3: Inherently Safer Design in the process design life cycle.

6.1 Selection of basic technology

There are many ways to disinfect water – for example, chlorination, ozone, ultraviolet light, radiation, and others. These various technologies have differing ISD characteristics relative to different hazards of concern. For example, chlorine is toxic and may produce hazardous chlorinated organic materials in water containing certain organic precursors. Ozone and ultraviolet light provide disinfection at the point of treatment, but do not have residual activity should the water be contaminated downstream of the treatment plant. To consider ISD for basic technology selection, the decision maker must understand all hazards of concern and the inherent safety characteristics of the available process options relative to those hazards.

6.2 Implementation of the selected technology

Once the basic technology has been selected, there may be many options available for actual implementation of that technology. Using the water treatment example, assume that chlorination technology has been selected. Now the process designer has to decide how chlorination will be implemented. Some options he might consider include elemental chlorine gas, sodium hypochlorite, or solid chlorinating agents. Each option has specific ISD characteristics relative to various hazards of concern. The designer must also consider other factors such as economics, feasibility of the technology, state of development (proven technology, a new process which has never been used, or somewhere in between), and other risk concerns (environmental, chronic health, etc.).

6.3 Plant design

At this point in the process life cycle, the designer must consider ISD for a specific plant design. Factors might include:

- Location of the plant relative to surrounding population, in-plant occupied areas, sensitive environmental areas, etc.
- general layout of the equipment on the selected plant site
- number of parallel systems and size of those systems (one big plant, or two or more smaller plants, for example)

If we assume that the designers of this particular water treatment system have decided that disinfection using gaseous chlorine is the optimum process, ISD should be considered when determining, for example, where the facility is located in the community, where the chlorine is stored and handled on the site once the site has been selected, the number and size of the water chlorination systems.

6.4 Detailed equipment design

The designer should consider ISD in the detailed design of each piece of equipment in the plant. There are many options in the design equipment such as heat exchangers, chlorine vaporizers, and other devices that might be included in the water treatment plant. Different equipment designs will have different ISD characteristics – for example, the inventory of hazardous material in the equipment. Also, the detailed layout of the equipment will impact things such as the length and diameter of piping containing hazardous materials. Consideration of human factors in the design of equipment, to minimize the potential for incorrect operation and human error, will also result in an inherently safer plant.

6.5 Operation

Once a plant is built, ISD should be considered in the development of operating procedures and maintenance procedures. These must be clear, logical, and consistent with actual human behavior. Ideally, the *easiest* way to operate plant equipment should also be the *right* way – and the ***safe*** way – to operate that equipment. Also, the plant should consider ISD options throughout the operational lifetime, particularly when modifications are made, or if new technology becomes available.

7. Some ISD Issues

ISD is not a magic bullet which will make all potential risk associated with chemical processing go away. For example, in many cases it is not possible to eliminate or reduce the hazard because the characteristic of a material or technology which makes it hazardous is the same as the characteristic which makes it useful.

- Jet airliners typically travel at about 600 miles per hour. This is what makes them useful – they can transport you half way around the world in less than a day. But that speed also makes them hazardous – an airplane traveling at 600 miles per hour has a large amount of kinetic energy which can cause major damage if it hits something, as well as likely killing all of the passengers.
- Gasoline is flammable, and has the potential for a major fire. But the flammability of gasoline is also why gasoline is useful – it stores a large amount of energy in a small mass of material making it a valuable transportation fuel.
- Chlorine is toxic. This makes it hazardous to most life, including people and animals. But this is also what makes it useful for killing pathogenic organisms in drinking water so that people can drink the water safely.

For these and other hazardous materials or technologies, the important factor in attaining the benefits of the technology and managing the hazard is control. In some cases there may be alternative technologies which are less hazardous or which are easier to control. But, for many technologies, there are no inherently safer technologies, those technologies are not economically feasible, or other risks (environmental, chronic health risks) are important enough that society chooses to use a technology which is less inherently safe. In these cases, we must rely on passive, active, and procedural safety strategies to manage the risk. These strategies can be highly effective – travel by airplane is extremely safe despite the significant inherent risks of flying. This is because of the highly effective safety management systems in place in the air transport system.

Every technology has multiple hazards. As an everyday example, consider automobile travel. Hazards include the speed of the car (kinetic energy), flammable fuel, toxicity of exhaust gases, hot surfaces in the engine, a pressurized cooling system for the engine, electricity, and others. For a chemical process, hazards might include acute toxicity, flammability, corrosiveness, chronic toxicity, reactivity, adverse environmental impacts, and others. The statement that a process is inherently safer can only be in the context of one hazard, or perhaps several specific, but not all, hazards. It is highly unlikely that any technology will ever be inherently safer with respect to all possible hazards. Any change to a technology designed to reduce one or more hazards will also impact other hazards, perhaps such that these hazards will be increased or new hazards will be introduced.

Chlorofluorocarbon (CFC) refrigerants provide an example of ISD conflicts. When first developed in the 1930s, CFCs were considered to be safer alternatives to existing refrigerants such as ammonia and light hydrocarbons (the term "inherently safer" was not in use at that time). CFCs have low acute toxicity and are not flammable. Toward the end of the 20th century, the adverse environmental impacts of CFCs were recognized, and many of them have been phased out. But, CFCs are still inherently safer than many alternatives with respect to flammability and acute toxicity hazards. Society has decided that the previously unknown hazard of adverse environmental impact is unacceptable, and is willing to apply passive, active, and procedural risk management strategies to manage the hazards associated with CFC replacements for refrigeration systems.

Different populations of potentially impacted people may perceive the inherent safety of technology options differently. For example, for a process which requires relatively small quantities of chlorine gas, a plant may have a choice between supply in 1 ton cylinders or 90 ton railroad tank cars. A neighbor located several miles away from the plant would consider the 1 ton cylinder supply to be inherently safer because it is unlikely that a leak would impact him at that distance. On the other hand, the plant operators would have to connect and disconnect cylinders 90 times for every one time they would have had to connect and disconnect a railroad car. They would consider the railroad car to be inherently safer because they would be impacted by any release, even a small one. The operators would have a much higher frequency of relatively high risk operations – connecting and disconnecting hoses which could potentially contain chlorine. Of course, these hazards can be managed with procedures, personal protective equipment, and other safety management systems, but these are not inherent. Both the neighbor and the operator are correct in their perception of the ISD characteristics of the chlorine supply options, but they are concerned about different kinds of incidents. The challenge for the designer of the system is to understand these conflicting requirements and make an intelligent choice, including consideration of the entire risk management system (inherent, passive, active, and procedural).

It is also important to consider whether an ISD option actually reduces risk or transfers it somewhere else, perhaps increasing overall risk. A plant might reduce the size of a hazardous material storage tank at the site, reducing inventory and site risk. But use of the smaller tank may require a change from shipment of the material to the plant from railroad tank cars (typically about 300,000 pound shipments for many materials) to trucks (typically about 30,000 pound shipments) because the smaller tank cannot contain more than a truck load of material. Now there will be 10 times as many shipments, and they will go by road rather than by rail. Depending on the specific characteristics of a particular plant location, road shipment may be more hazardous. While the site risk is reduced, the overall risk to society may actually be increased.

8. Implementing Inherently Safer Design

How do we incorporate inherently safer design philosophies into the design and operation of chemical processing plants? The best answer is to start early in the life cycle, and never stop (Figure 4). The greatest opportunities for fundamental changes to processes occur early in the process life cycle, during initial process conception and early development. At this stage, the researcher may have opportunities to select less hazardous raw materials and intermediates or less hazardous chemical synthesis paths from among the many options which might be available. However, it is never too late to consider inherent safety. There are many published reports of significant inherent safety improvements in plants which have been operating for many years.

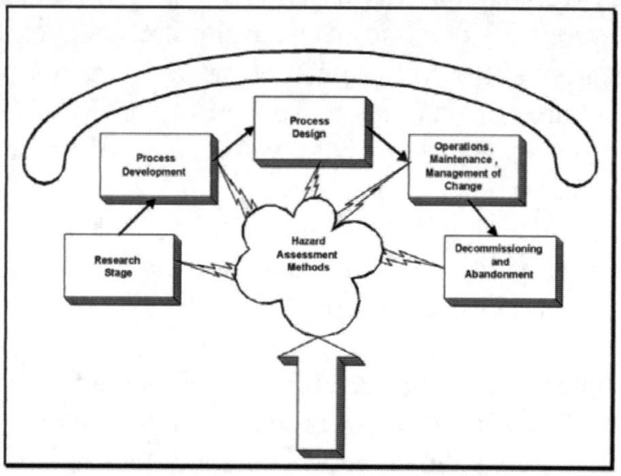

Figure 4: Hazard Assessment at various stages in the process life cycle.

An opportunity to consider inherently safer design at all stages in the process life cycle is to incorporate inherent safety considerations into the process safety management (PSM) activities that are normally done at every stage in the life cycle, from initial technology selection through detailed design and operation. These activities will vary in structure as the process moves through its life cycle, from general technology reviews at early stages through detailed design reviews and other PSM activities later in the life cycle. Some organizations prefer to do a separate inherent safety review at a relatively early stage of process or plant development, others incorporate inherently safer design considerations into the existing process safety review processes. Clearly the type of inherently safer design improvements will be different at different life cycle stages. Early on, the designer has opportunities to change things like the basic technology, process unit operations, plant location. Later on, in detailed design, inherently safer design opportunities will be more limited in impact – things like detailed piping configuration, user friendly operator interfaces, operating inventory in equipment. CCPS and others have published checklists which are useful in identifying ISD opportunities at various stages in the process life cycle. In the 2009 2nd edition of the CCPS book, *Inherently Safer Chemical Processes: A Life Cycle Approach*[2], these checklists are significantly improved.

Also, inherently safer design philosophy can be considered in determining how to respond to an identified hazard from any PSM activity, including Process Hazard Analysis (PHA), management of change (MOC), incident investigation, routine "walk around" safety inspections, and others. The team conducting the PSM activity is challenged to think about ways to eliminate or minimize hazards, rather than accepting that the hazard exists and focusing its efforts on controlling that hazard. The team should ask the following questions, in this order, once it has identified a hazard:

1. Can the hazard be eliminated?
2. If not, can the magnitude of the hazard be significantly reduced?
3. Do the alternatives identified in questions 1 and 2 increase the magnitude of any other existing hazards, or create new hazards? If so, consider all hazards in selecting the best alternative.

4. What technical and management systems are required to manage the hazards which inevitably will remain?

Designers and PSM activity teams often skip directly to the 4th question, identifying systems to manage hazards whose existence is accepted and believed to be unavoidable. This may be true in many cases, but no team will ever eliminate or reduce a hazard if it does not ask if this is possible. Teams involved in all PSM activities should challenge themselves to eliminate or reduce hazards, and then, only if this is not possible, shift to designing systems to manage risk from hazards which cannot be eliminated.

9. Some Inherent Safety "Myths"

There are a number of common misconceptions ("myths") that we often see in discussion of inherently safer design. Here are a few:

- ***Myth 1: Inherently safer design will eliminate all hazards.*** It is unlikely that any process or material will ever be completely non-hazardous. You can only speak of "ISD" in the context of a specific hazard. You can describe a process as "inherently safer" with respect to the hazard of exposure to a volatile and toxic material, but that says nothing about other hazards – thermal stability, runaway reaction potential, flammability, etc. Also, any change to a process or technology, even one intended to reduce a particular hazard of concern, has the potential to introduce a new hazard, or to increase the magnitude of a different hazard than the one which the change is intended to reduce. There are many examples of "no good deed goes unpunished" where a change intended to improve safety resulted in a new hazard or increased the risk of a different existing hazard.
- ***Myth 2: Because an inherently safer technology represents "the best" approach to managing a particular hazard, you must always implement that technology.*** This is not true because there may be other hazards and risks to be considered, and also because the societal benefits of a technology may justify the robust application of passive, active, and procedural risk management strategies. We have always known how to completely eliminate the risk of airplane crashes, whether caused by accident or by terrorist activity. If we stop flying airplanes, there will never be another airplane crash, so, if safety is the only objective, this would be "the best" approach. However, eliminating crashes is not our only objective. We also want to be able to get from New York to Los Angeles in 5 hours instead of 5 days, and at a reasonable price, along with many other design objectives of the air transportation system. With regard to safety, the objective is safety, not necessarily inherent safety. An appropriate combination of active, passive, and procedural safety management strategies can be extremely effective, as demonstrated by the extremely good safety record of commercial airlines.
- ***Myth 3: Inherently safer design is only applicable at early stages of process research and development and plant design***. ISD applies at any stage in a plant life cycle. While the greatest benefits accrue from selection of inherently safer basic technology, there are many examples of significant improvements in inherently safer operation of existing plants.
- ***Myth 4: Plant operating personnel have little to contribute to implementing inherently safer design.*** There are many examples of inherently safer design improvements in plants

which have been suggested by operating or maintenance personnel. While we don't expect plant operators to suggest a safer chemistry for the process, they are very familiar with the detailed mechanical functioning of the plant its equipment. Who is in a better position to identify issues with complex systems setting up operators for making errors than the people who use those systems every day?

- ***Myth 5: There is a "best technology" which is always inherently safer for the manufacture of a particular product.*** "Best" technology for inherent safety may be highly dependent on local factors – for example, plant location and environment, proximity of population, significance of other hazards in a particular plant environment, and practicality of other (passive, active, procedural) safety strategies at a particular location. As an example, single floor houses eliminate the risk of injury from falling down the steps, but, if you live in a flood plain, perhaps a second floor is a good idea!

10. Summary

The designer and operator of a chemical manufacturing process should consider ISD options throughout the process life cycle, from initial conception through research and development, plant design, construction, operation, modification, and eventual shutdown. Usually the best opportunity for implementing ISD is early in research and development before significant resources have been expended in process or product development, and before a plant has been built. The complete process and product life cycle also needs to be considered to avoid local optimization and transfer of risk from one sector of the overall economy to another. This means consideration of raw materials, the manufacturing process, transportation, storage at all stages in the supply chain, end use, and the safety consequences of changing technology – demolition of old facilities and construction of new ones.

All hazards must be considered so that informed decisions can be made about conflicting goals and impacts. Other factors must also be considered – economics, resource allocation (including capital, research and development resources, operating costs), the feasibility, reliability, and effectiveness of other process risk management features (passive, active, procedural). This may result in different choices for different situations, even for the same technology. In a different environment, the relative importance of different hazards or other factors may be different, leading to a different choice about the optimal technology.

References

[1] Center for Chemical Process Safety. *Inherently Safer Chemical Processes: A Life Cycle Approach*. New York, American Institute of Chemical Engineers, 1996.

[2] Center for Chemical Process Safety. *Inherently Safer Chemical Processes: A Life Cycle Approach*. 2nd Edition. Hoboken, NJ, John Wiley & Sons, 2009.

[3] *The American Heritage® Dictionary of the English Language*. 4th Edition. Houghton Mifflin Company. Boston, MA, 2000.

[4] Center for Chemical Process Safety. *Guidelines for Hazard Evaluation Procedures*. 3rd Edition. Hoboken, NJ, John Wiley & Sons, 2008.

(21b) The DHS Chemical Facility Anti-Terrorism Standards – A Risk-Based Approach to Chemical Facility Security

Larry Stanton, Infrastructure Security Compliance Division, Department of Homeland Security, Washington, DC

In October 2006, Congress granted DHS authority to regulate security at high-risk chemical facilities. The resulting program, CFATS, (1) prioritizes facilities that possess chemicals of interest to terrorists due to the release, sabotage or diversion hazards they present; (2) identifies the types of security performance measures that address these risks; and (3) adapts to different types of facilities, ranging from manufacturing plants to universities, hospitals and warehouses across various industries. The risk-based approach drives facilities to consider adoption of a range of risk-reduction measures, including inherently safer technologies, to achieve compliance with the standards.

See more of this Session: Inherently Safer Design/Inherently Safer Technologies (Part I)
See more of this Group/Topical: Topical 1: Global Congress on Process Safety

(21c) IST Trade-Offs

Jatin Shah, Baker Engineering and Risk Consultants, Inc., Chicago, IL

Inherently Safer Technologies should be a key consideration in any companies risk management effort. Often we are faced with making a choice between an inherently safer option and one that is more hazardous but provides better benefits. With proper safeguards and controls it is possible to manage the risk of the more hazardous option to levels that are comparable or even lower than the inherently safer option. It may cost more and involve a greater amount of oversight but the benefits may still outweigh the costs.

This paper presents three examples to illustrate the trade-offs that one should evaluate in deciding which option is the best for a given situation:

1. Comparison between travel by Automobile and Airplane 2. Comparison of HF vs. MHF (HF with additive)Transport 3. Comparison of HF Alkylation risk reduction options with Sulfuric Alkylation

While inherently safer technological options should be considered they may not always be the appropriate option to select, since it may require sacrificing desired benefits that an alternative option can provide at comparable or lower risk.

See more of this Session: Inherently Safer Design/Inherently Safer Technologies (Part I)
See more of this Group/Topical: Topical 1: Global Congress on Process Safety

HAZOP/LOPA/SIL

Be Careful What You Ask For!

Michael P. Broadribb
Baker Engineering and Risk Consultants, Inc.
3330 Oakwell Court, Suite 100, San Antonio, TX 78218-3024
mbroadribb@bakerrisk.com

Mervyn R. Currie
BP Exploration
1 Wellheads Avenue, Dyce, Aberdeen, AB21 7PB, UK
mervyn.currie@uk.bp.com

Prepared for Presentation at
American Institute of Chemical Engineers
2010 Spring Meeting
6th Global Congress on Process Safety
San Antonio, Texas
March 22-24, 2010

UNPUBLISHED

HAZOP/LOPA/SIL

Be Careful What You Ask For!

Michael P. Broadribb
Baker Engineering and Risk Consultants, Inc.
3330 Oakwell Court, Suite 100, San Antonio, TX 78218-3024
mbroadribb@bakerrisk.com

Mervyn Currie
BP Exploration

Keywords: Inherently Safer Design (ISD), Layer of Protection Analysis (LOPA), Safety Integrity Level (SIL), Safety Instrumented Function (SIF), Safety Instrumented System (SIS)

Abstract

In recent years Process Safety Engineers have promoted the use of HAZOP to identify hazards and assess safeguards, followed by LOPA to identify if additional protection layers are necessary and, if they are to be provided using safety instrumented functions, how reliable the SIFs need to be. Finally SIL verification (reliability analysis) is used to show that the selected SIF sensors, logic and final elements will achieve the required reliability performance for the selected testing regime.

Given that the addition of instrumented protection systems to a facility design is often viewed by engineering contractors and operating companies as easier to implement than changes to the overall process design, it has led to a proliferation of Safety Instrumented Functions/Safety Instrumented Systems (SIF/SIS), the implications of which industry has yet to fully understand. This paper will discuss some of these implications and makes a number of observations regarding approaches to improve the integrity of process plants.

1. Introduction

Sound design of process plants for the oil & gas, petrochemical and related industries requires implementation of various processes to identify hazards, understand the risks associated with those hazards, and to do whatever is necessary to manage the associated risks. This should start with implementation of the principles of inherently safer design (ISD) during concept selection, and progresses through Hazard and Operability Study (HAZOP) (and/or other Process Hazards Analysis (PHA) technique), Layer of Protection Analysis (LOPA), determination of Safety Integrity Level (SIL) and other methodologies to specify risk reduction measures in detailed engineering.

2. HAZOP/LOPA/SIL

For many years Process Safety Engineers have promoted the use of various PHA techniques to identify and qualitatively assess hazards in process plants. Techniques such as HAZID (a.k.a. Preliminary PHA), HAZOP, What If and checklists have been used at various stages in the development life cycle of a facility, and each methodology has its own strengths, weaknesses and applications. However many companies have adopted HAZOP as a powerful tool to identify potential hazards from P&IDs at the detailed engineering stage for new process units. HAZOP has also been used to identify hazards and assess the adequacy of any safeguards in the design of existing process units. Where safeguards are qualitatively judged to be insufficient, additional safeguards are often recommended by HAZOP teams.

In recent years Process Safety Engineers and others have also promoted the use of LOPA to assess in a semi-quantitative manner if further risk reduction is necessary. Often the LOPA has been conducted by the same team as the HAZOP and/or it follows immediately after the HAZOP. The key objectives are to determine if all safeguards (protection layers) are sufficient to adequately manage the hazards under consideration, and to help determine appropriate protective instrumentation systems.

If the LOPA determined that existing safeguards (post HAZOP) are insufficient to adequately manage the hazards under consideration, consideration is given to a range of options to close the gap. These options should encompass both instrumented and non-instrumented protective/safety systems. The former will include consideration of whether safety instrumented systems (SIS) may close the gap (fig.1). If so, the gap determines the minimum risk reduction and hence the reliability requirement for safety instrumented functions (SIF). This assumes that a SIF can functionally prevent the hazardous outcome in each case, i.e. operates quickly enough, and does not introduce other hazards that are intolerable, such as pressure surge.

In recent years Process Safety Engineers and others have also promoted the requirements of the IEC and ISA standards (refs 1 & 2) for SIL verification, which is reliability analysis, to ascertain if proposed safety instrumented functions (SIF) can achieve the required risk reduction at practical testing intervals.

Given the perception that the installation of SIS is easier than other more significant process design modifications to reduce risk, the result has been a significant increase in the specification and use of SIS in recent years in place of traditional means of hazard management. There has also been a corresponding increase in the use of SIL verification. The industry has yet to fully understand the implications of this trend.

Fig. 1 Example of Instrumented Safety System Classification (ref. 3)

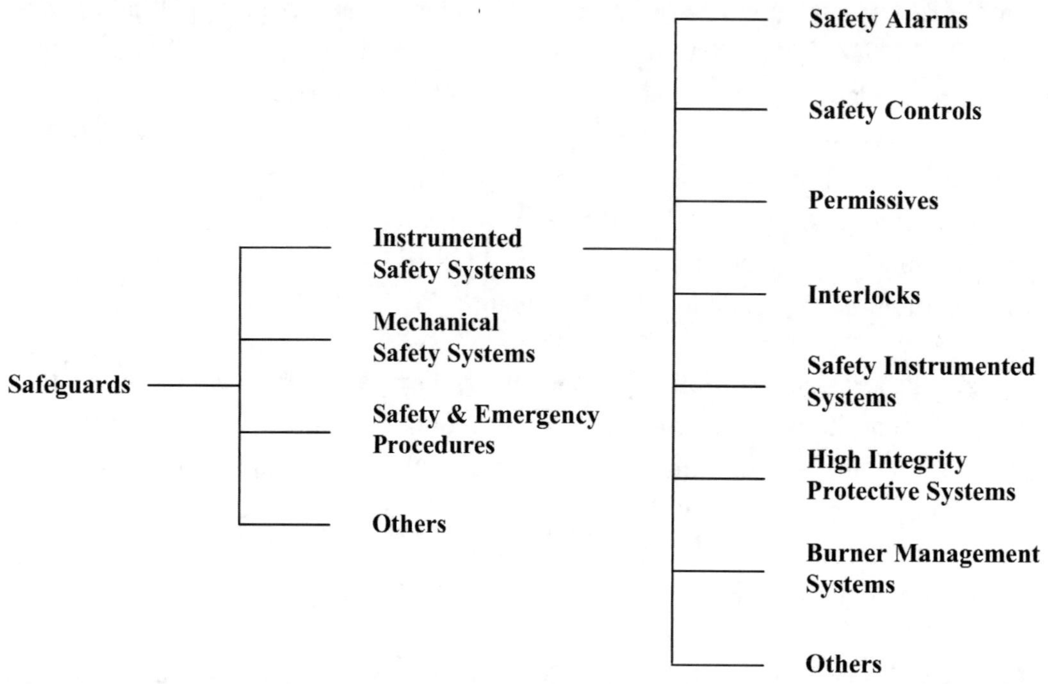

There are several reasons for this increase of SIS/SIFs. Here are a few:

- The inherently safer design (ISD) approach has not been fully embedded in new projects. One reason for this is that ISD is not fully understood and/or practiced by some major engineering contractors, and it is easy to "bolt on" protective systems. Another reason is that ISD has been applied at the front end in choosing design concepts, but "good design practice" is still used in detailed engineering, i.e. how we have always done something because that works and regulators would be suspicious of moving away from "good design practice".

- SIS/SIFs are desirable for operational or environmental reasons in some situations where they are not necessary for hazard management. An example is using a high pressure trip before a pressure safety valve (PSV) operates (assuming the SIF is not required as per LOPA). It is operationally more desirable to trip and hold pressure in the plant than relieve and have to shutdown to check the PSV has correctly reseated, and more desirable from an environmental perspective to avoid the PSV discharge.

And two examples from the oil/gas exploration and production (E&P) industry:

- The use of API RP14C (ref. 4) is mandated for US outer continental shelf in MMS regulations, and it is often applied elsewhere in the E&P industry including some onshore terminals. Process designers treat it as a basic design document, often without further thought. They like it because it is prescriptive and avoids the need to address more difficult questions, like:
  - Is it safe?
  - Are all the instrumented alarms and trips effective?
  - Is the design less safe because of a higher rate of unnecessary plant trips?

- The offshore E&P industry is starting to develop high pressure reservoirs in deep water that cannot be developed without either subsea or topsides HIPPS (high integrity pressure/pipeline protection system), which is usually a higher integrity level SIF. In some cases pipelines cannot be manufactured in the sizes and materials required - the pipe mills do not exist. Topsides facilities are impractical to design for 1000 barg (15,000 psig) shut in pressures.

This paper will discuss some of these implications and makes a number of observations regarding approaches to improve the integrity of process plants.

3. Inherently Safer Design (ISD)

As previously noted, an ISD approach has not always been fully embedded in new designs. Although properly installed and operated SIS will result in safe operations, operating companies and their engineering contractors need to put more emphasis upon 'elimination' and 'prevention' of hazards early in the project design process. It will be much too late if consideration of ISD is left to the technical specification, or even detailed design, stages of new developments. It is too easy for designers to copy previous designs and specify multiple additional protective systems. Companies need to hold their designers accountable for pursuing ISD harder.

When pursuing ISD, all alternative options for ISD should be considered and documented. The capital cost of some options may be higher than others, but consideration should be given to the lifecycle cost of the facility. Some lower capital cost options may initially appear attractive, but result in higher residual risks with attendant requirements for addition of more protective systems in detailed design and/or reliance on people to implement administrative controls and procedures. This may result in higher operating costs to maintain the integrity of protective systems and for additional manpower. After a thorough evaluation and selection of the optimum ISD concept, the residual risk should be reduced through use of multiple diverse protection layers. The instrumented route should never be seen as the easy way out and should be a last consideration after other measures are deemed not practical.

Fig. 2 Hazard Management Triangle

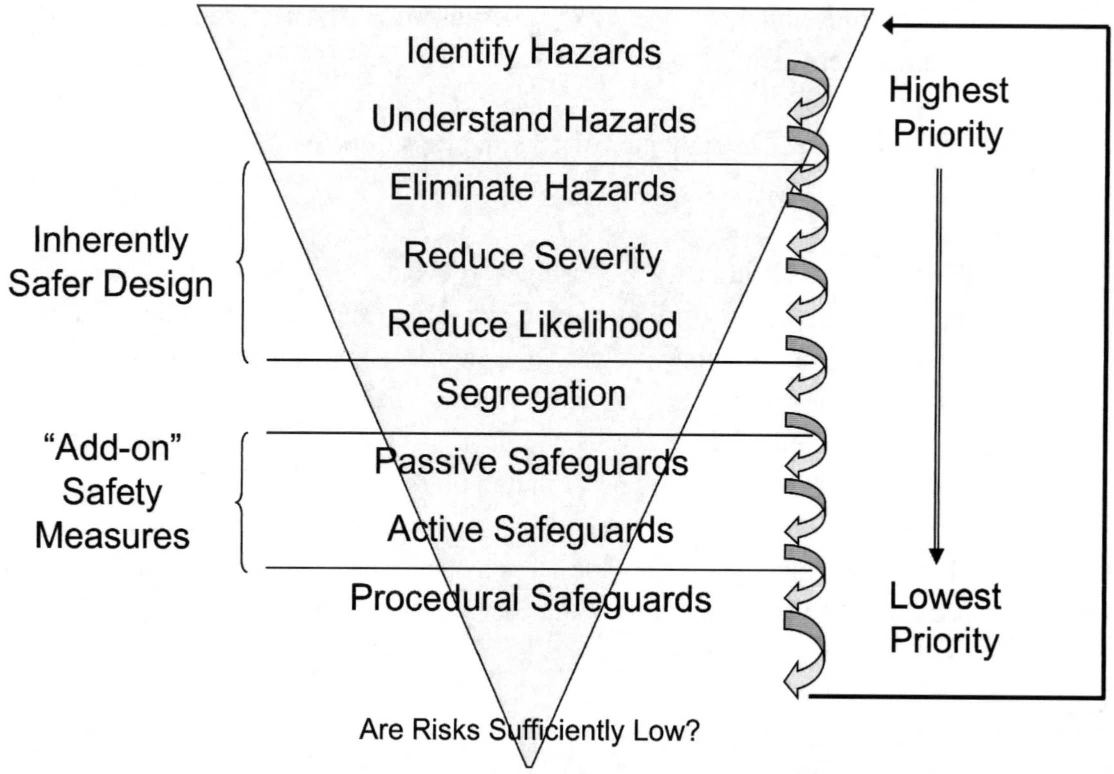

The use of SIF/SIS should be applied only after first considering and ruling out alternative approaches to eliminate the hazard, because, besides being "add on" safety measures, they are "active" systems, so come towards the bottom of the hazard management triangle (fig. 2). While projects and existing plants will inevitably find situations where SIS/SIFs are essential, consideration of non-SIS layers of protection, such as spacing and segregation, should be exhausted before determining any requirement for SIL 2/3 SIFs. The addition of multiple SIFs in detailed engineering, especially multiple SIL 2 and 3 SIFs, could be an indication that additional risk reduction via ISD should be implemented. Providing an ISD approach is fully implemented, new developments may still end up needing some high integrity SIL 2 and 3 SIFs, but hopefully fewer of them than in the past for a similar project scope.

Within the E&P industry, the trend to development of higher pressure oil/gas reservoirs is likely to require greater use of SIL 2 and 3 rated SIF or HIPPS as there is simply no alternative if these reservoirs are to be developed. Some new high pressure developments will likely require SIL 3 subsea HIPPS in deep water, because no pipe mills in the world can manufacture the pipelines in the line sizes and materials for the pressures required. Examples of some of these challenging reservoirs are:

- Rhum (North Sea),
- West Nile Delta (Egypt),
- Shah Deniz (Azerbaijan sector, Caspian Sea)

Policy cannot mandate that higher rated SIFs shall not be used unless the industry is willing to accept that it will not pursue development of these and similar reservoirs.

High integrity SIFs generally have high lifecycle costs requiring a lot of attention during both the project and operations stages. For subsea application costs are very high. During the project phase their design needs careful attention to ensure that the appropriate SIL is specified and verified to achieve the required risk reduction. During the operating phase regular testing and maintenance is necessary to preserve the intended integrity. These issues are discussed in more detail below.

4. SIS Design

The lifecycle, including design, of SIS is covered in detail in a number of standards (ref. 1, 2) and publications (ref. 3), and it is not the intent to repeat that content in this paper. Rather it is intended to highlight a few key steps and observations related to design.

The design process should encompass the following steps:

1. Confirm that a SIF is necessary to manage a hazard as a result of LOPA, operational or regulatory requirements.
2. Establish a functional specification - what does the SIF need to do and in what circumstances?
3. Establish integrity (reliability) specification from LOPA or by assertion if appropriate (e.g. a high pressure SIF is not necessary as a risk reduction measure but the PSV is and by assertion we may want the high pressure SIF to meet SIL 1 requirements) - how reliably does it need to operate on demand?
4. Establish a survivability specification if required. Critical valves may need to survive a defined fire and/or explosion overpressure to complete their safety function. Typically these valves provide isolation from large inventories such as in pipelines, hydrocarbon reservoirs and within process units. [The functional, integrity and any survivability specification form the performance standard for the SIF].

5. Select sensors, logic and final elements and any voting necessary between them (e.g. 2 out of 3 (2oo3) sensors, two series valves effectively voting 1 out of 2) to meet the functional and reliability specification with a proof-testing regime agreed with the Operations team
6. Design for robustness to minimize unnecessary facility trips and maximize production efficiency. For example, 2oo3 sensors can have one of them fail in safe manner and not result in a 'spurious' trip. This arrangement can also have one sensor fail in a dangerous manner without impairing the safety function.

Table 1 provides definitions of SILs in terms of probability of failure on demand (PFD) and risk reduction factors from refs 1 and 2.

Table 1 SIL Definitions

SIL	PFD range	Risk Reduction Factor
1	0.1 - 0.01	10 - 100
2	0.01 - 0.001	100 - 1,000
3	0.001 - 0.0001	1,000 - 10,000
4	0.0001 - 0.00001	10,000 - 100,000

To meet the SIL rating, and probability of failure on demand (PFD) within the SIL range (because LOPA gives the PFD requirement), testing is required at intervals as determined by reliability analysis. The purpose of this 'proof' or 'function' testing is to detect any dangerous or hidden failures which otherwise would only become apparent when the SIF failed to function on a genuine demand. If testing is not carried out at the calculated test interval the average PFD will be higher than the requirement and so the facility will be running at increased risk. Obviously, any faults found during periodic proof testing need to be rectified in a timely manner. The higher specification equipment that is used (lower dangerous failure rate) the longer the allowable interval between proof-testing for a given PFD. Proof-testing of SIFs during operations needs to test for all potential dangerous failure modes.

Some SIF/SIS are designed so that they cannot be tested on line, and require a production shutdown to test. Unplanned shutdowns will not test all input sensors/devices that initiate the SIF/SIS so they must be tested during planned production shutdowns. Many SIF/SIS are designed with overrides (a.k.a. bypasses) to enable testing on line without disrupting production. Referring to Fig. 3, if there is a single sensor, there would generally be an override/bypass switch, either a hardwired key switch or a soft switch applied through an operator screen, to inhibit the sensor so the logic always sees a healthy signal while the sensor is calibrated and function tested which takes it to the trip condition. Installing sensor bypass switches may not be necessary if double or triple redundant sensors are installed. (This is more common on SIL-2 or

SIL-3 systems when more frequent calibration would be required). Sometimes overrides/bypasses are provided between the logic and valve to allow sensor and logic to be tested without shutting the valve, but this is not considered good practice by many operating companies. A bypass valve in parallel with the SIF final element valve may be provided, though typically not offshore for weight reasons, which can be opened to allow function testing of the SIF valve without affecting production.

Fig. 3 Safety Instrumented Function

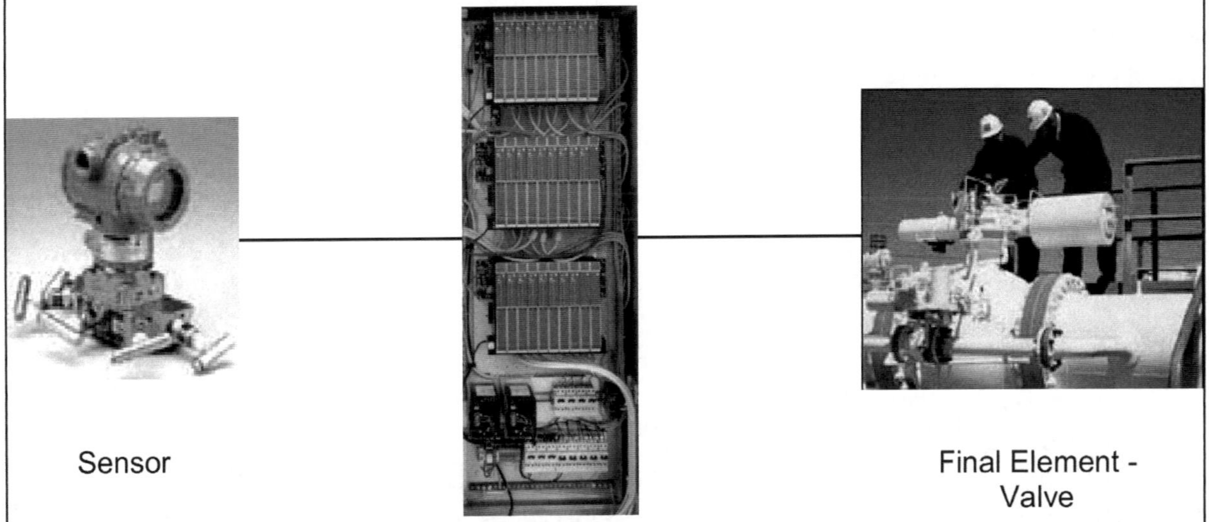

When an override/bypass is applied to sensor (if only single sensor), logic or final element that SIF is not available to perform its protective function. Hence strong procedural controls need to be in place during these periods and overrides/bypasses should be applied for the shortest possible times. Compensating protective measures should be applied to manage the temporary loss of automatic protective systems.

This raises questions about the overall reliability of the system. The original requirement for a SIF was presumably because a certain (high?) reliability was appropriate. However by incorporating an override/bypass with procedure, the overall reliability of the system will be lower as, even with some compensating protective measures in place, there may not be sufficient alternative means of managing the increased risk. There has been much research on human factors, specifically human error, demonstrating that the reliability of an operator correctly implementing a procedure is in the range 90 – 99% depending upon a number of factors such as:

- number of tasks to be performed,
- task complexity,

- pressure on the individual (work and personal life),
- time available to perform tasks,
- etc.

Let's assume that the operator correctly (i) applies the override/bypass, (ii) tests the SIF element e.g. sensor, (iii) resets the SIF, and (iv) removes the bypass, with a reliability of 95%, i.e. performed correctly 19 times out of 20 on average. If the SIF has been designed to a required reliability of correct operation (1 - PFD) of 0.90 to 0.99, i.e. SIL 1, then provision and use of the override/bypass means a significant deterioration to the overall reliability for two reasons. Firstly the required risk reduction of 10 to 100 was removed during the period when the override/bypass was in place. Secondly, the potential for human error discussed above means this planned risk reduction may not be correctly reinstated after testing is completed. There is also potential due to human error of causing an unplanned plant trip when the override/bypass is removed because the SIF element was not left in the correct state.

5. Operating Discipline

All SIF/SIS require testing at intervals during operations as determined by reliability analysis to meet the SIL/PFD rating. This proof-testing is needed to test for all potential dangerous/hidden failure modes. As indicated above, some SIF/SIS are designed so that they cannot be tested on line, and require production to be shutdown. Operations need input to the design process and agreement/buy-in on the testing regime. Will Operations always shutdown to test them at the required frequency? If Operations do not have the necessary discipline to shutdown at the required frequency, then the original integrity and risk reduction will not be met. Worse still there may be a false sense of security that could lead to complacency with regard to plant operation and hazard management.

Many SIF/SIS are designed so that they may be tested on line without impacting production, such as via the installation of redundant sensors. Even under these circumstances where overrides/bypasses are provided, Operations teams need strong discipline and practices to ensure that procedures are followed correctly ***every time*** that testing is required. The demands on testing and maintaining SIL 2/3 SIFs will be significant in terms of operating cost and manpower. Will Operations teams have the operating discipline to maintain these SIF/SIS over the full lifecycle (30+ years) of the plant? If not, these SIF/SIS should not be specified in the design.

Operations teams should put in place a commensurate alternative means of risk reduction when they are bypassing, overriding or inhibiting SIFs. Overriding a SIL 2 SIF is equivalent to removing 10 times the risk reduction provided by a SIL 1 SIF. Plant managers would not operate a hazardous operating facility without a PSV in place. In LOPA it is typical to credit a fully rated PSV with risk reduction of 100, i.e. on the border between SIL 1 and SIL 2. Why do we tolerate overriding/bypassing SIL 2 and SIL 3 SIFs?

Periodically during operations the SIF proof test results should be analyzed, after a statistically significant number of results are available, to ensure that required reliability levels are being achieved in practice. If not, more frequent testing will be required or more reliable equipment needs to be used. If required reliability is more than achieved less frequent testing can be implemented.

There is a huge burden placed upon Operations teams to have the operating discipline to manage their SIF/SIS correctly every time without fail for the lifecycle of the plant. How many Operations teams recognize and fully understand this? How good is the transfer of design intent from Project to Operations? Even if today's Operations team fully understands it, will next year's team? Or next decade's team? How good is the transfer of design information from Operations to new employees (management, supervision and workforce) when organizational change occurs?

6. Conclusion

By pursuing an inherently safer design approach harder, and mitigating risks using multiple diverse protection layers, reliance on SIF/SIS will be reduced. SIL 2 and SIL 3 systems should be considered as a last resort after other measures are considered and the required risk reduction cannot be achieved. Avoid especially SIL 3 and SIL 4 systems - whenever possible, as they should be the exception rather than the rule. There may not be any alternatives to SIL 2 & 3 rated SIF or HIPPS for high pressure oil/gas reservoirs or some chemical processes handling highly hazardous chemicals.

The lifecycle cost of SIFs is high, particularly at higher SILs, due to the significant attention required during both Project and Operations phases. Operations teams need a strong operating discipline to maintain the integrity of SIF/SIS over the full lifecycle of the asset. Projects need to exercise care during design to ensure that that the potential impact of human factors, especially human error, is minimized so that it does not impair the required integrity/reliability.

While there are several standards and numerous associated technical reports available that cover engineering design aspects of SIS and address in high level terms the operations and maintenance requirements, perhaps there is an opportunity for CCPS to take the lead and develop specific guidance (or white paper?) to emphasize the operational aspects of SIF/SIS addressed in this paper?

7. References

[1] IEC, "IEC 61511, Functional Safety: Safety Instrumented Systems for the Process Industry Sector", Geneva, Switzerland, 2003

[2] ISA, "ANSI/ISA 84.00.01 – 2004 (IEC 61511 modified), Functional Safety: Safety Instrumented Systems for the Process Industry Sector", Research Triangle Park, NC, 2004

[3] CCPS, "Guidelines for Safe and Reliable Instrumented Protective Systems", New York, 2007

[4] API, "RP 14C, Recommended Practice for Analysis, Design, Installation, and Testing of Basic Surface Safety Systems for Offshore Production Platforms", Washington DC, 2001

LOPA:
Going Down the Wrong Path

Robert F. Wasileski
Sr. Process Safety Engineer
NOVA Chemicals, Inc.
U.S. Commercial Center
1550 Coraopolis Heights Road
Moon Township, PA, USA 15108

Fred Henselwood
Process Safety Leader
NOVA Chemicals Corporation
Canadian Operating Centre
1000 7th Avenue SW
Calgary, AB, Canada T2P 5L5

Prepared for Presentation at

American Institute of Chemical Engineers

2010 Spring Meeting

6th Global Congress on Process Safety

San Antonio, Texas

March 22-24, 2010

UNPUBLISHED

LOPA:
Going Down the Wrong Path

Robert F. Wasileski
Sr. Process Safety Engineer
NOVA Chemicals, Inc.
U.S. Commercial Center
1550 Coraopolis Heights Road
Moon Township, PA, USA 15108

Fred Henselwood
Process Safety Leader
NOVA Chemicals Corporation
Canadian Operating Centre
1000 7th Avenue SW
Calgary, AB, Canada T2P 5L5

Keywords: LOPA, Layer of Protection Analysis, Mitigation, Risk

Abstract

Layer of Protection Analysis (LOPA) has quickly gained acceptance in the Chemical Processing Industries (CPI), and has risen to be one of the leading risk assessment techniques used for process safety studies. LOPA generally employs more rigor and science than what is encountered with qualitative risk assessments, while still not becoming overly onerous when compared to detailed Quantitative Risk Assessments (QRA). In the interest of balancing time and resources against science and accuracy, certain tradeoffs and assumptions are made within the LOPA assessment. In turn, these tradeoffs and assumptions can lead to inaccurate conclusions.

One of these tradeoffs is that a LOPA assessment is based on only a single outcome, rather than an evaluation of the full spectrum of possible outcomes which would be assessed in a QRA. Generally within LOPA the approach is based on selecting what is perceived to be the most significant event sequence, with respect to the overall risk contribution. Failure to correctly identify the most significant event sequence can, however, result in the risk being understated.

For example, this selection issue arises in the treatment of protection layers associated with mitigation of consequences. LOPA teams have a choice to account for mitigation layers in the consequence assignment, or alternatively treat these layers as Independent Protection Layers (IPL). While this may appear to be an inconsequential decision, it can in fact result in very different conclusions. In the course of treating mitigation layers as Independent Protection Layers, organizations must ensure the necessary Inspection, Testing, and Preventive Maintenance (ITPM) practices are in place for these layers. Furthermore, recognizing this dichotomy in treatment, one can also show that these mitigation layers should be designed so as to achieve a balance between consequence reduction and desired reliability.

This paper discusses alternative treatments of risk mitigation layers that are commonly applied by LOPA teams, and demonstrates their impacts through case studies.

1. Introduction

LOPA as a risk tool differs from many other risk measurement methodologies in that LOPA is designed to assess only a single cause-consequence pair, whereas risk is typically expressed as a measure that is reflective of all potential cause-consequence pairs. As such, LOPA does not measure the full risk associated with a situation but rather attempts to focus on what is believed to be the dominant cause-consequence pair, likely representing a majority of the overall risk. The cause-consequence pair selected for analysis in the LOPA can generally be likened to one path within an event tree, with an associated unique outcome (Figure 1).

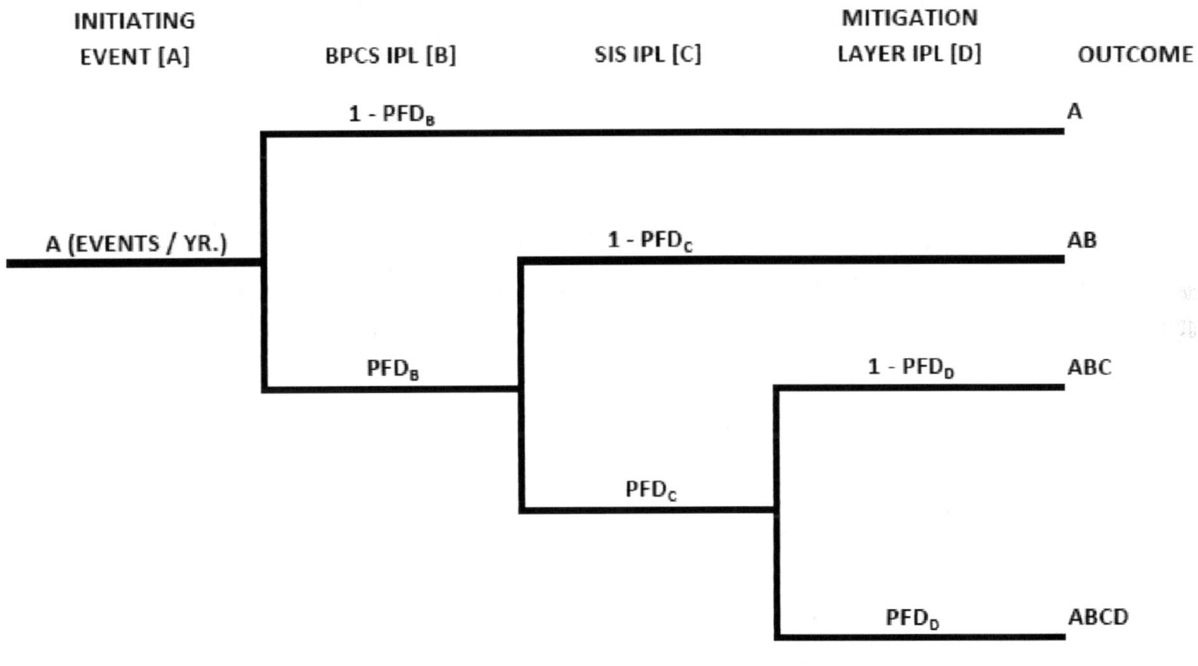

Figure 1. Example Event Tree with Four (4) Unique Outcomes

Further, each outcome in the event tree will have a unique frequency of occurrence. Using the event tree shown in Figure 1 as an example, the frequency of outcome ABCD can be calculated as follows:

$$f_{ABCD} = f_A \times PFD_B \times PFD_C \times PFD_D \qquad \text{[Eq. 1]}$$

Where f_{ABCD} = Frequency of Outcome ABCD, yr^{-1}

f_A = Frequency of Initiating Event A, yr^{-1}

PFD_n = Probability of Failure on Demand of the nth Independent Protection Layer, dimensionless

As LOPA is not a cumulative measure of risk the selection of cause-consequence pairs for analysis becomes critical for the proper application of this tool and the application of appropriate risk criteria.

From a risk evaluation perspective, the selection of the appropriate cause-consequence pair is vital to ensuring that potentially unacceptable risks are identified and then managed. Also of importance is that this cause-consequence pair path selection issue defines a relationship between the optimum effectiveness of a risk mitigation layer and the likelihood of success for that risk mitigation layer (i.e., the effectiveness of a risk mitigation layer should align with the reliability of that risk mitigation layer).

2. Background

Within LOPA, the Independent Protection Layers (IPL) that are allocated to the mitigation layer (i.e., post-release protection) are typically passive devices or features in a plant. For example, secondary-containment dikes, fireproofing, blast walls, and underground drainage systems fall into this category.

When assessing a risk mitigation layer of protection, two possible cause-consequence pairs can be readily identified; the first option being the risk mitigation layer working successfully, resulting in a smaller (mitigated) anticipated consequence. The second option is where the risk mitigation layer fails, resulting in a larger (unmitigated) anticipated consequence. On a relative basis within the LOPA approach, the first option is likely to be associated with a higher frequency, lower severity cause-consequence pair, while the second option is likely to be associated with a lower frequency, higher severity cause-consequence pair. As such, depending on the relative change in frequency versus the relative change in consequences, one cause-consequence pair is likely to result in a higher risk and therefore be the cause-consequence pair that should be assessed.

3. Approaches

In the first approach where the risk mitigation layer is viewed as working successfully, the LOPA team may or may not make this decision deliberately. For example, the LOPA team may intuitively assume that because post-release protection is in place, the scenario with the higher severity consequence is not credible. In doing so, the risk mitigation layer has effectively been assigned a Probability of Failure on Demand (PFD) equal to zero. Furthermore, the worst-credible scenario has likely been overlooked by assuming the post-release protection is completely effective 100% of the time. A similar approach is often taken with Inherently Safer Design (ISD) features, such as a pressure vessel that has been designed and constructed to withstand the maximum pressure that can be created by the system. In the case of ISD features, it may be a valid approach to treat certain scenarios as "not credible" and thus eliminate them given the inherent nature of the protection layer. However, the same conclusion cannot be made for post-release protection layers as these layers will always have a nonzero PFD.

An organization may also make a deliberate decision to consistently and uniformly treat mitigation layers as working successfully 100% of the time. When this approach is taken it is common to account for the mitigation layers through the assignment of consequence severity. While the worst-credible scenario may in fact be discussed by the LOPA team, it will not necessarily be examined as a unique cause-consequence pair. Rather, the risk mitigation layer(s) will be assumed to work successfully, and the severity of consequence will typically be adjusted downward to account for this assumption. For example, when considering a large spill resulting from a tank overfill, this approach would assume the secondary-containment dike is 100% effective at preventing the release from spreading beyond the dike walls. As a result, the consequence selected for analysis would be that associated with a large material spill contained inside the dike.

Whether or not the decision to treat post-release protection layers as 100% effective is deliberate, there are impacts associated with this approach. First, lower frequency, higher severity cause-consequence pairs (scenarios) may not be explicitly evaluated, or may not be considered at all. Second, the organization will have erroneously assumed a PFD equal to zero for these layers. Finally, the opportunity to understand the optimal balance between required effectiveness and desired reliability will have been missed.

The second approach recognizes the imperfect reality of these post-release protection layers, and evaluates them against the LOPA rules used for Independent Protection Layers. In this approach the mitigation layers are tested against the rules for Effectiveness, Independence, and Auditability [1]. When the rules are found to be met, the mitigation layer (i.e. the safeguard) can be treated as an Independent Protection Layer in the LOPA, and assigned an appropriate PFD value.

This second approach – of evaluating mitigation safeguards for eligibility as an IPL - can benefit an organization in a number of ways. First, when mitigation safeguards are tested against the LOPA rules for an IPL, deficiencies in design, physical condition, and testing practices will often become apparent and understood. Second, post-release protection facilities in a plant, such as dikes and fireproofing, are upheld to the Inspection, Testing, and Preventive Maintenance (ITPM) practices required to maintain the PFD value claimed in the LOPA study. In so doing, the mitigation layer becomes an IPL in the organization's LOPA database and must be periodically audited. Lastly, this approach gives an organization a way to objectively compare similar scenarios among multiple plants that may have been designed using different standards or practices.

4. Case Example #1: Styrene Monomer (Flammable Material) Pool Fire

In this scenario, transfer pump P-101 is used to transfer Styrene Monomer from a storage area within the Polymerization Unit to the reactor train. The LOPA team desires to evaluate the risk associated with a pool fire arising from a pump seal failure.

The pump operates on a continuous basis, transferring Styrene at a controlled temperature of 50 degrees F. The pump is also equipped with a double-mechanical seal fitted with a failure alarm that alarms to the Board Operator in the Central Control Room (CCR). Since the pump is located in a fire hazard area it is protected by a pilot-operated waterspray system. Further, structural steel members in this area have been fireproofed in accordance with API 2218 [2] and company standards, and drainage is present to prevent excessive pooling of firewater.

4.1 EXAMPLE 1A: Localized Pool Fire with Minor-to-Moderate Equipment Damage

Using the first approach, the LOPA team has assumed that, because fireproofing and drainage are present, the consequence-of-interest resulting from a major pump seal failure (Initiating Event, 1×10^{-1}/year) is a localized process fire. This results in minor equipment damage and business interruption, with a Tolerable Risk Criteria (TRC) frequency of 1×10^{-2}/year. The LOPA team identifies one IPL in this scenario, and assigns it a Risk Reduction Factor (RRF) of "10" (i.e., a PFD = 0.1). The Risk Gap is further calculated with the following equation:

$$Risk\ Gap = \frac{\text{MEF}}{\text{TRC}} \qquad \text{[Eq.2]}$$

Where MEF = Mitigated Event Frequency, yr^{-1}
TRC = Tolerable Risk Criteria, yr^{-1}

Note: MEF = $f_{\text{IE}} \times P_{\text{EE}} \times P_{\text{CM}} \times PFD_{\text{IPL}}$ [Eq.3]

Where f_{IE} = Frequency of the Initiating Event, yr^{-1}

$P_{\text{EE},\,n}$ = Probability of Occurrence of the nth Enabling Event, dimensionless

$P_{\text{CM},\,n}$ = Probability of Occurrence of the nth Conditional Modifier, dimensionless

$PFD_{\text{IPL},n}$ = Probability of Failure on Demand of the nth Independent Protection Layer, dimensionless

The calculated Risk Gap in this scenario is equal to "1", and thus the scenario risk is determined to be acceptable (Figure 2).

4.2 EXAMPLE 1B: Large Pool Fire with Widespread Damage

Using the second approach, the LOPA team has chosen to evaluate a higher severity consequence, and judge the area fireproofing and drainage system against the rules for an IPL. The LOPA team agrees that if the fireproofing and underground drainage system meets the requirements for an IPL, the fireproofing would be eligible to receive a RRF of "100" (i.e., a PFD = 0.01) and similarly the drainage system would be eligible for a RRF of "10" (i.e., a PFD = 0.1). Specifically, the fireproofing must meet the following requirements to be considered an IPL:

1) ***Effectiveness.*** Structural steel supports located within the fire hazard area envelope must be fireproofed commensurate with the guidelines of API 2218. The fireproofing must be applied up to the support level on all structural members of vessels and pipe racks. Fireproofing must have a two-and-a-half hour (2.5 hours) rating as per UL-1709 [3].
2) ***Independence.*** The fireproofing may not share any devices or common-cause failures with the Initiating Event (Pump seal failure) or the other IPL's in this scenario (Pump seal failure alarm and the underground drainage system).
3) ***Auditability.*** A visual inspection of the condition of the fireproofing in the unit must be conducted every quarter. In addition, a civil inspection of the fireproofing must be performed every 3 years. The results of these inspections must be documented and maintained on file.

In similar fashion, the underground drainage system must meet the following requirements to be considered an IPL:

1) ***Effectiveness.*** The design capacity of the underground drainage system must be equal to or greater than the combination of: (i) the deluge/sprinkler system flow in the immediate area of concern, plus (ii) the two adjacent deluge/sprinkler systems, plus (iii) two hose streams, plus (iv) the anticipated quantity of spilled process material [4]. The ground

within the fire hazard area must be sloped or otherwise graded (1-2%, depending on surface material) towards a process catch basin. Catch basins must be equipped with valves that remain operable under fire conditions. Drain ditches in the area must be designed such that ignited flammables may be safety consumed (e.g., a wicking trench design; see NFPA 15, Annex A [5]).

2) ***Independence.*** The underground drainage system may not share any devices or common-cause failures with the Initiating Event (Pump seal failure) or the other IPL's in this scenario (Pump seal failure alarm and the fireproofing).
3) ***Auditability.*** A visual inspection of the catch basins and drain ditches for debris and other accumulations that can hinder performance must be conducted every quarter. In addition, a flow test must be performed every 3 years, using the firewater capacities described in the requirements for *Effectiveness*. The results of these flow tests must be documented and maintained on file.

During the assessment of the ability (*Effectiveness*) of the fireproofing to prolong the structural strength and integrity of steel members during a fire event, it is noted that the fireproofing has either deteriorated over the years, or has been intentionally removed in many areas for maintenance reasons but not replaced (Figures 3 and 4). Additionally, while reviewing the results of the most recent flow (proof) test (*Auditability*) conducted on the underground drainage system, the LOPA team discovered that the system failed the proof test. During the test, large quantities of firewater pooled throughout the unit and encompassed buildings and pipe racks beyond the fire hazard area envelope (Figure 5).

As a result, the LOPA team agrees that neither the fireproofing in the area or the underground drainage system meets the requirements for an IPL. Accordingly, the LOPA team assigns each a RRF of "1" (i.e., a PFD = 1). Moreover, given the catastrophic nature associated with a large pool fire in the Unit, the TRC frequency for this scenario is 1 x 10^{-5}/year.

The calculated Risk Gap in this scenario is equal to "500", and thus the scenario risk is determined to be unacceptable (Figure 6).

LOPA/SIL Analysis Hazardous Event Description				Event Frequency Analysis						Independent Protection Layers				LOPA		Is remaining risk acceptable?
				Initiating Event		Enabling Events		Unmit. Event Freq (Events/yr)	Min Reqd RRF							
Hazard Scenario Title	Hazard Scenario Description	TRC/y	Safeguards	Initiating Event	Events/yr	Enabling Event	Prob.			Independent Protection Layer Description	IPL Type	IPL RRF	IPL Total RRF	Mitigat. Event Freq (Events/yr)	RRF Gap	
1. EXAMPLE 1A - Localized Pool Fire with minor-to-moderate damage	Loss of Process Containment: Styrene Monomer (Flammable Liquid; syn. Vinyl Benzene) spill in Polymerization Unit, leading to formation of a liquid pool and subsequent ignition, resulting in minor-to-moderate equipment damage	1E-02	1. Fixed Firewater Protection Systems - Pump Waterspray and Area Deluge Systems	Mechanical Failure - Loss of Process Containment due to major seal failure (double mechanical seal) on Primary Transfer Pump P-101	1.E-01	1. Probability of Ignition - Release of Styrene Monomer expected to be below the material Flash Point (88 deg. F closed-cup; 98 deg. F open-cup)	0.5	5.E-02	5	1.1. BPCS ALARM + OPERATING PROCEDURE: Pump seal-failure alarm to Central Control Room (CCR) with Operator response to incipient events	ALM+OP	10	10	5.E-03	1	Yes
			2. Emergency response plan (ERP)													

Figure 2. LOPA Worksheet for Localized Pool Fire with Minor-to-Moderate Equipment Damage

Figure 3. Fireproofing in need of Repair

Figure 4. Fireproofing Deficiency

Figure 5. Firewater Pooling Around Pipe Supports and Structures During Proof Test

LOPA/SIL Analysis Hazardous Event Description				Event Frequency Analysis						Independent Protection Layers				LOPA		Is remaining risk acceptable?
				Initiating Event		Enabling Events										
Hazard Scenario Title	Hazard Scenario Description	TRC/y	Safeguards	Initiating Event	Events/yr	Enabling Event	Prob.	Unmit. Event Freq (Events/yr)	Min Reqd RRF	Independent Protection Layer Description	IPL Type	IPL RRF	IPL Total RRF	Mitigat. Event Freq (Events/yr)	RRF Gap	
1. EXAMPLE 1B - Large Pool Fire with widespread damage to Unit	Loss of Process Containment: Styrene Monomer (Flammable Liquid; syn. Vinyl Benzene) spill in Polymerization Unit, leading to formation of a liquid pool and subsequent ignition, resulting in widespread damage to surrounding equipment in the Unit (from accumulation and pooling of firewater, and substantial fire spread)	1E-05	1. Fixed Firewater Protection Systems - Pump Waterspray and Area Deluge Systems	Mechanical Failure - Loss of Process Containment due to major seal failure (double mechanical seal) on Primary Transfer Pump P-101	1.E-01	1. Probability of Ignition - Release of Styrene Monomer expected to be below the material Flash Point (88 deg. F closed-cup; 98 deg. F open-cup)	0.5	5.E-02	5000	1.1. BPCS ALARM + OPERATING PROCEDURE: Pump seal-failure alarm to Central Control Room (CCR) with Operator response to incipient events	ALM+OP	10	10	5.E-03	500	No
			2. Emergency response plan (ERP)							1.2. FIREPROOFING: Extensive fireproofing present in Unit on pipe supports and structural members. This is NOT an IPL due to numerous instances of missing, damaged, or detoriated fireproofing in the Unit	Not an IPL	1				
										1.3. PLANT SEWER SYSTEM: Drainage systems in place to remove firewater. This is NOT an IPL since this system has recently failed a proof test	Not an IPL	1				

Figure 6. LOPA Worksheet for Large Pool Fire with Widespread Damage to Unit

4.3 Analysis of Case Example #1

Several potential shortcomings have been highlighted in the previous case example, by illustrating two different treatments of the risk mitigation layer. Where the mitigation layers have been accounted for in the consequence assignment (Example 1A), the analysis failed to identify flaws in these layers, yet still concluded the risk was acceptable. This oversight was due in part to the erroneous assumption that the mitigation layers have a Probability of Failure on Demand equal to zero. Further, the analysis did not explicitly evaluate the lower frequency, higher severity cause-consequence pair associated with the failure of these two mitigation layers.

Conversely, where the mitigation layers were evaluated as Independent Protection Layers (Example 1B), the analysis identified deficiencies in these layers, resulting in a substantial risk gap. Clearly, an organization benefits from understanding where these exposures exist. However, this example also underscores the level of conservatism found in the LOPA technique. Since the mitigation layers did not meet the criteria for an IPL, they did not contribute a risk reduction factor to the analysis (i.e., their assigned PFD = 1). While this is mathematically consistent with the basic (order-of-magnitude) LOPA approach, it is in all likelihood overly conservative and an overstatement of the risk.

5. The Relationship between Reliability and Effectiveness

In situations where a mitigation safeguard meets the requirements for an IPL, a limit on the PFD value for that mitigation layer can be proposed based on the degree of consequence reduction provided by that layer. This relationship can be established through assessing both LOPA scenarios: the path where the mitigation layer functions and the path where the mitigation layer fails. Through assessing both paths it may be demonstrated that there is little benefit to managing a mitigation layer at a PFD value which goes beyond the predicted consequence reduction to be provided by that layer. Further, in cases where the PFD value is small relative to the consequence reduction, the path associated with success of the mitigation layer may prove to dominate the overall risk contribution.

6. Case Example #2: Plant Wastewater Spill

In this scenario, Tank 1000 is routinely filled with a plant wastewater from a larger storage tank. The wastewater is a mixture of numerous effluent streams from the main plant, containing trace quantities of suspended solids, soluble organics, and insoluble organics. The LOPA team desires to evaluate the risk associated with overfilling the tank.

Prior to beginning the transfer operation, a Field Operator manually gauges the level in Tank 1000, estimates the time required to complete the transfer, and then starts the pumping operation.

The entire transfer typically requires 3 to 4 hours to complete, and as such is not attended to continuously by the Field Operator.

Tank 1000 is equipped with two independent level instruments. The first, a continuous level instrument, provides both a local readout to the Field Operator at the perimeter of the dike, and a remote readout with a high level alarm (at 90% tank capacity) to the Board Operator in the Central Control Room (CCR). The second instrument is a point level device that is used for a Safety Instrumented Function (SIF) to shutdown the transfer pump at 95% of tank capacity.

Tank 1000 is located inside a concrete secondary-containment dike. The catch basins inside the dike are interconnected with the plant's contaminated sewer system (CSS). All materials entering the CSS are routed to the plant's wastewater treatment facility for further treatment. Just outside of the dike are additional catch basins that collect storm water, steam condensate, and other non-contact water. This storm-water sewer (SWS) system is not integrated with the wastewater treatment system, but rather goes through a detention basin before being discharged directly into a nearby river.

6.1 EXAMPLE 2A: Wastewater Spill Inside the Dike

Using the first approach, the LOPA team has assumed that because the secondary-containment dike is present, the consequence-of-interest resulting from overfilling Tank 1000 is a spill inside the dike (Initiating Event of Human Error by the Field Operator, 1 x 10^{-1}/year). This results in substantial clean-up and business interruption costs, with a Tolerable Risk Criteria (TRC) frequency of 1 x 10^{-4}/year. The LOPA team identifies two IPL's in this scenario and assigns a Risk Reduction Factor (RRF) of "10" (i.e., a PFD = 0.1) to each IPL.

The calculated Risk Gap in this scenario is equal to "10", and thus the scenario risk is determined to be unacceptable (Figure 7).

6.2 EXAMPLE 2B: Wastewater Spill Outside the Dike

Using the second approach, the LOPA team has chosen to evaluate the higher severity consequence, and judge the dike against the rules for an IPL. The LOPA team agrees that if the dike meets the requirements for an IPL it would be eligible to receive a RRF of "100" (i.e., a PFD = 0.01). Specifically, the dike must meet the following requirements to be considered an IPL:

1) ***Effectiveness.*** The available capacity of the dike must exceed the volume equivalent released from the tank between rounds by the Field Operator. Further, effective administrative controls over drain valves inside catch basins must be in place. Drain

valves for passive containment must be car-sealed in the correct position and subjected to a monthly car seal inspection program [6].

2) ***Independence.*** The dike may not share any devices or common-cause failures with the Initiating Event (Human Error by the Field Operator) or the other IPL's in this scenario (BPCS Alarm + Operating Procedure, and Safety Instrumented Function).
3) ***Auditability.*** A visual inspection of the secondary-containment system and car seals must be conducted every month. In addition, a civil inspection of the dike must be performed every 5 years. The results of these inspections must be documented and maintained on file.

During the assessment of the capability (*Effectiveness*) of the dike to contain a large spill, it is determined that the dike meets the requirements for an Independent Protection Layer. As a result, the LOPA team assigns it a RRF of "100" (i.e., a PFD = 0.01). Further, given the greater severity associated with a release of wastewater outside the dike and the subsequent entry of untreated wastes into the local watershed, the TRC frequency for this scenario is 1 x 10^{-5}/year.

The calculated Risk Gap for this scenario is equal to "1" and thus the scenario risk is determined to be acceptable (Figure 8).

LOPA/SIL Analysis Hazardous Event Description				Event Frequency Analysis						Independent Protection Layers				LOPA		Is remaining risk acceptable?
				Initiating Event		Enabling Events										
Hazard Scenario Title	Hazard Scenario Description	TRC/y	Safeguards	Initiating Event	Events/yr	Enabling Event	Prob.	Unmit. Event Freq (Events/yr)	Min Reqd RRF	Independent Protection Layer Description	IPL Type	IPL RRF	IPL Total RRF	Mitigat. Event Freq (Events/yr)	RRF Gap	
1. EXAMPLE 2A - Wastewater Spill Inside Dike	Tank 1000 overfilled, leading to a large spill (greater than 10,000 Gallons) of wastewater inside the secondary-containment dike, resulting in substantial clean-up and business interruption costs	1E-04	1. Emergency response plan (ERP)	Human Error - Field Operator incorrectly gauges the level of the receiving tank (Tank 1000) prior to the filling process	1.E-01	1. None used - Operation occurs regularly	1	1.E-01	1000	1.1. BPCS ALARM + OPERATING PROCEDURE: High Level Alarm LAH-1000-A with setpoint at 90% tank capacity, alarms to Board Operator to shutdown filling operation	ALM+OP	10	100	1.E-03	10	No
										1.2. SAFETY INSTRUMENTED FUNCTION: High-High Level Instrumented Shutoff System LAHH-1000-B, with setpoint at 95% tank capacity (this SIF is completely independent from LAH-1000-A)	SIF	10				

Figure 7. LOPA Worksheet for Wastewater Spill Inside the Dike

LOPA/SIL Analysis Hazardous Event Description				Event Frequency Analysis						Independent Protection Layers				LOPA		Is remaining risk acceptable?
				Initiating Event		Enabling Events		Unmit. Event Freq (Events/yr)	Min Reqd RRF							
Hazard Scenario Title	Hazard Scenario Description	TRC/y	Safeguards	Initiating Event	Events/yr	Enabling Event	Prob.			Independent Protection Layer Description	IPL Type	IPL RRF	IPL Total RRF	Mitigat. Event Freq (Events/yr)	RRF Gap	
1. EXAMPLE 2B - Wastewater Spill Outside Of Dike	Tank 1000 overfilled, leading to a large spill (greater than 10,000 Gallons) of wastewater inside the secondary-containment dike, with a reportable quantity breaching the dike and entering the plant Storm-Water Sewer (SWS) system, resulting in contamination of the local waterway and regulatory penalties	1E-05	1. Emergency response plan (ERP)	Human Error - Field Operator incorrectly gauges the level of the receiving tank (Tank 1000) prior to the filling process	1.E-01	1. None used - Operation occurs regularly	1	1.E-01	10000	1.1. BPCS ALARM + OPERATING PROCEDURE: High Level Alarm LAH-1000-A with setpoint at 90% tank capacity, alarms to Board Operator to shutdown filling operation	ALM+OP	10	10000	1.E-05	1	Yes
										1.2. SAFETY INSTRUMENTED FUNCTION: High-High Level Instrumented Shutoff System LAHH-1000-B, with setpoint at 95% tank capacity (this SIF is completely independent from LAH-1000-A)	SIF	10				
										1.3. DIKE: Secondary Containment Dike to contain the spill, thus preventing it from entering the plant SWS.	ERRF	100				

Figure 8. LOPA Worksheet for Wastewater Spill Outside the Dike

6.3 Analysis of Case Example #2

In this example it was proposed that the dike had a RRF of "100" (i.e., a PFD = 0.01) and that the dike would reduce the resulting consequences from a 1 x 10^{-5}/year event to a 1 x 10^{-4}/year event (one order of magnitude). In this case the reliability of the dike would now exceed the one order of magnitude reduction in consequences which is expected. As such the LOPA analysis considering the possibility of the dike failing does not show a risk gap, while the LOPA analysis assuming success of the dike does show a risk gap. Therefore, although the preferred approach is to treat the mitigation layer as an IPL, care has to be taken to ensure the scenario associated with success of the mitigation layer is also evaluated for acceptability.

7. Conclusions

Layer of Protection Analysis provides organizations with a practical risk assessment technique that attempts to bridge the gap between purely qualitative methods and precise quantitative risk assessment. In doing so, the LOPA technique exhibits a number of shortcomings, created by the inherently conservative rules for application, the focus on single cause-consequence pairs, and the selection alternatives faced by LOPA analysts. Organizations must have a full understanding of these limitations when applying the LOPA technique.

In conclusion, organizations applying the LOPA technique must consider these dichotomies and seek to implement policies that result in consistent application of the technique. Moreover, LOPA analysts must recognize the spectrum of outcomes associated with a given initiating event, and evaluate all of the cause-consequence pairs that provide a substantial contribution to the overall risk.

8. References

[1] Center for Chemical Process Safety, *Layer Of Protection Analysis: Simplified Process Risk Assessment*, ISBN 0-8169-0811-7, American Institute of Chemical Engineers, New York, NY, 2001.

[2] "Fireproofing Practices in Petroleum and Petrochemical Processing Plants", **American Petroleum Institute (API) Publication 2218**, Second Edition, August 1999.

[3] "Rapid Rise Fire Test of Protection Materials for Structural Steel", **UL-1709**, Underwriters Laboratories Inc.

[4] "Sewers and Drains", **NOVA Chemicals Loss Prevention Standard 6.12**, Rev. No. 5, December 2006.

[5] "Standard for Water Spray Fixed Systems for Fire Protection", **NFPA 15**, 2007 Edition.

[6] Center for Chemical Process Safety, *Independent Protection Layers and Initiating Events,* American Institute of Chemical Engineers, New York, NY, pending 2010.

6TH GLOBAL CONGRESS ON PROCESS SAFETY

Chronic Pain Resulting from Your Relief Device Management Systems

David A. Moore
President & CEO
AcuTech Consulting
dmoore@acutech-consulting.com

Prepared for Presentation at
American Institute of Chemical Engineers
2010 Spring Meeting
6th Global Congress on Process Safety
San Antonio, Texas
March 22-24, 2010

UNPUBLISHED

Chronic Pain Resulting from Your Relief Device Management Systems

David A. Moore
President & CEO
AcuTech Consulting
dmoore@acutech-consulting.com

Keywords: layer of protection analysis, LOPA, process hazard analysis, PHA, relief device, safeguard

Abstract

With the increasing use of Layer of Protection Analysis (LOPA) as a semi-quantitative method of assessing process risk, LOPA practitioners are asked to assign credit to numerous safeguards including relief devices. Relief devices are only as strong as the initial design of the component and the systems that manage subsequent operation and maintenance of the devices. If the management systems for Process Safety Information, Management of Change, PSSR, Operating Procedures, and Training are not robust enough to ensure diligent management of relief devices, then credit may be applied liberally during a LOPA study resulting in a residual risk lower than the actual risk. This paper analyzes the effects that various PSM management systems have on relief devices, incorporates examples from LOPA studies that reinforce how relief device credits can be liberally applied, and suggests questions that LOPA practitioners can ask about management systems to apply appropriate credits.

1. Introduction

Several Chemical Safety Board (CSB)-investigated incidents from the last decade, including Synthron, T2, and Catalyst Systems Inc., highlight management system failures associated with relief devices that contributed to catastrophic losses. In addition to these incidents, there are likely countless other incidents and near-misses that have occurred in industry that identify inadequate relief device management systems as a contributing factor. If investigated further, back to the specific node and hazard identified in the PHA or LOPA, a relief device may have been identified as a safeguard with credits assigned to it resulting in a reduction of the scenario risk. Many PHA facilitators and LOPA practitioners may question the appropriateness of a relief device for a certain scenario, but others may not due to their training or experience. Some may begin a study by questioning the relief devices, but may soon stop based on the confidence exuded by the team participating in the study. Process safety information about the chemistry of the process or actual design basis for the relief device may not be known or well understood by the PHA or LOPA team. The relief device in question may be robust enough for the specified scenario discussed at the time of the hazard study, but the device may not be maintained appropriately after the team has disbanded. Or, perhaps, process changes were made after the hazard study and applicable hazard scenarios were not reviewed for the impact of the change. Relief devices, when appropriately sized, designed and maintained through the application of robust management systems are safeguards which can provide a high degree of protection against overpressure [1].

Over application of LOPA safeguards can occur when one or more of the relief device management systems fails. The management system failure is not likely to be a single event, such as one missed inspection, but a result of a more chronic, system failure to design and implement robust processes to design, operate, and maintain relief devices.

2. Process Safety Information

Designing a relief device can be as straightforward as following methodology outlined in the American Petroleum Institute Recommended Practice 520 or it can be more complex for reactive chemistries that would include thermodynamic analysis and modeling of two-phase flow using the Design Institute for Emergency Relief Systems (DIERS) methodology. Whichever design methodology used, a facility needs to identify design cases for the relief device, including the governing case that determines relief device flow area, set point, etc. That is, those hazard scenarios that could result in a challenge to the relief device resulting in activation of that device. Frequently, these hazard scenarios may include external fire, exothermic reaction, loss of cooling, instrument failure, or others (see API RP 520) but may not always include foreseeable, credible scenarios that should be evaluated.

During a PHA that is followed by a LOPA, hazard scenarios and safeguards are many times identified by the team and fed into a LOPA with little or no review of the integrity of the device design, including:

- identification of inappropriate safeguards given other design parameters of the process
- hazard scenarios identified during the PHA that have not been evaluated when considering the design of a relief device;
- a lack of full understanding of the consequence of the scenario, specifically with respect to reactive hazards

So what is a PHA and/or LOPA team to do in these cases? While relief device design and design basis is included as part of a facility's process safety information, sometimes these records have been lost or not maintained since the original equipment was designed and installed. Where records are available, quite often the team will not realize that a scenario identified in the PHA has not been considered in the documented design basis and will take credit for the relief device. If the relief device design were reviewed for the scenario, perhaps the result would indicate that the relief device is properly sized for the scenario in question. Conversely, it may be found that the relief device is undersized for the overpressure scenario identified by a PHA team. Allowing credit to be given for this safeguard may result in continued residual risk in the field and a missed opportunity during the analysis to flag this residual risk and initiate corrective actions. A few examples are provided to highlight the opportunities that PHA and LOPA teams have when evaluating relief devices safeguards to ensure that the integrity of the design is maintained.

In a first example, a PHA team identified a hazard scenario where a pressure controller malfunctions and reads lower than the actual pressure, resulting in increased pressure in the vessel and an overpressure hazard. The updated P&IDs that the team was using indicated that the rupture disk is set at 73 psig. Seeing a rupture disk in place on the drawing, the team was in favor of documenting this safeguard in the study without evaluating the maximum allowable working pressure (MAWP) of the vessel, information that was also on the P&ID. Upon review, the team realized that the MAWP of the vessel is 65 psig – lower than the rupture disk burst setting. In this case, because the rupture disk is set at a pressure higher than the MAWP of the vessel, no credit should be taken for the relief device and a recommendation should be made to investigate the discrepancy – is there an error on the P&ID? What is actually installed in the field? What do the design documents specify should be installed in the field and on the P&ID? Had the team not had the MAWP readily available or had they not compared the MAWP versus the burst pressure, it is likely that they would have taken credit for this device.

In a similar scenario, the PHA team identified a hazard scenario where a pressure controller malfunctions and reads lower than the actual pressure, resulting in increased pressure in the vessel and an overpressure hazard. The MAWP of the vessel is 65 psig and the rupture disk is set at 63 psig. After discussion, the team is in favor of documenting the relief device as a safeguard without evaluating the specific design scenarios for the relief device. If the team or team leader had reviewed the documented process safety information, it would have been noted that the device was only designed for an external fire scenario. In this case, no credit should be taken for the relief device because the hazard scenario identified – pressure controller malfunction – was not considered as a credible case in the design of the relief device. A recommendation should be made to evaluate this new scenario and review the design case. Perhaps this scenario will not become the worst credible case, but until it is reviewed, it cannot be known for sure. Had the team or team leader not reviewed the documented process knowledge, it is likely that they would have taken credit for this device as a safeguard.

The first two examples identify simple documentation reviews that PHA team leaders and team members can conduct to ensure appropriate credits are taken for relief devices during a PHA or LOPA. When it comes to reactive hazards, simple documentation reviews can be conducted, but the analysis and understanding can also be more complex. In 2002, the Chemical Safety Board published the report "Improving Reactive Hazard Management" [2] in which they cited inadequate hazard evaluation as a contributing factor in more than 40 percent of 167 incidents involving reactive materials. In some of these cases, the hazard was known but the potential magnitude and/or the potential severity was not. In other cases, the initiating event was not properly identified.

In a third example, the PHA team lacked of full understanding of the consequence of the hazard scenario, specifically with respect to reactive hazards. On January 31, 2006, there was a runaway chemical reaction and vapor cloud explosion at the Synthron LLC facility in Morganton, North Carolina that resulted in one fatality and fourteen injuries. According to the CSB report, the facility had received an order for an acrylic monomer that was 12% larger than the normal batch size [2]. While there are other management system failures that contributed to this incident and that will be reviewed further, the focus in this example will be on the lack of complete process safety information. It should also be noted that the CSB report indicated that Synthron had not fully implemented a PSM program.

In this example, a novice team leader or team member, not fully aware of the hazards associated with reactive chemistries, may not identify a 12% increase in batch size, along with adding all the monomer to the initial charge, as a new hazard scenario and questioned the changes in reaction thermodynamics. Had the facility conducted testing to evaluate the thermodynamics of this new batch size, they would have seen, according to the CSB, that the increased temperature in the process was not linear, but orders of magnitude greater (Figure 1). The increased temperature in the modified process exceeded both the cooling capacity of a clean condenser and the combined cooling capacity of a fouled condenser and emergency jacket. Although the PHA

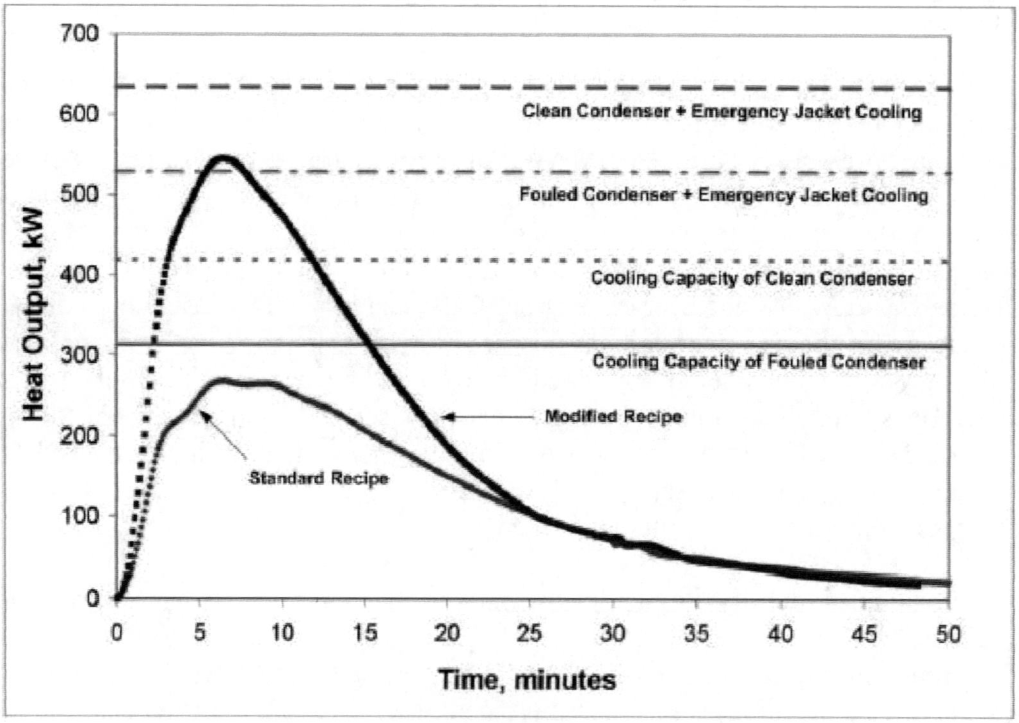

Figure 1 – Synthron reaction calorimetry heating curves for standard (lower) and modified (upper) batch recipe [2]

team may not have the technical expertise to understand all of the ramifications of a change such as described from this incident, PHA teams (and specifically PHA team leaders) should be sensitive to changes such as this and ask questions to team members and others to ensure that safeguards are still appropriate for the scenario given the change. In this case, the PHA team's technical knowledge may have been limited to simply making a recommendation (with an appropriate risk ranking or other priority) that required that the appropriate thermodynamic analysis and/or testing be performed to ascertain whether the relief device is adequate for the 12% increase in batch size.

Documentation of process safety information is the first step in evaluating hazard scenarios and appropriate safeguards during a PHA or LOPA. Team leaders and team members, while well intentioned and often technically educated, may often overlook information that is in front of them or not ask questions of safeguards to ensure that they are appropriate for the hazard identified. The three examples in this section identify some additional questions that team members and team leaders may ask while identifying safeguards to ensure that they are appropriate for the hazard scenarios. Additional questions to consider include:

- Is the set pressure of the relief device lower than the MAWP?
- Is the device installed in clean, fouling, or corrosive service? Has this been taken into account in the design of the relief device?
- Have block valves been installed under the relief device? Are they car-sealed open? What is the operating discipline for respecting car-seals at the facility?
- Where is the relief device discharged to and are there subsequent hazards presented that will require additional review?
- Has the hazard scenario identified in the study been specifically evaluated as part of the relief device design and design basis?
- Are the P&IDs consistent with the documented relief device design basis?
- Are the relief devices installed in the field consistent with the requirements of the documented relief device design basis?
- Are the reactive chemistry hazards known and understood, including but not limited to [2], [3]:
    - Normal and abnormal process conditions, including high/low temperature, too much/too little concentration, too much/too little agitation, batch sequence, reaction time, utility and instrumentation failure, etc.
    - How variations in reactant and catalyst quantity affect the process
    - How variations in concentrations affect the process
    - Rate of heat generation
    - Maximum operating temperature to avoid decomposition, including the acute and accumulated thermostability characteristics of the reactants?
    - Generation of byproducts, including evolution of toxic gas?
    - Contaminants?
    - Available heat removal capacity?

3. Management of Change and Pre-Startup Safety Reviews

The old saying goes "the only thing constant in life is change." In the process industries, changes occur throughout the lifecycle of a process - from the initial conceptual review, during commissioning and through operation. A robust management of change process will identify the safety and health impacts of a change, including any impacts to the relief device design or installation. Frequently, hazard reviews are conducted at various stages of this design process and will review the relief device design and design basis. A robust MOC review process will continue reviewing the relief device designs as the process evolves through commissioning.

When conducting a PHA or LOPA for existing processes, changes made to the process since the last revalidation must be incorporated into the study. While this is often done, a full evaluation of the impact of those changes during a study and how they would affect a relief device safeguard is not conducted since this was supposed to have been done during the MOC process. The following examples highlight relevant technical issues that PHA and LOPA teams have when evaluating relief device safeguards in the context of change.

The first example continues an evaluation of the Synthron incident introduced in the previous section. Let's assume that the MOC process was followed at the facility for the 12% increase in batch size. Upon reviewing the MOC documentation available to the PHA team, the team determined that they can take credit for the relief device safeguard because the MOC had been completed and there was no indication of the change impacting the relief device. Had the team reviewed the relief device PSI along with the MOC, they would have noticed that the last calculations completed for the device were done before the change was made. This means that there is not documentation of calculations having being completed to verify whether or not the 12% increase in batch size has an effect on the relief device design. Only when the calculations are completed can the PHA team really determine whether or not the relief device can be credited as a safeguard.

The second example is related to how current and accurate the P&ID is that is used in a PHA or LOPA. Since the PHA team cannot bring a scale model of the unit into the conference during the study and since it is not practical to bring the team out to the field to conduct the study next to the process, PHA and LOPA teams rely on the P&IDs as their blueprint to the process. In this example, a distillation column has a pressure safety valve installed. Since the last PHA, the unit was shut down for maintenance activities and during the shutdown, while the relief valve was removed from the vessel for bench testing, an isolation valve was installed. The well-intentioned maintenance worker that completed the work did not follow the MOC process and the isolation valve never made it on to the drawing. Since the drawing doesn't match the current field conditions, the LOPA team may take credit for that relief device without considering the potential impact of the isolation valve.

In a third example, formal hazard reviews were not completed during design and installation of a benzoyl peroxide (BPO) drying system. On January 2, 2003, there was an explosion and fire at the Catalyst Systems, Inc. facility in Gnadenhutten, Ohio resulting in one injury and significant damage to the processing building. The CSB concluded that the explosion occurred due to the 200 pound BPO batch operating at or near the thermal decomposition temperature. The CSB noted that Catalyst Systems, Inc. had neither completed a safety review during various phases of the design, commissioning, and operation of the BPO vacuum dryer system, nor a PSSR prior to start-up. Had formal MOC and PSSR systems been implemented for this capital project, the PHA and LOPA team could have evaluated causes of high temperature in the dryer, the consequences of a decomposition reaction in the dryer and whether or not there is adequate venting to alleviate pressure buildup [4].

In the previous section, documentation of the relief device design and design basis was highlighted. With respect to management of change, documentation of modifications to the process, to process conditions, to process chemistry, etc. that have a potential to impact the relief device must be considered and documented accordingly. PHA team leaders and members trust the process safety information presented to them during a PHA or LOPA and may not challenge the conclusions documented in the MOC or challenge the accuracy of other process safety information, including P&IDs. The examples in this section identify some additional questions that team members and team leaders may ask while identifying safeguards to ensure that they are appropriate for the hazard scenarios:

- Were P&IDs verified after major shutdown activities?
- Does the facility or unit have a good operating discipline to red-line or otherwise update P&IDs after changes?
- For reactive processes, have there been any changes since the last PHA that may resulted in [2], [3]:
  - Increases or decreases in process parameters including temperature, pressure, batch size, process time, etc.
  - Quantities, proportions, and sequencing of reactor feeds
  - Raw material charge rates
  - Changes in raw material specifications
  - Changes in raw material concentration
  - Conditions that could cause initiator or monomer accumulation
  - Conditions that could affect the deactivation of monomer inhibitors or stabilizers
  - Modified reliability of short stop or reaction inhibiting chemicals
- Where changes have been made to a reactive process, has chemical testing been completed in an effort to identify and evaluate new hazards from process changes?
- Have the modifications to process chemistry been evaluated for their impact on relief devices, including:
  - Runaway reaction potential
  - Thermal stability and thermal decomposition
  - Potential forseeable contaminates
  - Gas evolution

4. Operating Procedures & Training

Detailed procedures and knowledgeable employees are critical to operating and controlling the process and are important to the PHA and LOPA. The depth of knowledge that an employee working in a process unit has is dependent on numerous factors including their formal education, on the job training, tenure on the job and in the process industries, and continuing education. In the 2002 CSB Report "Improving Reactive Hazard Management" [2], the CSB indicated that "more than half of the reactive incidents, where some causal information is available, are attributed to inadequate operating procedures and training".

Personnel working in the process industries must have an understanding of the hazards of the process and actions they can take to ensure safety. Training is often conducted at levels from the operators and maintenance workers to the process engineers and chemists to establish a baseline understanding of process hazards. However, this training may not necessarily focus on the design, operation, and maintenance of relief devices or of the pressure related hazards of the process. Certainly, it is not suggested that every person in the plant become experts in API RP 521 or the DIERS methodology. However, during the PHA and LOPA it is necessary to have someone that is able to the team who understands and is able to interpret the relief device design and design basis, as well as the process chemistry for reactive hazards. In addition, providing a broader, high level understanding of relief devices to technical, operations, and maintenance personnel could possibly enhance the implementation of other management systems associated with relief devices, including process safety information and management of change.

Operating procedures are another key component to providing hazard awareness to facility personnel. Procedures that include the following components are useful not only as training tools, but also as documents for use during operations and PHA studies [2]:

- appropriate cautions and warnings for pressure related hazards
- consequences of deviations and abnormal operations
- appropriate responses to control or mitigate their effects

On December 19, 2007, an explosion – equivalent to 1,400 pounds of TNT – at the T2 Laboratories, Inc. facility in Jacksonville, Florida resulted in four fatalities and thirty-two injuries, twenty-eight of which were members of the public. Debris from the explosion was found a mile away and buildings within one-quarter mile of the facility were damaged [5]. The CSB concluded that the root cause of the incident was that T2 did not recognize the runaway reaction hazard associated with the chemical being produced. In the CSB video for the T2 explosion, process safety expert Dennis Hendershot stated that "a large portion of the incidents have occurred because the people who are actually running the process were not aware either of the existence of the reactive chemistry hazard or of the potential magnitude of the consequences associated with the chemical reaction hazard…a lot of the problem is education and awareness." Had T2 been aware of the reactive chemistry hazard and the magnitude of the consequence, they may have identified – possibly during a PHA or LOPA – that the relief device was incapable of relieving the pressure from a runaway reaction [5].

Personnel that work within process units that have pressure relieving devices installed must understand the hazards and how the pressure relief device can act as a safeguard to ensure safety. This understanding is completed through training of both technical personnel, including process engineers and chemists, and of operations and maintenance personnel. For those process units with reactive chemistry, the training and understanding extends to needing to have a working knowledge of the potential for runaway reactions and of the complex intersection of chemical properties and process-specific conditions [3]. This training is essential for day to day operations and maintenance of the process, and will provide value during the PHA and LOPA study as the participants will be able to better understand the hazards and when credits are appropriate to apply.

The T2 incident, along with the Synthron incident described earlier, highlight the importance of understanding the pressure related hazards associated with the process, including that of reactive chemistry. While some of this knowledge may be known and documented as part of a facility's process chemistry information in their process safety information, it is necessary that someone be trained to a level to understand and interpret the information contained in that documentation. The example in this section identified some additional questions that team members and team leaders may ask while identifying safeguards to ensure that they are appropriate for the hazard scenarios:

- Is there a thorough knowledge and understanding of the process chemistry, including:
    - Runaway reaction potential
    - Thermal decomposition
    - Thermal stability
    - Contaminates
    - Gas evolution potential
    - Presence of and reliability of short stop vessels
- Has formal reactive hazards training been provided to PHA and LOPA team members? [2]
- Is there recognition of the potential change in process hazards when modifications to product recipes, batch sizes, or reaction conditions are made? [2]
- Have PHA and LOPA team members been trained in the nature of the reactive hazards, including the process safety margins? [2]
- Do operating procedures include the following and has this information been used to develop hazard scenarios:
    - appropriate cautions and warnings for pressure related hazards
    - consequences of deviations and abnormal operations
    - appropriate responses to control or mitigate their effects

7. Conclusion

Numerous incidents in industry have occurred due in part to inadequate relief device management systems. While a robust process safety program provides several opportunities to identify and correct deficiencies in a relief device management system, including incident investigations, compliance audits, and mechanical integrity programs, these opportunities are often missed prior to completing a PHA and LOPA. The net result is overconfidence in the designed, installed, and maintained relief device that provides a residual risk higher than expected.

Relief device management systems, including process safety information, management of change, pre-startup safety review, operating procedures, and training were evaluated to highlight some of these missed opportunities and suggest questions that PHA facilitators and LOPA practitioners can ask during a study to increase confidence in an individual device prior to taking credit.

It is often the case that the PHA team will not know the answer to these questions, such as what the design basis was for the relief device, during the study. In this case, the PHA team leader has several options:

1) assign a team member to go find out the information and report back to the team
2) if the information is not available, a recommendation should be entered into the PHA to find it or create it, and then evaluate it with respect to it being an effective safeguard.

If the team knows the information isn't available – such as from previous engineering work or an audit – the team should skip directly to option 2). If the PHA team makes regular recommendations regarding relief device design documentation during the PHA or LOPA, then the recommendation may be generalized for the whole unit under the study.

When relief devices have been appropriately sized, installed, and maintained, and when facility personnel have an understanding of the process-related hazards that may challenge the device, relief devices can provide a high degree of protection against overpressure. The PHA and LOPA provide opportunities to verify this information prior to taking credit for the device and having conclusions drawn of the residual process risk.

8. References

[1] CCPS. (2001). *Layer of Protection Analysis - Simplified Process Risk Assessment.* New York.

[2] CSB. (2007). *Case Study - Runaway Chemical Reaction and Vapor Cloud Explosion - Synthron, LLC.* Washington D.C.

[3] CSB. (2002). *Improving Reactive Hazard Management Report.* Washington D.C.

[4] CSB. (2003). *Case Study – Fire and Explosion: Hazards of Benzoyl Peroxide.* Washington D.C.

[5] CSB. (2009). *Investigation Report – T2 Laboratories, Inc. Runaway Reaction.* Washington D.C.

Quantitative Risk Assessment Case Study for Organic Acid Processes

R Wayne Garland
Technical Associate
Corporate Health, Safety, Environment and Security
Eastman Chemical Company
P O Box 511 Bldg 54D,
Kingsport, TN 37662,
423-229-5518,
wgarland@eastman.com

Prepared for Presentation at
American Institute of Chemical Engineers
2010 Spring Meeting
6th Global Congress on Process Safety
San Antonio, Texas
March 22-24, 2010

UNPUBLISHED

Quantitative Risk Assessment Case Study for Organic Acid Processes

R Wayne Garland
Technical Associate
Corporate Health, Safety, Environment and Security
Eastman Chemical Company
P O Box 511 Bldg 54D
Kingsport, TN 37662
423-229-5518
wgarland@eastman.com

Keywords: risk assessment, quantitative risk assessment, case history, acetic acid, organic acids

Abstract

Acetic acid is not a chemical covered by the OSHA PSM regulation, as its flash point is above the 100 F threshold for flammables; however, it is by no means a non-hazardous chemical. Eastman Chemical Company has a large number of processes containing acetic acid and related organic acids, and as a corporation, we wanted to assure that risk for these processes was at an acceptable level. To facilitate these risk assessments, flash point data was generated and a qualitative analysis for a large number of process vessels was performed. Selected Quantitative Risk Assessments (QRA) were then performed using fault tree/event tree methodology, comparing the level of risk to industry and government benchmarks as well as the risk of common daily activities. The highest contributors to risk were identified, analyzed with respect to the current protective measures, and practical improvements recommended which further reduced risk to a low level.

Background and Objectives

At Eastman Chemical Company, there are a number of production systems dating back to the 1930's that utilize acetic acid and related organic acids as solvents. These processes were constructed according to engineering standards and guidelines of their time. They are not covered by the OSHA PSM regulation. These systems include several hundred vessels of varying size, type and pressure rating. Routine safety studies per company policy require confirmation that the process vent systems for these vessels are working satisfactorily to prevent flammable concentrations of chemical vapors from escaping into building areas. They also require that systems meet our company criteria for an acceptable level of risk.

However, flash point and flammability data was limited. Nearly all existing published flash point data was for pure components only. The vessels studied contained mixtures of acetic, propionic and butyric acids with either water or the corresponding anhydride. Very little data existed about flash points or flammability limits for these mixtures and the data that did exist was inconsistent. For example 90% acetic acid:10% water had reported flash points 126-142 F and 80% acetic acid:20% water had reported flash points 145-159 F. Using pure component data alone would have been overly conservative.

A significant part of this assessment was to determine whether adding nitrogen blanketing (inerting) to any of our existing vessels that do not currently have it was appropriate. As a part of this work, our company guideline for inerting a vessel (if the liquid is likely to be above its flash point) was revisited in order to ensure consistency with other, external engineering practices (e.g. NFPA 69).

The objectives of this study were:

1. Survey all processes where organic acids (acetic, Propionic and butyric) are used as solvents
2. Determine vessel operating temperatures versus flash point
3. Analyze current protective measures
4. Determine risk
5. Recommend changes as needed to reduce risk to a very low level

Approach

Factors influencing t he selection of hazard evaluation techniques are discussed in *Guidelines for Hazard Evaluation Procedures (ref 1)*. This reference lists the following factors for selection of hazard evaluation techniques as follows:

- Motivation for the study
- Type of results needed
- Type of information available to perform the study
- Characteristics of the analysis problem
- Perceived risk associated with the subject process or activity
- Resource availability and analyst/management preference

Our study was on a process already in operation and was not driven by regulatory requirements, but rather to get a general hazards list. Ultimately a list of safety improvement opportunities was generated, allowing for a ranking of hazards and ultimately, a more detailed hazard review based on this ranking.
Criteria for risk assessments are discussed in *Guidelines for Developing Quantitative Safety Risk Criteria*, (ref. 2). This includes a comparison of qualitative and quantitative approaches. Based on predicted consequences, a hazard identification or qualitative analysis is adequate for low to moderate consequences but a detailed quantitative risk assessment (QRA) may be justified for higher consequences. This reference states that "each layer provides more detailed information and reduces the uncertainty associated with results, but requires greater resources. Analysis it

typically performed at the level necessary to obtain the appropriate information for risk-based decisions." There is also a discussion of a combined strategy (a tiered approach) utilizing multiple layers of assessment. This was the approach taken. We focused initially on potential consequences for each group of vessels, and then used this information to determine whether a qualitative or quantitative risk assessment was appropriate.

Generally, a quantitative risk assessment (QRA) is most useful for:
1. Situations where there is a large capital cost for potential safety improvements and some type of cost/benefit analysis needs to done. The QRA can help provide an estimate of benefits.
2. Situations where decisions must be made among a number of alternatives.
3. Highly hazardous processes. For less hazardous processes it is more difficult to justify the time commitment required for a QRA study.
4. Situations where the emphasis is on avoidance of major incidents.
5. Situations where the objective is to improve the safety of a process.

Based on these criteria we felt that a QRA could be useful for some subset of our process vessels.

To execute this approach, a team was formed, consisting of knowledgeable operations, engineering and lab personnel. To aid in the analysis, ABS Consulting (formerly JBF Associates) was contracted for guidance in the earlier stages and to perform the detailed QRA calculations in the last stage.

Flash Point Models

The team felt that an accurate assessment required experimentally validated data on chemical flash points. Available data was limited primarily to pure components and there were significant discrepancies among various sources. Production systems contained mostly mixtures - organic acids with water or anhydrides. Assessments using pure component data would have been overly conservative and would have resulted in major process changes. To generate more accurate data we ran a four component designed experiment for organic acid: water mixtures. Two different models were required to describe the results due to boundary issues. Excellent correlations were achieved. Results from this study helped us achieve our safety objectives with fewer process changes and lower capital expenditures.

The first step was to generate flash point data for the mixture compositions that we had in our processes. To generate more accurate data we ran a four component designed experiment for organic acid: water mixtures. Two different models were required to describe the results due to boundary issues. Excellent correlations were achieved. We developed models for flash point boundary and flash point values for mixtures of acetic acid, propionic acid, butyric acid and water. This work has been previously reported in *Evaluating Vent Manifold Inerting Requirements: Flash Point Modeling for Organic Acid-Water Mixtures*, (ref. 3)

Methods used for flash point were Setaflash, Tag closed cup (TCC) and Pensky-Marten. Primarily the Pensky-Marten method was used due to the viscosity range of our samples. The flash point boundaries [above the stated water concentration no flash was observed up to the limit of the test (225 F)] were:

Acid Mixture Flash limits
- Acetic/water - 26% (wt) water
- Propionic/water - 23% (wt) water
- Butyric/water - 4% (wt) water

Since all our processes operate in this temperature range, this effectively ruled out the possibility of flashing for all process vessels containing water concentrations above these levels. Vessels that met these conditions were therefore identified as lower risk. We also developed data for acid-anhydride mixtures and were able to similarly identify vessels containing these mixtures as lower risk.

Since the flash models were developed using laboratory mixtures, we verified results by testing actual production samples which in some vessels contain additional components in low concentrations, such as cellulose acetate and salts. We found that these additional components depressed the flash points as much as 10 degrees F. Even taking this flash point depression into consideration, we were able to classify a large number of vessels as lower risk due to no flash under our operating temperatures and concentrations.

We also utilized this new flash point data in a review of our electrical classifications.

Experimental Data

The team felt that we needed additional data and understanding for several other critical parameters. We investigated several experimental data issues for organic acid (primarily acetic acid) mixtures with water. First, in order to assure that we were taking meaningful measurements, we needed to know the amount of time required for an acid mixture to reach liquid-vapor equilibrium. Most processes studied contain at least some water and we didn't know whether they were operating in an equilibrium state or a dynamic state. Data plotted in Figure 1 indicated that about 30 minutes is required to reach equilibrium vapor concentrations. These results were utilized to design the remainder of the experiments, along with a typical range of temperatures and concentrations.

Further investigate liquid-vapor equilibrium investigations were attempted by measuring the composition in the vapor spaces of actual process vessels with an Industrial Scientific gas monitor; however technical problems were encountered (in particular, condensation in the measurement system) that we could not solve in the short term.

Ignition energies were then explored. Figure 2 illustrates the expected impact of water on Minimum Ignition Energy (MIE) of the mixtures, demonstrating the expected positive influence with higher water concentrations in the vapor space resulting in higher MIEs. Temperature and agitation were also included as variables (many of the process vessels studied are agitated). Experimental results for 100% acetic acid are listed in Table 1 and results for acid-water mixtures in Table 2. They are plotted together in Figure 3. As expected, there is a positive correlation with water concentration and the data shows that the effect is very pronounced. For 100% acetic acid, a negative correlation with temperature was observed; however for mixtures with water a positive correlation with temperature (based on our limited data) was seen, presumably due to the lower volatility of water. Water concentration in the vapor phase would be expected to be higher than in the liquid phase. It appears that in our range of operating temperatures and concentrations the effect of water is larger than the effect of temperature.

These organic acids and their mixtures with water have higher MIEs than common solvents. High concentration acetic acid has the lowest MIE and is in the borderline region where static electricity cannot be ruled out as a potential ignition source, although the nearly 80 year operating history of these units, there has never been an incident where static electricity was identified as the ignition source.

We also wanted to know the energy that we could expect in a deflagration (Kg) expressed as rate of pressure rise and maximum pressure. Data for 100% acetic acid is listed in Table 3 and plotted in Figure 4. There is no data on acid-water mixtures, but a negative correlation with water concentration would be expected, since any water present would reduce the percentage of vapor space within the explosive range. This is illustrated in Figure 5. Therefore 100% acetic acid was used as the worst case.

Figure 1

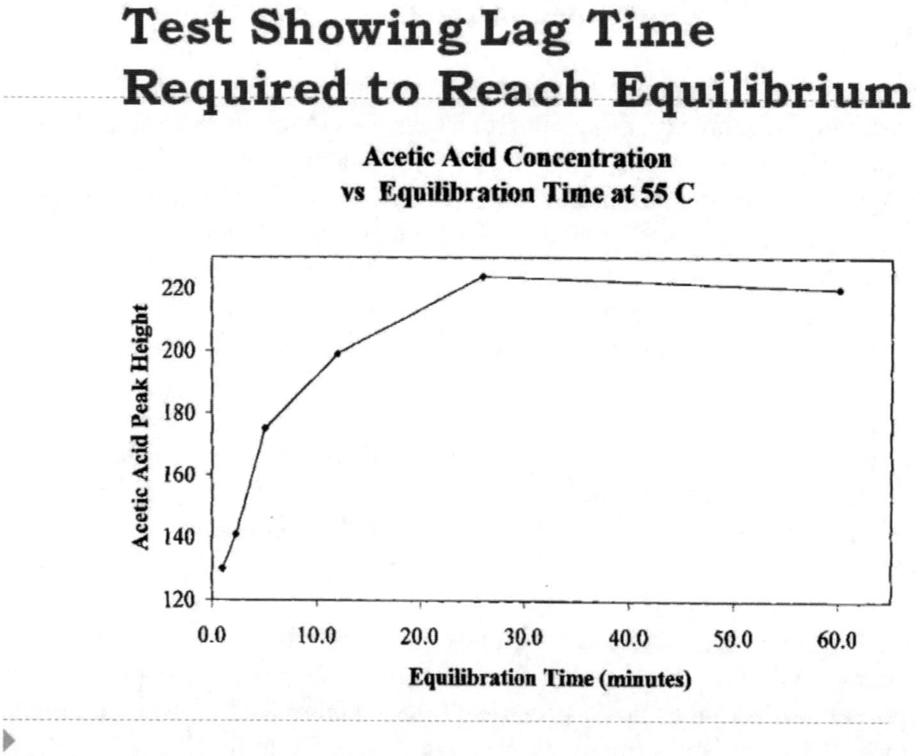

Figure 2

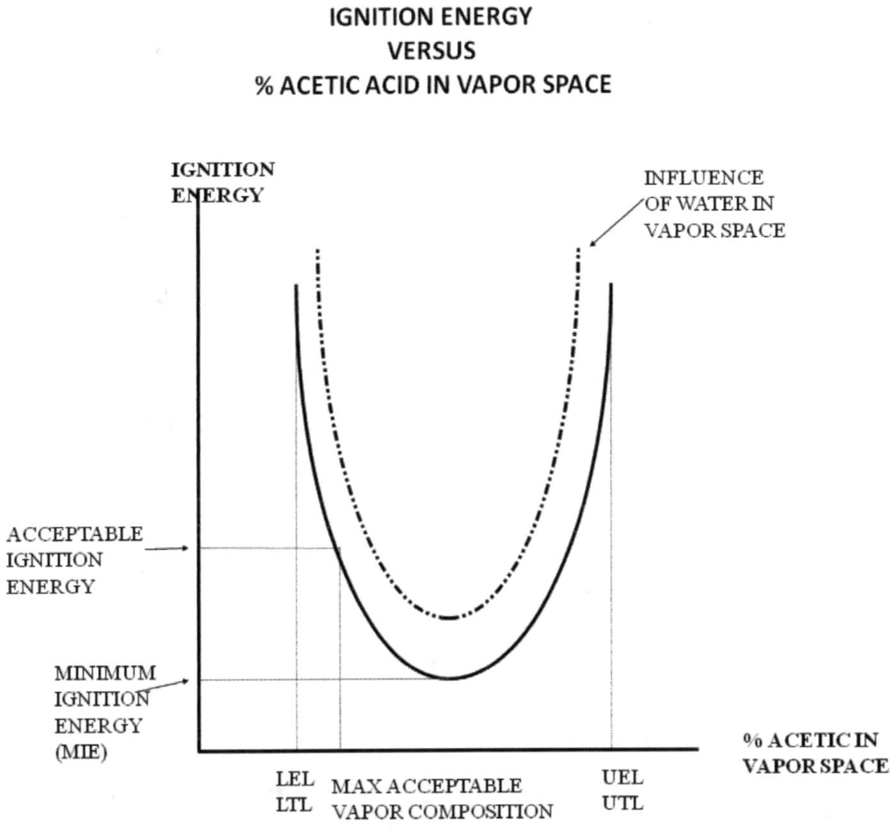
IGNITION ENERGY
VERSUS
% ACETIC ACID IN VAPOR SPACE
IGNITION ENERGY
INFLUENCE OF WATER IN VAPOR SPACE
ACCEPTABLE IGNITION ENERGY
MINIMUM IGNITION ENERGY (MIE)
% ACETIC IN VAPOR SPACE
LEL
LTL
MAX ACCEPTABLE VAPOR COMPOSITION
UEL
UTL

Table 1 – Ignition Energy Tests for Acetic Acid

Material	Test Temperature, C	Agitation Time, min	Ignition Energy, mj
Acetic Acid, 100%	37	None	>500
Acetic Acid, 100%	37	5	>500
Acetic Acid, 100%	37	30	>500
Acetic Acid, 100%	48	None	>500
Acetic Acid, 100%	48	5	>500
Acetic Acid, 100%	48	30	400-500
Acetic Acid, 100%	55	None	>500
Acetic Acid, 100%	55	5	25-50
Acetic Acid, 100%	55	30	20-25
Acetic Acid, 100%	60	10	15-20
Acetic Acid, 100%	60	30	5-10
Acetic Acid, 100%	65	30	5-10
Acetic Acid, 100%	70	30	5-10

Table 2 – Ignition Energy Tests for Acetic Acid Mixtures

Material	Test Temperature, C	Agitation Time, min	Ignition Energy, mj
Acetic Acid + 10% Water	46	None	>500
Acetic Acid + 10% Water	55	None	>500
Acetic Acid + 20% Water	55	None	>500
Acetic Acid + 20% Water	60	None	>500
Acetic Acid + 20% Water	60	30	5-10
Acetic Acid + 20% Water	70	30	25-50
Acetic Acid + 10% Anhydride	60	30	10-25
Acetic Acid + 10% Anhydride	60	30	5-10
Acetic Acid + 10% Anhydride	70	30	50-100

Figure 3 MIE of Acetic Acid and Acetic Acid Mixtures

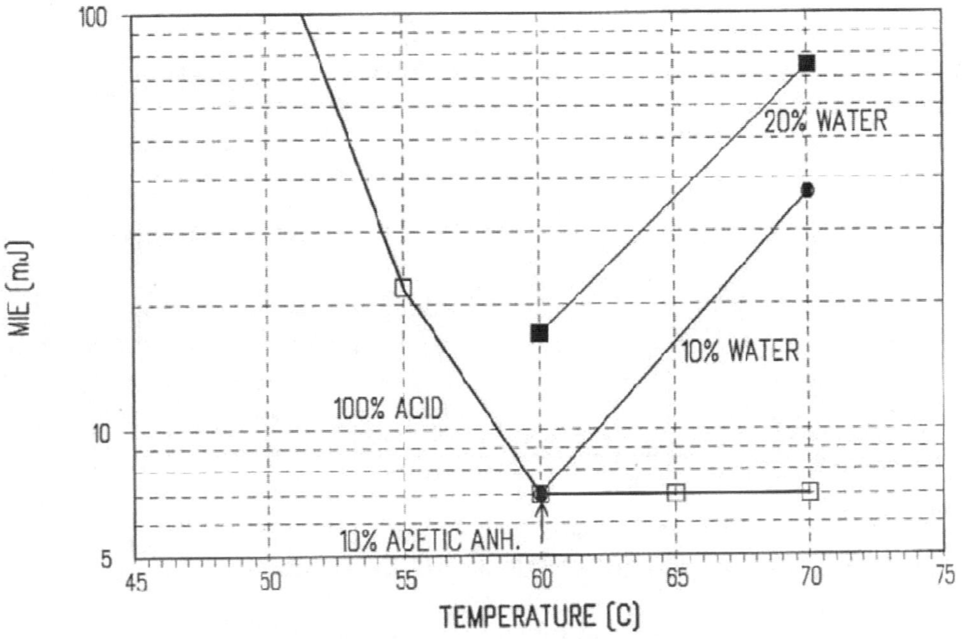

Table 3 – Explosion Test Data

Material	Test Temp, C	Hold Time, Min	Pmax, bar	dP/dt, bar/sec
Acetic Acid, 100%	55	30	2.4	16
Acetic Acid, 100%	55	30	0	0
Acetic Acid, 100%	55	45	0	0
Acetic Acid, 100%	55	45	2.1	2
Acetic Acid, 100%	55	60	0	0
Acetic Acid, 100%	55	60	3.3	30
Acetic Acid, 100%	55	60	0.2	1
Acetic Acid, 100%	55	60	3.7	36
Acetic Acid, 100%	60	60	3.9	18
Acetic Acid, 100%	70	60	4.7	45

Figure 4 – Maximum Pressure and dP/dt for 100% Acetic Acid

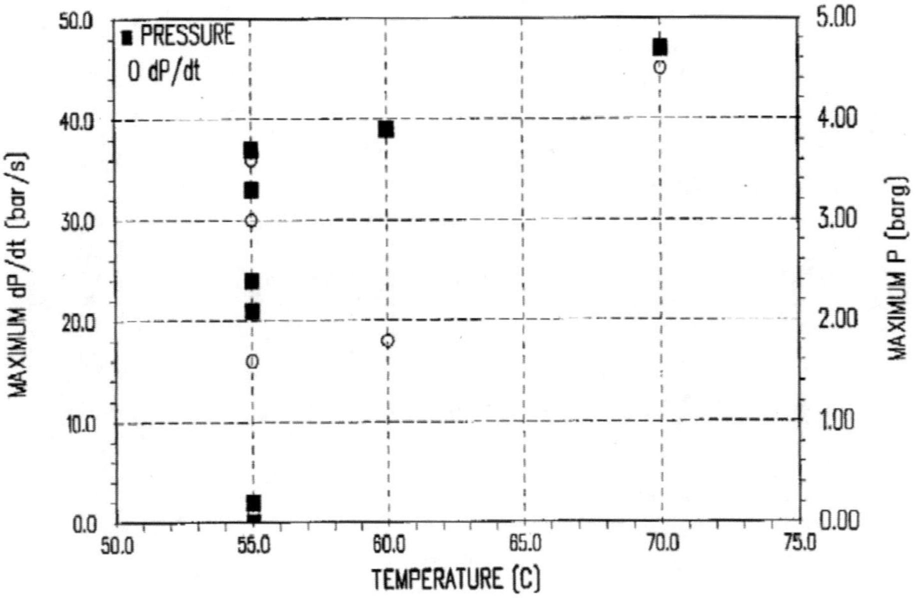

Figure 5

THE GRAPHIC ON THIS PAGE WAS NOT AVAILABLE ON THE E-MEDIA VERSION OF THIS PAPER.

Qualitative Study

Utilizing all of the experimental data, a qualitative study based on hazard identification/evaluation and consequence analysis was performed. This included all vessels that had not already been identified as lower risk due to absence of flash in the flash point testing data.

A number of factors were included:

- Size of Vapor Space
- Vessel Operating Temperature
- Vessel Operating Pressure
- Susceptibility of People (how often people are in the area)
- Accident History
- Sources of Ignition
- Control Levels and Reliability
- % of Time in Higher Risk State
- Business Interruption Potential (BI)

We used the following parameter weightings.

Parameter	Weight	High	Medium	Low
Size of Vapor Space	4	>2 cubic meters		<2 cubic meters
Temperature	5	>(LTL+10F)	LTL to (LTL + 10F)	(LTL-20F) to LTL
Pressure, P	3	>1.5 atm	0.75 atm < P < 1.5 atm	<0.75 atm
People	5	continuous	frequent	infrequent
Business Interruption	4	>$10 M	$1-10 M	<$1 M
% of Time in Higher Risk	4	>90%	5-90%	<5%
Accident History	3	more than one	one	none
Control Levels	3	none	one	2 or more
Sources of Static Ignition	2	bearings & rotating parts	rotating parts	none or MIE > 30 mj

This qualitative study was a look at potential fire and deflagration consequences and some other risk parameters. It was used to help determine if a QRA was needed for a group of vessels. The steps used in the qualitative study were as follows:

1. Group vessels by function
   Grouping by function means vessels in essentially identical service with virtually the same physical configuration. This also allowed us to consider the context of their location and all nearby vessels, piping, personnel, etc. Operator assignments were also considered when doing the grouping since the ultimate objective was to calculate the risk for individual operators.

2. Perform qualitative study based on weighted parameters

3. Perform limited numerical analysis based on ignition sources.

Based on this analysis a number of additional vessels were placed in the lower or medium relative risk categories. Remaining vessels proceeded through a detailed QRA.

QRA Benefits and Limitations

The primary benefits of a QRA are:

- An in depth study utilizing generally accepted industry "best practice" methodology.
- If a consultant is used, certification by an outside expert.
- A logical, structured approach (usually utilizing fault trees/event trees).
- A more objective, more quantitative estimation of risk associated with equipment and job assignments compared to some methods. These estimates can be compared to generally accepted government and industry benchmarks as well as risk levels for some common activities.
- Improved prioritization of potential hazards so that efforts can be concentrated on areas where they would have the largest benefit. Often a small number of items will be identified which are contributing a large percentage of the risk. Identifying the major sources of risk in a process is perhaps the most important benefit of a QRA.
- A list of recommendations for improving safety of processes. Implementation of some of these recommendations may result in less hazardous electrical classifications and smaller relief devices in certain areas, potentially saving large capital expenditures. These recommendations are one of the major benefits of a QRA.
- Determine processes for which a reduced level of safety review would be appropriate, therefore saving labor.

The primary costs and limitations of a QRA are:

- The QRA process itself is labor intensive and requires a major time commitment.
- Specific data may not be available concerning the probability of some initiating events; estimates may have to be used based on the best expert judgment available. Likewise specific data or adequate modeling capabilities may not exist for credible consequences. Therefore ultimately there may still be a significant amount of uncertainty and some degree of subjectivity in the results. Unless complete specific data is available, order of magnitude accuracy is probably the best that can be achieved. Numbers should be viewed more as relative rather than as absolute.
- A QRA should not be considered a substitute for Best Engineering Practices but rather a supplement to them.
- A study leader well trained in QRA methodology is recommended to assure that the analysis is done properly and in a consistent manner with other QRA studies.

QRA Process

To validate the QRA approach used in this work, a pilot study on a typical system (as shown in Figure 6) was conducted. Particular consideration was given to various ignition sources such as mechanical friction, static electricity and maintenance operations. The pilot QRA was deemed a success because the relative sources of risk were able to be ranked reliably, and therefore used to develop targeted recommendations for reducing risk.

Figure 6

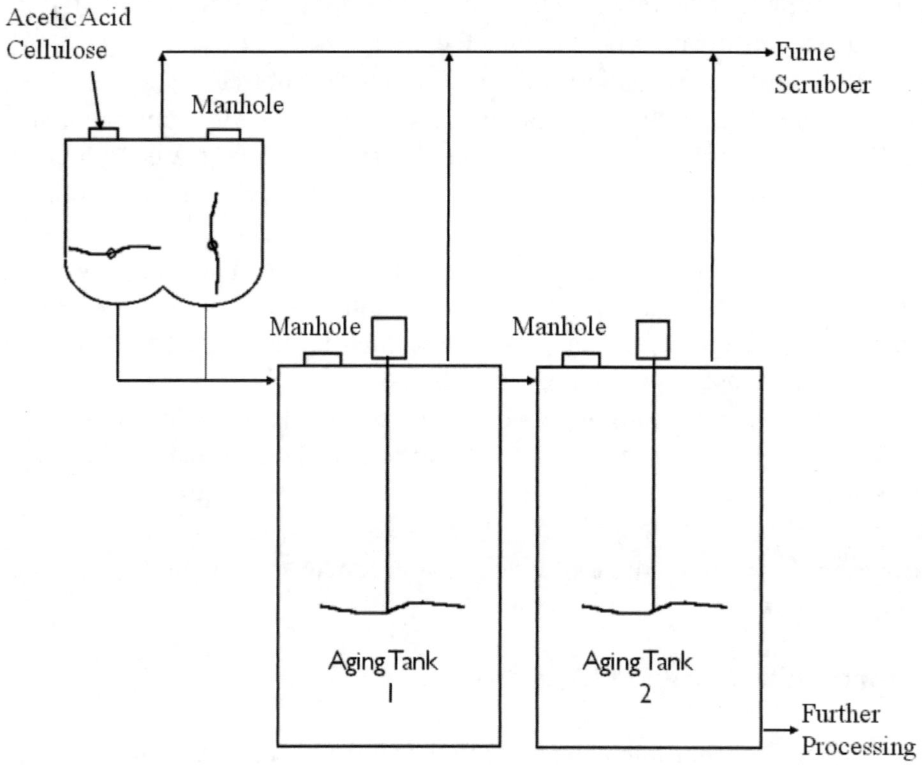

A detailed quantitative risk assessment (QRA) for all vessels that had not been placed into the lower or medium risk categories were then performed using fault tree/event tree methodology with three basic steps:

1. Determine Event Tree Leading To An Undesirable Consequence
2. Consider Severity Of Consequences And Probability Event May Occur
3. Do Relative Comparisons Of Risk

Expected results from the QRA were:

1. Calculated Risk (C x P)

2. Risk Relative To Other Industrial Risk
3. Risk Relative To Public Sector Risk
4. Which Factors Most Influence Risk
5. What Changes Can Be Made That Would Most Influence The Overall Risk

Due to similarities in some of our process vessels, templates were generated and used for multiple studies, which increased efficiency of the analysis.

A contract firm, ABS Consulting, was used to do the QRA calculations. For more specific information about the QRA method see "QRA of Chemical Reaction Systems: The State of the Practice (ref 4.)

Per standard practice QRA results were expressed in terms of total risk to an individual person (generally the person who is most at risk in a particular operation), which was then compared the level of risk to common daily activities as well as to known government and industry benchmarks. One example is the guidelines used by Rohm and Haas Company. They consider a probability of fatality risk $>1x10^{-5}$ to be unacceptable and a risk $<1x10^{-7}$ to be negligible (see "Risk Guidelines as a Management Tool", ref. 5). There are many other published guidelines that have been developed by various companies and government agencies. Hendershot suggests that "A proposed risk guideline should insure that facilities tested against the guideline and passing will be at least as safe as the organization's past safety record. Furthermore, the risk guideline should provide a driving force for continuing improvement of that safety record."

Process Vent System Studies

Due to issues around vent manifold systems raised during the studies, the team felt that we needed experimental data for this as well. Studies were performed, and calculated vent system capacities developed using measurements of flow and pressure at various locations in the header manifolds as well as flow velocities at vessel lids (some of the vessels have lids that can be opened for inspection). Some additional data on the agitator shaft seals was also collected, as well as optimum scrubber system operating parameters.

Conclusions and Recommendations

The risk assessment was deemed a success because utilizing existing and new data, the relative sources of risk were able to be ranked reliably and therefore used to develop targeted recommendations for reducing risk.

Conclusions were as follows:

1. Mixtures of acetic acid and water require about 30 minutes to achieve equilibrium in the vapor space.

2. Flash points of acetic, propionic, and butyric acid mixtures are strongly influenced by the presence of water. Above a critical water concentration they will not flash (up to 225 F) by the Penske-Martens test method.
3. Flash points are depressed as much as 10 F by components such as salts and cellulose acetate in small percentages.
4. Mixtures of acetic acid, propionic acid, butyric acid and water have higher Minimum Ignition Energies (MIE) than many common solvents.
5. Flash point data and a quantitative risk assessment results supports external fire case not being credible for some mixtures containing propionic and butyric acids, as well as most mixtures containing cellulose acetate ("acid dopes").
6. Individual vessels should not be evaluated in isolation but always in the context of their location and all nearby vessels, piping, personnel, etc.
7. At stated operating conditions, flash point data and a quantitative risk assessment results do not support the need for nitrogen blanketing of most vessels containing mixtures of acetic acid with propionic acid, and butyric acid with water. The data also does not support the need for nitrogen blanketing of vessels containing mixtures of acids and anhydrides.
8. Even though the risk assessment done was resource intensive, it was justified by the significant results achieved.
9. Company risk acceptance criteria were met for all the processes studied.

Several improvement opportunities to further reduce risk were identified:

1. Improvements in electrical classifications based on the new data
2. Cooling acid mixtures to lower temperatures in some processes
3. Modifications to process vent manifolds to improve venting of some processes
4. Increased layers of protection against process vent scrubber failure by adding additional interlocks.
5. Enhanced hot work procedures in some specific areas
6. Improved MI program for mechanical seals in one area
7. Inerting of some vessels
8. Reduction in fire hazards and improved fire detection in some areas
9. Agitation interlocks to reduce probability of spark generation when vessels are empty
10. Improved procedures to provide another layer of protection for certain chemical reactivity issues
11. Improvements to grounding in some specific areas
12. Increased layers of protection against overfill for some specific process vessels

Many of these improvements required only administrative changes; some required capital work, especially those involving electrical changes, acid cooling, inerting, process ventilation changes, and fire detection.

References

1. *Guidelines for Hazard Evaluation Procedures (3rd Edition),* Center for Chemical Process Safety, New York, 2008, pp. 176-194

2. *Guidelines for Developing Quantitative Safety Risk Criteria*, Center for Chemical Process Safety, New York, 2009, pp. 20-24

3. *Evaluating Vent Manifold Inerting Requirements: Flash Point Modeling for Organic Acid-Water Mixtures*, R. Wayne Garland and Michael O. Malcolm, Process Safety Progress, Vol. 21, No. 3, September 2002

4. *QRA of Chemical Reaction Systems: The State of the Practice* by Farquharson et.al., JBF Associates, Process Safety Progress, Vol. 16, No. 4, Winter 1997

5. *Risk Guidelines as a Management Tool*, Dennis C. Hendershot, Process Safety Progress, Vol. 15, No. 5, Winter 1996

6TH GLOBAL CONGRESS ON PROCESS SAFETY

Simulating the Consequences of an HF Release and Evaluating the Effectiveness of Safeguards to Reduce those Consequences

Randall J. Hawkins
RRS/Schirmer
6455 South Shore Blvd., Suite 400
League City, Texas
Randy_Hawkins@SchirmerEng.com

Daniel Sheahan
RRS/Schirmer
6455 South Shore Blvd., Suite 400
League City, Texas
Daniel_Sheahan@SchirmerEng.com

Prepared for Presentation at
American Institute of Chemical Engineers
2010 Spring Meeting
6th Global Congress on Process Safety
San Antonio, Texas
March 22-24, 2010

UNPUBLISHED

Simulating the Consequences of an HF Release and Evaluating the Effectiveness of Safeguards to Reduce those Consequences

Randall J. Hawkins
RRS/Schirmer
6455 South Shore Blvd., Suite 400
League City, Texas
randy@rrseng.com

Daniel Sheahan
RRS/Schirmer

Keywords: hydrogen fluoride, quantitative risk assessment, QRA, semi-QRA, consequence, modeling, probit analysis, mitigation

Abstract

A semi-quantitative risk assessment (semi-QRA) is less rigorous than a full-blown QRA, but it necessarily involves some or all of the same elements as that more comprehensive study. A semi-QRA may not deliver a full and complete picture of risk, but it can be quite useful in its own right. A semi-QRA study was recently completed for a portion of a refinery's alkylation process containing hydrogen fluoride (HF), and that study serves as a good illustration of the value of a semi-QRA.

1. Introduction

A fully rigorous Quantitative Risk Assessment (QRA) is often used as the tool of last resort to help bring clarity and objectivity to complex safety issues and to assist in making informed risk-based decisions. A QRA for a process unit necessarily requires the analyst to identify all of the hazardous materials present in that unit, to develop a comprehensive list of postulated release scenarios of those hazardous materials from the unit, to assess the consequences and the frequencies associated with each of those release scenarios, and to blend all of that information together with a number of other factors (particularly meteorology, mitigation, and occupancy) to arrive at an estimate of the overall risk for that unit.

A completely rigorous QRA can be very time consuming and rather expensive. As a result, in many instances, risk-based decisions are made based on the results of less rigorous studies, often referred to as semi-quantitative risk assessments (or semi-QRAs). A semi-QRA may not deliver a full and complete picture of risk, but it necessarily involves some or all of the same elements used to conduct a full-blown QRA … and it can be quite useful in its own right.

2. Conducting a Semi-Quantitative Risk Assessment

A semi-QRA study was recently completed for a portion of a refinery's alkylation process containing hydrogen fluoride (HF), and that study serves as a good illustration of the value of a semi-QRA. In fact, the study highlights the importance of several questions that must be answered in both semi-QRA and full-blown QRA studies, including:

- What types of release scenarios should be considered?
- What range of meteorological conditions should be considered?
- How should one predict the downwind consequences associated with a release of HF?
- How can one translate predictions of HF-concentration-vs.-time into probabilities of fatality for individuals located at downwind points of interest?
- How can one assess the benefits of mitigation (e.g., water spray or water curtain) on the postulated release scenarios?

2.1 Release Scenarios

What types of release scenarios should be considered? Each item of process equipment (vessels, lines, valves, pumps, etc.) in contact with HF could fail and lead to a release of hazardous material, and each failure could yield events ranging from pinhole leaks to catastrophic ruptures. Some type of strategy must be employed to reduce the number of scenarios to be investigated to a manageable level while still producing a representative and credible spectrum of scenarios. A technique to accomplish that goal has been developed using failure frequency information published by the Exploration and Production Forum, and the use of that technique will be illustrated using the HF study as an example.

2.2 Meteorological Conditions

What range of meteorological conditions should be considered? Meteorology is an important parameter in risk assessment, since weather conditions can have a significant impact on the dispersion of emissions of hazardous materials. Meteorology (involving wind speed, stability class, ambient temperature, and other variables) is quite complex, with an almost infinite number of permutations possible. Once again, some type of strategy must be employed to reduce the number of conditions to be investigated to a manageable level. A technique to accomplish that goal has been developed using data collected by the National Weather Service, readily available from an EPA-sponsored web site.

2.3 Predicting Consequences

How should one predict the downwind consequences associated with an HF release? A number of dispersion modeling programs are available. Some are in the public domain and are available at little or no cost (e.g., ALOHA, DEGADIS, or HGSYSTEM), while others are only available commercially (e.g., PHAST, TRACE, or CANARY). All of those programs work reasonably well for almost all chemical species. However, HF is one possible exception. In dry conditions, gaseous HF forms dimers, tetramers, and hexamers upon release to the atmosphere, and unless a dispersion model is equipped to simulate the effects of oligomerization, its predictions may be in error. This semi-QRA study was conducted using PHAST, and that program does contain a special thermodynamics package that is supposed to be especially well-suited for releases of HF.

2.4 Probabilities of Fatality

PHAST can be used to predict an HF-concentration-vs.-time profile for any downwind point of interest. Given that information, what is the best way to assess the consequences that might be delivered to a receptor located at that point? A number of toxic threshold criteria have been published (e.g., TLV, PEL, TWA, STEL, IDLH, and ERPG), but none of those criteria are ideally suited to a semi-QRA study. They are all based on exposure to a constant concentration of material, they are all associated with rather long exposure durations, and the penalties associated with exceeding the criteria are not clear. The consequences assessed in this semi-QRA study involving HF were conducted using the probit method of analysis – a rather sophisticated toxic assessment method that is very well suited to this type of study.

2.5 Benefits of Mitigation

This semi-QRA study was conducted for a refinery with "fire monitors" (water cannons) that could be used to spray water onto the source of an HF release. The owner/operator of the facility is also considering the installation of a water curtain to contain any potential spills. One of the purposes of this semi-QRA study was to quantify the benefits of both mitigation strategies. All of the modeling tools and techniques described above were combined with published information on the effectiveness of water sprays and water curtains to assess the benefits of those mitigation options.

3. Conclusion

The HF study performed must be classified as a semi-QRA because it did not include any frequency assessments, and as a result, it did not actually provide a quantified estimate of risk. Nevertheless, it did provide a very valuable initial characterization of the risk associated with the unit, and it did identify some weaknesses in the current mitigation scheme. In fact, the results of the semi-QRA have been used to develop several ideas for even more effective mitigation strategies.

6TH GLOBAL CONGRESS ON PROCESS SAFETY

Modeling the National Chlorinated Hydrocarbon Supply Chain and Effects of Disruption

Dr. Margaret E. Welk
Sandia National Laboratories
PO Box 5800, Albuquerque, NM 87185-0734
mewelk@sandia.gov

Dr. Amy Sun
Sandia National Laboratories
PO Box 5800, Albuquerque, NM 87185
acsun@sandia.gov

P. Sue Downes
Sandia National Laboratories
PO Box 5800, Albuquerque, NM 87185
psdowne@sandia.gov

Prepared for Presentation at
American Institute of Chemical Engineers
2010 Spring Meeting
6th Global Congress on Process Safety
San Antonio, Texas
March 22-24, 2010

UNPUBLISHED

Modeling the National Chlorinated Hydrocarbon Supply Chain and Effects of Disruption

Dr. Margaret E. Welk
Sandia National Laboratories
PO Box 5800, Albuquerque, NM 87185-0734
mewelk@sandia.gov

Dr. Amy Sun
Sandia National Laboratories
PO Box 5800, Albuquerque, NM 87185
acsun@sandia.gov

P. Sue Downes
Sandia National Laboratories
PO Box 5800, Albuquerque, NM 87185
psdowne@sandia.gov

Keywords: supply chain, modeling, ethylene dichloride, vinyl chloride monomer, polyvinyl chloride

Abstract

Chlorinated hydrocarbons represent the precursors for products ranging from polyvinyl chloride (PVC) and refrigerants to pharmaceuticals. Natural or manmade disruptions that affect the availability of these products nationally have the potential to affect a wide range of markets, from healthcare to construction. Analysis of chemical supply chains is an inherently complex task, given the dependence of these supply chains on multiple infrastructure systems (e.g. transportation and energy). This effort requires data and information at various levels of resolution, ranging from network-level distribution systems to individual chemical reactions. The U.S. Department of Homeland Security (DHS) has tasked the National Infrastructure Simulation and Analysis Center (NISAC) with developing a chemical infrastructure analytical capability to assess interdependencies and complexities of the nation's critical infrastructure, including the chemical sector. To address this need, the Sandia National Laboratories (Sandia)[1] component of

---

[1] Sandia is a multiprogram laboratory operated by Sandia Corporation, a Lockheed Martin Company, for the United States Department of Energy's National Nuclear Security Administration under contract DE-AC04-94AL85000.

NISAC has integrated its existing simulation and infrastructure analysis capabilities with various chemical industry datasets to create a capability to analyze and estimate the supply chain economic impacts resulting from large-scale disruptions to the chemical sector. This development effort is ongoing and is currently being funded by the DHS's Science and Technology Directorate. This paper describes the methodology being used to create the capability and the types of data necessary to exercise the capability, and it presents an example analysis focusing on the chlorinated hydrocarbon portion of the chemical supply chain.

1 Introduction

The U.S. Department of Homeland Security (DHS) has tasked the National Infrastructure Simulation and Analysis Center (NISAC) with developing a chemical infrastructure analytical capability to assess interdependencies and complexities of the nation's critical infrastructure, including the chemical sector. To address this need, the Sandia National Laboratories (Sandia) component of NISAC has integrated its existing simulation and infrastructure analysis capabilities with various chemical industry datasets. The intent of this capability development is to provide consequence analysis of natural and manmade events with respect to the chemical industry and chemical-dependent sectors of the economy [1]. This paper reports on the results of exercising the capability, using a disruption scenario similar to the 2008 Hurricane Ike. The portion of the chlorinated hydrocarbon supply chain evaluated in this study is the ethylene dichloride (EDC) to vinyl chloride monomer (VCM) to polyvinyl chloride (PVC) chemical supply chain. Chlorinated hydrocarbons are used in a wide range of products affecting many industries including PVC polymers, solvents, polystyrene, herbicides, fuel additives, bactericides, cosmetics, sanitizers, anesthetic, refrigerants, and more.

NISAC has developed expertise in the modeling and simulation of very difficult, complex problems for national security over the past decade, and this expertise is being used to provide estimates regarding the disruption consequences due to interdependencies among the critical infrastructures. Figure 1 outlines the types of tools used in the chemical analysis capability development. Each of these analytical methods can be used independently or in combination depending on the analytical needs.

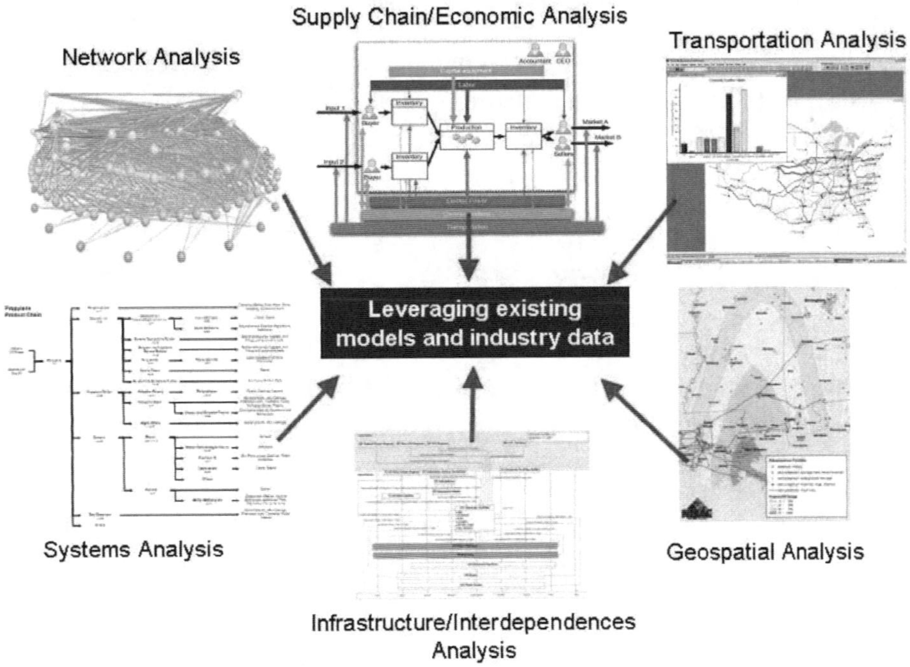

Figure 1: Range of analytical capabilities within NISAC

These capabilities are described as follows:

- The geospatial analysis component is conducted using a tool called FASTMap. FastMap is a geospatial information system (GIS)-based tool that creates maps of chemical plants and other critical infrastructures in disrupted areas and provides data on the names and number of facilities.

- The infrastructure dependency analysis is conducted within a tool called the Fast Analysis Infrastructure Tool (FAIT). FAIT provides data on the dependencies of specific chemical sector components (plants and pipelines) on other infrastructure (such as electric power, telecommunications, and transportation).

- Network analysis is conducted using a tool called Loki. The network model and analysis tool is designed to quickly estimate potential production losses among chemical manufacturing processes (does not trace to final use, just chemical uses).

- Rail transportation analysis is accomplished through a network tool called the Railroad-Network Analysis System (R-NAS). R-NAS represents the U.S. national rail network and estimates the impact to national rail commodity flows given a single disruption or multiple disruptions to the rail system (bridges, rail yards, and so forth).

- The supply chain analysis capability is conducted using the NISAC Agent-Based Laboratory for Economics™ (N-ABLE™), a large-scale supply chain model and tool that allows for the analysis of the impacts to individual firms (production, sales, transportation, and inventories) and the broader supply chain over time (output, shipments, and inventories) resulting from disruptions to firm operations and disruptions to the transportation networks that connect them.

Given all the variety of models that are used in the analytical process, it is imperative that the data used by each tool are consistent. To ensure data integrity across these models, Sandia has created and maintains the NISAC Chemical Data Model (CDM).

2 Analysis Tools

2.1 NISAC Chemical Data Model (CDM)

The foundation of the chemical analysis capability is the NISAC CDM. The effort to build the CDM began in 2007 and is ongoing. The NISAC CDM is composed of many datasets from a variety of vendors. Table 1 lists the various data sources. They are updated annually, where possible, or as often as updates become available.

Table 1: Chemical data sources within CDM

Dataset Name	Provider
World Petrochemicals Program	SRI Consulting
Chemical Economics Handbook	SRI Consulting
Directory of Chemical Producers	SRI Consulting
Oil & Gas Pipelines	NGA HSIP Gold (PennWell)
Oil & Gas Facilities	NGA HSIP Gold (PennWell)
United States Census	U.S. Census Bureau
County Business Patterns 2	U.S. Census Bureau
County Business Patterns Employees Estimation 2007	U.S. Census Bureau
Geographic Names Information System	U.S. Geological Survey
IMPLAN States Summary	Minnesota IMPLAN Group
International Trade Statistics	U.S. Department of Commerce
Refinery Location Data	Argonne National Laboratory*
Commodity Flow Survey, Department of Transportation	Waybill Sample, Surface Transportation Board
Class I Railroad Statistics, Association of American Railroads	Producer Price Index, Department of Labor
E-Plan Emergency Response Information System	U.S. Environmental Protection Agency/U.S. Department of Homeland Security

* Argonne data were updated in 2007 domestic data from the Energy Information Administration (EIA) and foreign data from SRI Consulting.

The data sources provide chemical information, pipeline network information, transportation network information, labor and economic statistics, and geospatial information. The 2008 release of the NISAC CDM was used for the example analysis within this paper.

Figure 2 represents a simplified schematic of how the information within CDM is organized, merged, and stored. Each plant in the database has attributes that identify where it is located, what chemicals are produced and stored, the associated capacities for production and storage, and what production technologies are used at the plant (if known).

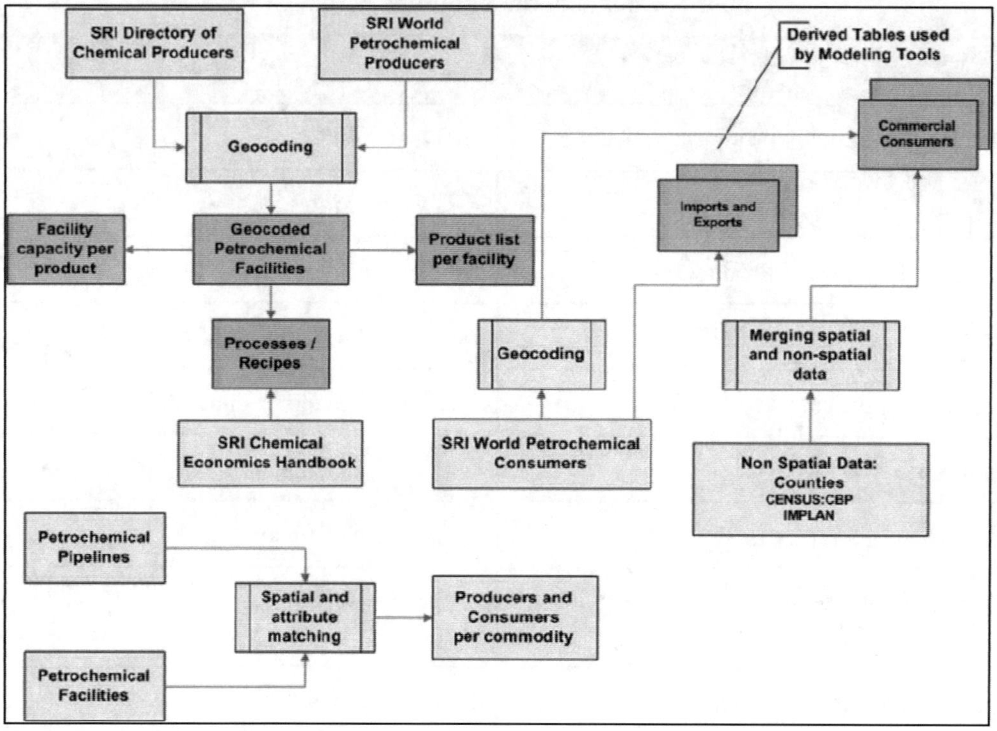

Figure 2: The NISAC Chemical Data Model

Multiple modeling tools use the CDM as their data input. Static modeling tools, including FASTMap, FAIT, Loki, and R-NAS, provide rapid analysis with respect to the area affected, the direct/initial impact to the transportation sector, and dependencies among facilities within an area. The focus here is on the analysis of the supply chain, which includes measures of the above tools and provides the microeconomic assessment and system recovery estimate.

2.2 The N-ABLE™ Supply Chain Representation

The chemical supply chain analysis capabilities are accomplished using the N-ABLE™ microeconomics simulation tool ([2], [3]) to simulate disruptions to the chemical sector from various types of disturbances. N-ABLE™ uses an agent-based enterprise firm structure to simulate production, storage, buying, selling, and consumption behavior. A schematic is shown in Figure 3. Transportation representation is also included (although it is not shown in Figure 3).

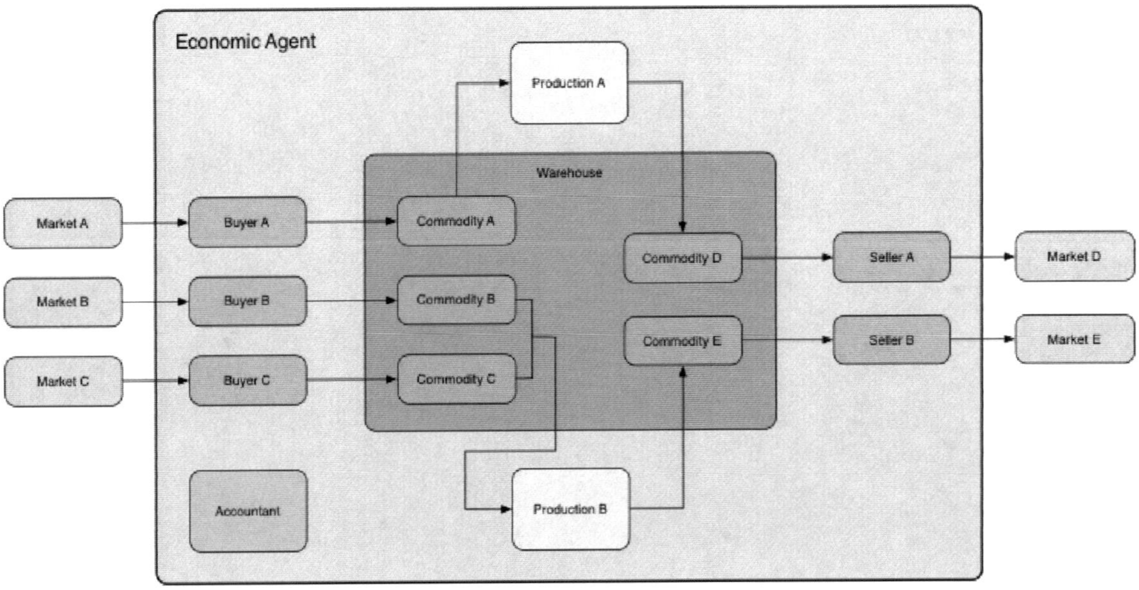

Figure 3: The N-ABLE™ Enterprise Model of an Economic Firm

N-ABLE™ simulation results provide quantitative and qualitative information necessary for consequence analyses. For example, if a hurricane temporarily shuts down a set of chemical production facilities, N-ABLE™ estimates economic impacts resulting from a decreased chemical supply to downstream facilities (e.g., customers of the closed facilities, the customers of the customers, etc.). N-ABLE™ also estimates losses resulting from a decreased demand of input chemicals used by the closed production plants to upstream facilities (e.g., suppliers to the closed plants, suppliers of the suppliers, etc.). These economic impact and loss estimates can be used to measure the systemic impacts to the chemical supply chain from a hurricane.

In addition, N-ABLE™ estimates the time necessary for the system to recover from a disruption. In the case of chemical supply chains, supply interruptions can cascade through many other sectors at different rates. Some downstream consumers will feel the impact of interrupted production immediately, some will not feel the impact until days or weeks later, and some will not feel it at all. Inherent in N-ABLE™, is the capability to represent the search for other supply sources when losses occur and any changes in transportation costs associated with the need to use alternate suppliers. The cost estimates associated with the recovery and adaptation processes are crucial to calculating the total recovery effort of the chemical supply chain following a hurricane.

For the purposes of this paper, N-ABLE™ has been applied to the portions of the petrochemical and chlor-alkali supply chains that feed the chlorinated hydrocarbon industries. Figure 4 is map showing the general location of the ethylene dichloride (red-markers) and VCM/PVC (blue-markers) facilities within the United States.

Figure 4: General Locations of the Selected Primary Chlorinated Hydrocarbon Facilities

2.3 Example Hurricane Disruption Scenario

This study analyzes the impacts on the chlorinated hydrocarbon section of the chemical industry of a Category 2 hurricane making landfall in the Gulf Coast. The scenario hurricane was patterned after the actual Hurricane Ike (2008), which developed during the early part of September and made its landfall over coastal Texas on September 13, 2008. The storm moved at a projected forward speed of 12 miles per hour (mph), carrying maximum sustained winds of over 100 mph. The storm size was approximately 230 miles, and it was the third most destructive hurricane in U.S. history. Figure 5 shows the projected path of Hurricane Ike early on September 11, 2008.

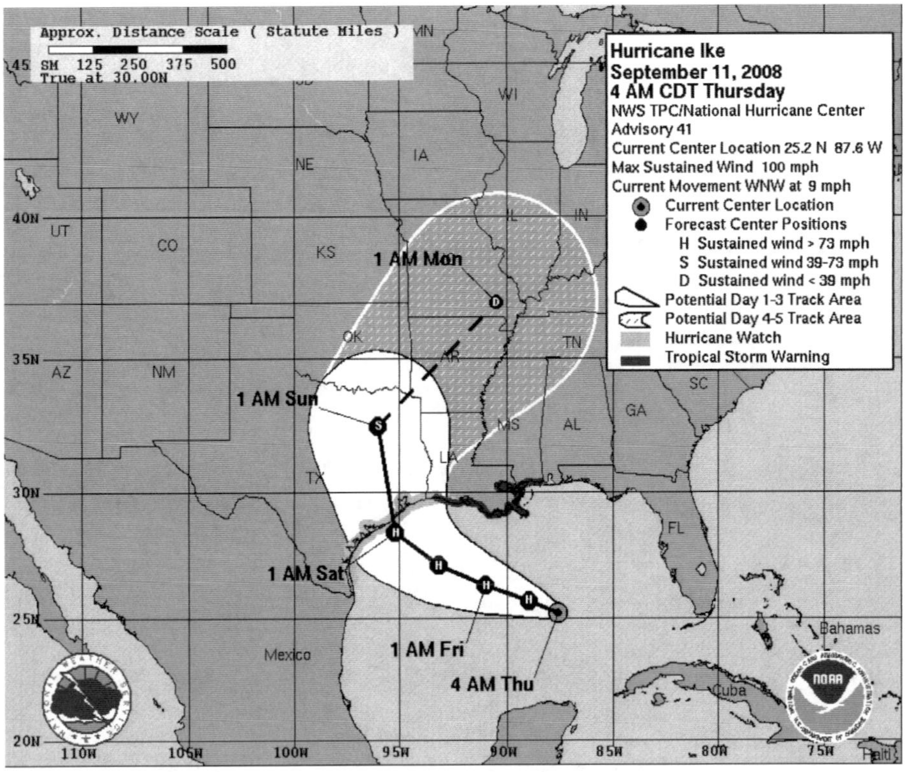

Figure 5: Projected Path of Hurricane Ike as of Early September 11, 2008

We used the scenario storm parameters (trajectory and category) in this analysis to first estimate the damages from wind and surge waters. These damage estimates are translated into areas of probable electric power outage and inland flooding depths. Based on these results, Sandia analysts assessed potential impacts to chemical facilities, petroleum refineries, and the natural gas network for elements directly affected by this scenario storm. The analysis was then extended to those facilities and infrastructures that are not in the path of the hurricane, but are dependent on facilities within the disruption area. Cascading impacts to the chemical industry and petrochemical supply chain were then calculated at a regional, national, and global perspective. Figure 6 shows the estimated disruption area for the scenario hurricane.

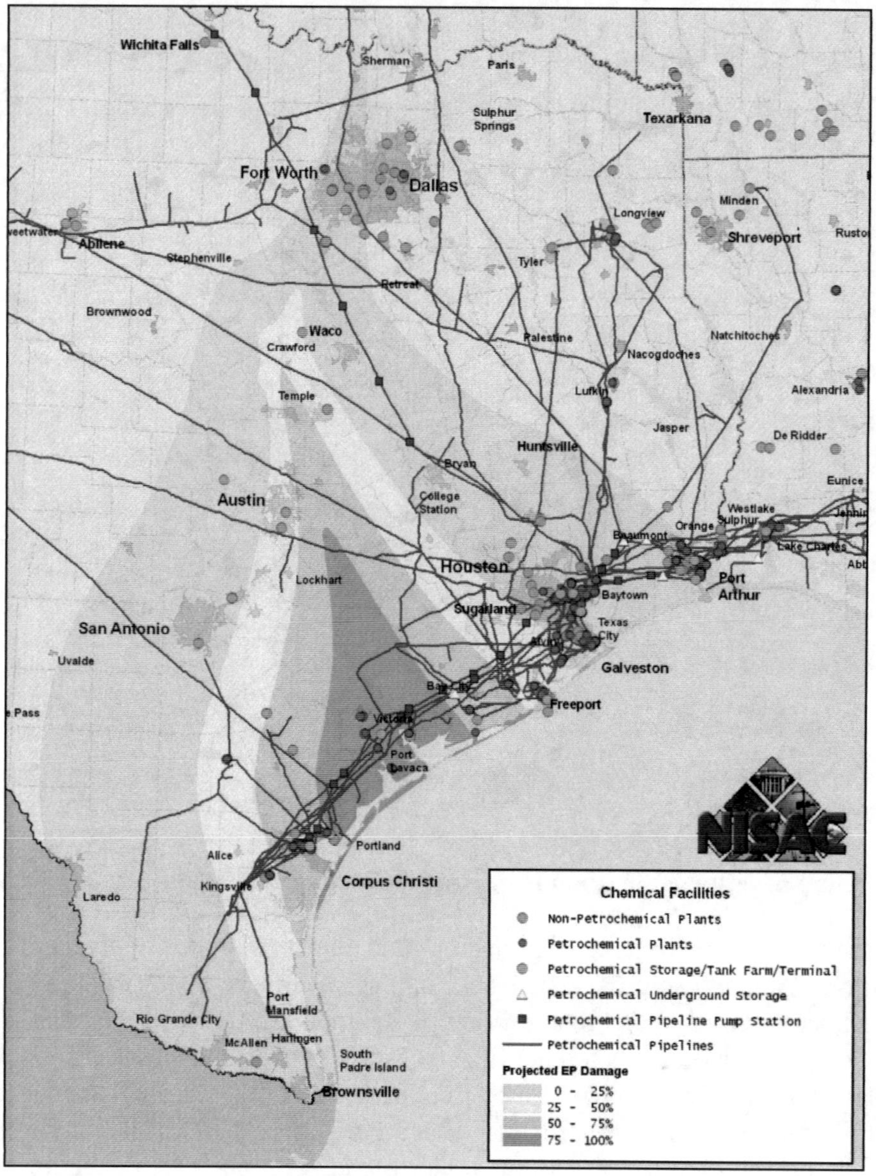

Figure 6: Estimated Disruption Area of the Scenario Hurricane

3 Analysis Results and Discussion

3.1 Ethylene Dichloride, Vinyl Chloride Monomer, and Polyvinyl Chloride

EDC, the VCM, and PVC form a sequential supply chain within the chemical industry. The VCM is the major reactant in this process. Industry data for 2008 indicate 95 percent of EDC was used to produce VCM, and 97 percent of VCM was used in the manufacture of PVC resins. Because of the coupled nature of this supply chain, one would expect a disruption in the EDC supply chain to cascade to VCM and PVC production.

Based on prior feedback from petrochemical industry representatives, hurricane disruptions are assumed to be a minimum of 2 weeks with no physical damage, or 3 weeks to months, if damage occurs [5]. For purposes of this analysis, the disruption time was assumed to be 25 days. The region of the country directly affected by the scenario hurricane includes one or more chemical production plants for the 63 major petrochemical groupings within SRI Consulting's World Petrochemical Program dataset, including the chlorinated hydrocarbons, EDC, VCM, and PVC. Within the estimated disruption region, there are 13 EDC producers and 10 VCM/PVC production plants, which account for a significant contribution to U.S. market. For purposes of this example, all of the facilities were assumed to be down for the entire 25-day disruption.

To evaluate the cascading disruption, we began with EDC and followed the supply chain through to the end consumers. EDC production was approximately 10,640,000 metric tons in 2008 in the United States. [4] The N-ABLE™-simulated capacity utilization of EDC, VCM, and PVC, with respect to time, is shown in Figures 7, 8, and 9, where capacity utilization is defined as the ratio of production level over nameplate capacity. The baseline scenario (non-disruption) and the disrupted scenario are shown on each graph.

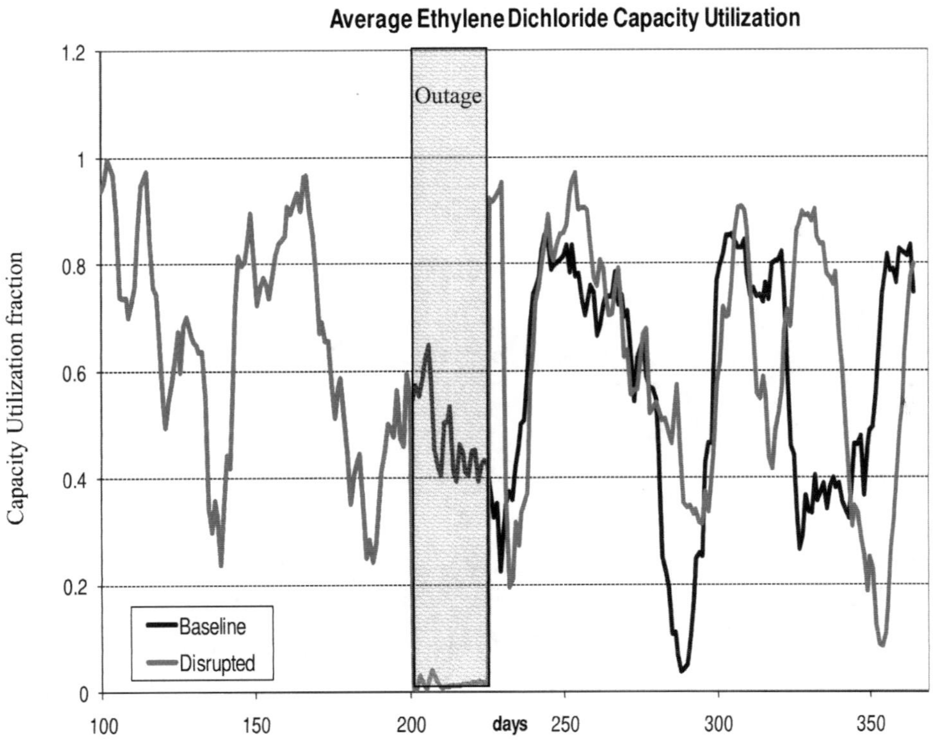

Figure 7: Simulated Results for EDC Capacity Utilization

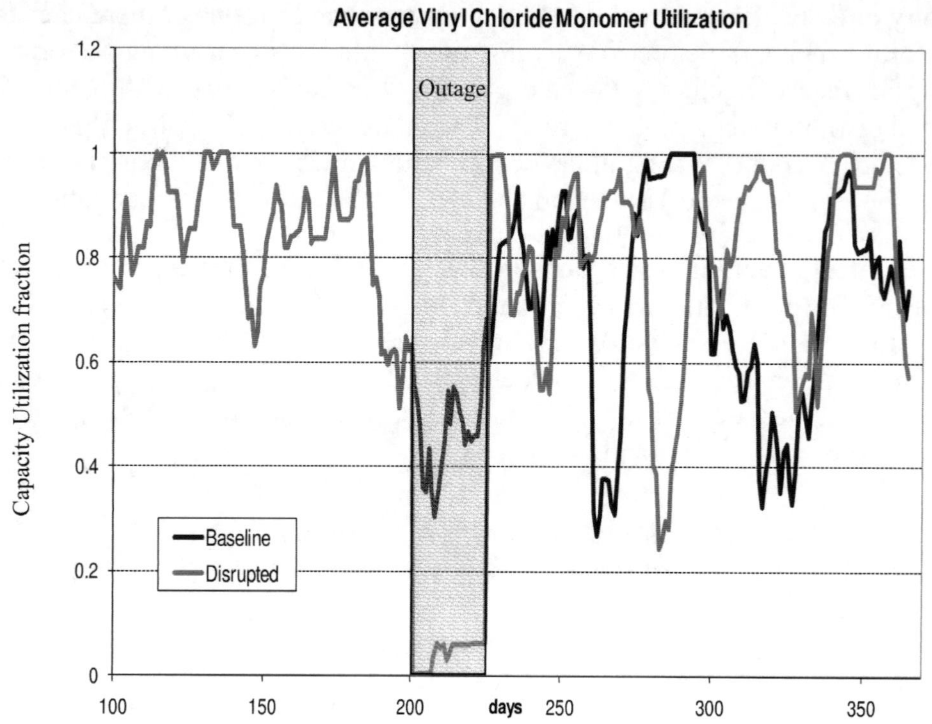

Figure 8: Simulated Results for VCM Capacity Utilization

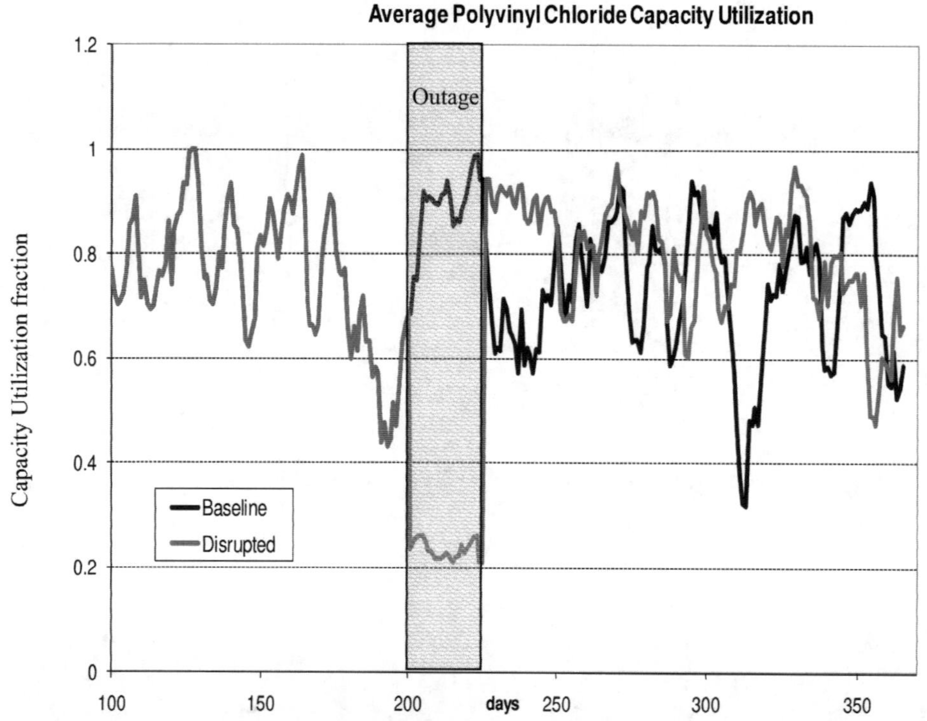

Figure 9: Simulated Results for PVC Capacity Utilization

The baseline representation of Sandia's capability closely matches the known production and consumption of EDC for the United States. Published data report a supply-demand ratio for the PVC sector of the 87 percent. The Sandia simulation results indicate an 85-percent supply-demand ratio, indicating that the production mechanisms in the simulation results are reasonable, as are the consumption results.

Perhaps the most useful capacity utilization graphic is the analysis of the ability to meet demand for the PVC consumer products. There are 10 categories of PVC consumer goods within the simulation:

- Rigid siding/profiles
- Rigid blow molding
- Rigid pipe and fittings
- Rigid film and sheet
- Rigid-other
- Flexible wire and cable
- Flexible synthetic leather
- Flexible floor coverings
- Flexible film and sheet
- Flexible-other

Figure 10 compares the capacity utilization between the baseline and the disruption results for these downstream consumptions.

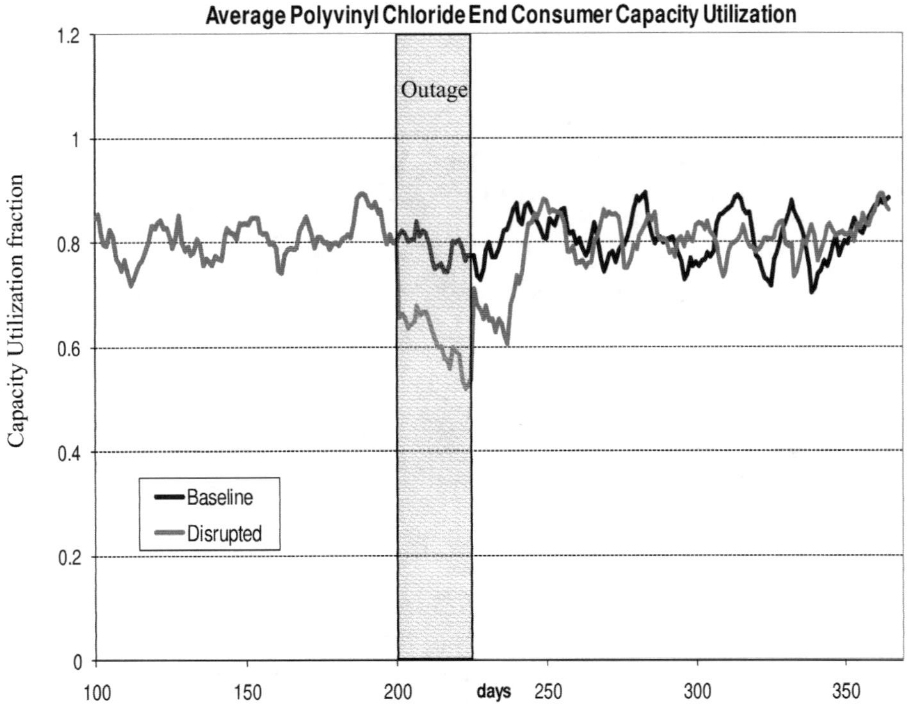

Figure 10: Simulated Results for PVC Capacity Utilization

Once production resumes, the bull-whip shape of the capacity utilization reflects the time lag for downstream production to resume and stabilize. Factors such as storage, disrupted transportation, and transportation capacity cause this non-baseline behavior. Figure 11 shows the sale of PVC production before and after the disruption. The simulated results obtained from N-ABLE™ provide key insights toward how the physical supply chain responds as well as the monetary impacts associated with a prolonged disruption. The sales remain at fairly high level for an extended period of time post disruption.

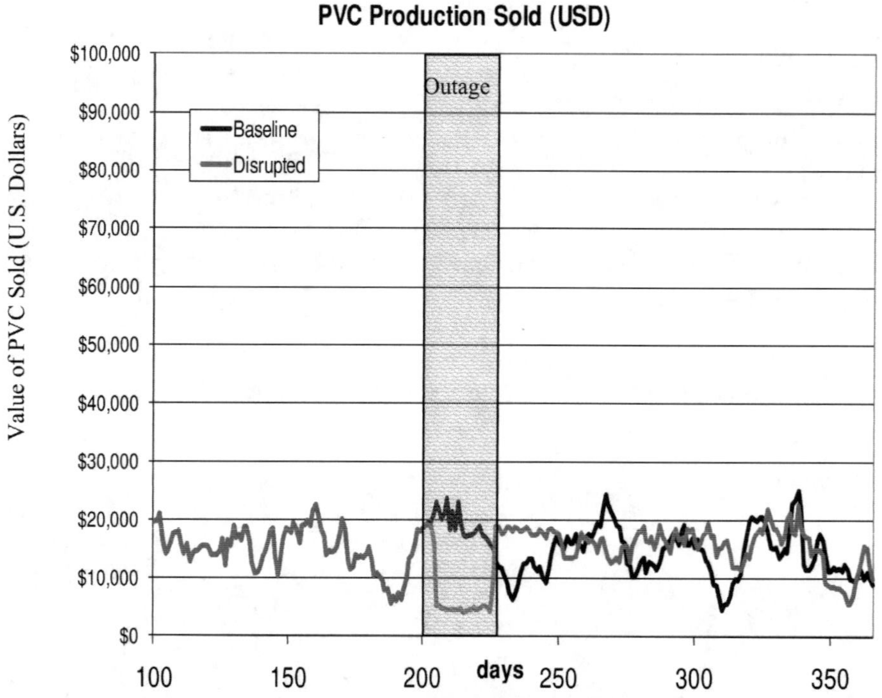

Figure 11: PVC production in terms of US dollars generated

4 Summary

Analysis of chemical supply chains is an inherently complex task, given the dependence of these supply chains on multiple infrastructure systems (e.g. transportation and energy). The capability developed at Sandia is intended to provide information to the DHS with respect to the consequences of large-scale disruptions to the chemical sector, including interrupted supply and resulting economic impacts to the nation. The ongoing development effort includes the development of several tools along with a comprehensive database that feeds the tools. The database is constructed by merging many datasets in combination to provide a high degree of resolution within the data so that individual plants can be uniquely represented.

The hurricane disruption scenario presented herein shows that large-scale disruptions to the EDC supply chain affect many supply chains and, consequently, take considerable time to recover (Figures 10 and 11). Supporting this result in the scenario analysis, information reported in *Chemical Week* showed that

> Several Texas Gulf Coast chemical plants began to restart operations after shutting down ahead of Hurricane Ike's landfall on September 13, 2008. However, producers claim that the ready availability of utilities, raw materials, and logistics, and the damage at some customer sites negatively affect their effort to restart operations. [6]

The disruption to chemical plants cascade both up and down the supply chain, affecting recovery efforts.

5 Acknowledgement

The authors would like to thank Mark Ehlen and John Masciantoni for providing N-ABLE™ assistance. This work is funded by the U.S. Department of Homeland Security, Science and Technology Directorate.

6 References

[1] For a description of the 18 infrastructures, see U.S Department of Homeland Security, "Infrastructure Protection Plan: Partnering to enhance protection and resiliency," 2009, http://www.dhs.gov/xlibrary/assets/NIPP_Plan.pdf; and "Sector-Specific Plans," http://www.dhs.gov/xprevprot/programs/gc_1179866197607.shtm

[2] Eidson, E. D., and M. A. Ehlen, 2005. "NISAC Agent-Based Laboratory for Economics (N-ABLETM): Overview of Agent and Simulation Architectures." Sandia National Laboratories, SAND Report SAND2005-0263, February 2005.

[3] Ehlen, M. A., A. J. Scholand, and K. L. Stamber, 2007, "The Effects of Residential Real-Time Pricing Contracts on Transco Loads, Pricing, and Profitability: Simulations using the N-ABLE™ Agent-Based Model," *Energy Economics*, 29:2, pp. 211-227, 2007, http://www.sandia.gov/nisac/docs/CERTS.pdf

[4] For SRI Consulting, "World Petrochemical Program 2008," access http://www.sriconsulting.com/WP/Public/ProgramContents.html

[5] NISAC Chemical Supply Chain Industry Workshop 2007: Working papers and notes

[6] Coons, Rebecca, "Hurricane Ike Batters Texas Gulf Coast Production," *Chemical Week*, September 22, 2008, http://media.web.britannica.com/ebsco/pdf/34/34625778.pdf

A Guide to Developing and Implementing Safety Checklists: Plant Steam Utilities

Mark Fecke, P.E.
Exponent, Inc.
1011 Warrenville Road, Suite 215, Lisle, Illinois, 60532
630-743-7714, mfecke@exponent.com

John Martens, Ph.D., P.E.; Joel Cowells, P.E.; and Delmar "Trey" Morrison, Ph.D., P.E.
Exponent, Inc.
jmartens@exponent.com, jcowells@exponent.com, and tmorrison@exponent.com

Prepared for Presentation at
American Institute of Chemical Engineers
2010 Spring Meeting
6th Global Congress on Process Safety
San Antonio, Texas
March 22-24, 2010

UNPUBLISHED

Guide to Developing and Implementing Safety Checklists: Plant Steam Utilities

Mark Fecke, P.E.
Exponent, Inc.
1011 Warrenville Road, Suite 215, Lisle, Illinois, 60532
630-743-7714, mfecke@exponent.com

John Martens, Ph.D., P.E.; Joel Cowells, P.E.; and Delmar "Trey" Morrison, Ph.D., P.E.
Exponent, Inc.

Keywords: boiler, steam, checklist, process hazard analysis, burner management system, combustion control system

Abstract

Steam generation is an integral part of most chemical process plants; however, the steam plant often is or can be overlooked in the area of hazard analysis. The reasons for this oversight are obvious: steam generation is considered to be an old and well understood process, and steam boiler systems are often not considered to pose the same hazards as other plant units. However, modern steam boiler systems are fueled with natural gas, pulverized coal, and/or fuel oil; each of which poses significant fire and explosion hazards. For example, a moderately sized chemical plant's boiler house may have two or three boilers operating at 240 MMBTU/hr, with each using approximately 11 tons/hr of subbituminous pulverized coal feed. Chemical plants rely upon equipment design and installation, maintenance practices, operating procedures, and programmable logic controllers (PLCs) to operate safely. Steam plants rely on these same safety mechanisms. Thus, the same process safety management objectives may be applied to these systems as to the balance of the chemical plant.

Several hazard analysis techniques can be applied to existing designs, management of change, and new designs of steam power plants in chemical facilities. A method that we have found useful to apply to these systems is that of a safety checklist based upon relevant standards and good engineering practice. More than just identifying the hazards and required safeguards for such a system, if implemented effectively, the checklist will enhance training and staff knowledge of boiler operation and safety systems. This paper aims to provide guidelines for when it is appropriate to implement a checklist and for reasonable alternatives for developing such a checklist. We will also discuss typical hurdles in implementing such a hazard analysis program for existing boiler systems.

1. Introduction

The modern integrated chemical processing plant, electrical power utility plants, and other industrial facilities rely upon onsite steam generation. The onsite "steam plant" is often the oldest and most central processing unit in these facilities. While the surrounding processing units have been built, demolished, rebuilt, and redesigned over the decades, the steam plant has remained largely unchanged. Modern steam plants may be fired on natural gas, pulverized coal, and/or fuel oil; each of which may pose significant fire and explosion hazards if uncontrolled. Chemical plants rely upon equipment design and installation, maintenance practices, operating procedures, and programmable logic controller (PLC) systems to operate safely. Steam plants rely on these same safety systems due to the hazards. Thus, the process safety management objectives may equally apply to both the boiler and the balance of the chemical plant. Steam is a utility, not a product that directly generates revenue; therefore, it may be difficult to justify significant upgrades to the mechanical systems, operations, and safety systems without a direct financial payback. However, when considering costs, risk is considered as part of the justification process. Upgrades to the safety of the steam plant should generally keep pace with the improving safety objectives for the balance of the plant.

Several formal hazard analysis techniques can be applied to the risk management of existing boiler systems, management of change, and new system designs in steam power plants within chemical process facilities. An analytical method that we have found particularly useful to evaluate the safety of these systems is that of a Checklist Analysis based upon relevant standards and good engineering practice. More than just identifying the hazards and required safeguards for such a system, if implemented effectively, the checklist may be used to enhance training objectives and personnel knowledge of boiler operation and safety systems. This paper aims to communicate guidelines for when it is appropriate to implement a checklist and for reasonable alternatives to consider when developing such a checklist. We will also discuss typical hurdles in implementing such a hazard analysis program for existing boiler systems.

2. Background

To understand the importance and evolution of safety through the years, it is important to understand the history of steam generation. For nearly 300 years, steam generation has been used in industrial settings. In those 300 years, the three basic components of any steam generator have not changed: air, water, and fuel. Early steam generators in the 1700's were simple kettle boilers that operated at very low pressures and were used to help drive pumps. In the early 1800's, Richard Trevithick designed one of the first fire tube boilers. This was an improvement over the previous generation of boilers because it operated at a greater pressure. This operating pressure increase was due primarily to the cylindrical design of the boiler. This boiler was known as a fire tube because the fire (or hot gases) was inside a tube with water surrounding the tube.

The Trevithick single fire tube was improved by adding several fire tubes. As time passed and the demand for higher pressures and greater steam capacities grew, the limitations and safety issues with these types of boilers became significant. Engineers recognized the need for a new design that would increase capacity while reducing the consequences of pressure part ruptures. This led to the development of the water tube boiler in which a small amount of water was now contained inside tubes while the fire or hot gas surrounded the tubes. In practice, the water tubes enclose a firebox where the heat radiated by flames from one or more burners heats the water (tube) walls. Modern boilers are also equipped with different subsystems to aid in economic steam generation such as a superheater section, steam drum, and mud drum. A thorough discussion of modern steam boiler systems can be found in the Steam Handbook by Babcock and Wilcox [1].

Most industrial boilers today are the water tube type. They can be designed to generate anywhere from 200,000 lbs of steam per hour to as much as 10,000,000 lbs of steam per hour. These boilers generally are fired with coal, natural gas, and/or oil. While these boilers may be fired with several different types of fuel the basic systems for the boiler is the same. A common steam boiler uses a mixture of fuel and air to generate heat that is transferred to the walls, roof, and pendants. The water that is flowing through these tubes is heated to or above the given saturation temperature producing steam. The hazards posed by the pressurized water and steam are contained by the mechanical systems due to current design practices [2].

2.1 Relevant Guidance Documents

Early on, safety from pressure boiler failures was accomplished largely by making the walls and tubes of a boiler thicker. However, a number of serious boiler accidents still occurred. From 1893 to 1903 more than 1200 people were killed in the US in 1900 separate boiler explosions [1]. These deaths led to the first version of the ASME Boiler and Pressure Vessel Code (ASME Code) entitled "*Boiler Construction Code*", which was published in 1915 [1]. This code established safety rules governing the design, fabrication and inspection during construction of boilers and pressure vessels. This document continues to provide guidance on these issues today as it is updated every three years. The ASME Code is currently used as a Federal, State, and Local regulation for the construction and maintenance of steam boilers [1].

While the development of the ASME Code helped increase the structural quality and mechanical safety of steam boilers, it did not provide guidelines on how to operate a boiler. This deficiency led to the development of a 1924 document entitled NBFU 60 *Regulation of the National Board of Fire Underwriters for the Installation of Pulverized Fuel Systems as Recommended by the National Fire Protection Association* [3]. This document changed from a NBFU/NFPA document to an NFPA standard in 1946. The evolution of several NFPA standards on operations of different types of boilers were eventually compiled into a single document in 2001 entitled NFPA 85 *Boiler and Combustion System Hazards Code.* Like the ASME Code, NFPA 85 is updated every 3 years and continually incorporates new technology that will improve safety and reliability. NFPA 85 is subdivided into chapters describing the following industrial boiler types:

- Single burner boilers
- Multiple burner boilers
- Fluidized Bed Boilers
- Heat Recovery Steam Generator and other combustion turbine exhaust systems
- Stoker boilers
- Pulverized Fuel System

Within these chapters, the standard also addresses the types of fuel (solid, liquid, or gas), necessary design features, operations, maintenance, and safeguards. Many safeguards are provided by trips and interlocks between specific control functions (e.g., burner firing) and deviations from desired operating states (e.g., loss of fan). Perhaps the most complex hazards posed by industrial boiler systems involve controlling the burner operation to maintain safe operation throughout start-up, shut-down, normal operation, and abnormal operations.

Most conventional burners are controlled by two main types of systems. The first system, called the Burner Management System (BMS), is the main safety system dedicated to maintaining safe burner functions during start-up and shut-down of the burner and boiler. This system monitors a number of sensors that may indicate misoperation of the boiler. The system will then initiate a Master Fuel Trip which shuts off fuel and purges the boiler. See Figure 1 for an example list of trips. The second system, called the Combustion Control System (CCS), regulates the furnace fuel and air inputs in response to steam demand and maintains the proper ratio through the use of cross limits. This system takes primary control after the boiler and/or burners have started and the BMS safe-start permissives have been met. While the CCS maintains primary control to maintain firing rate and the fuel-air ratio, the BMS continuously monitors several critical boiler states to shut-down (i.e., trip) the boiler if potentially dangerous operating states are achieved.

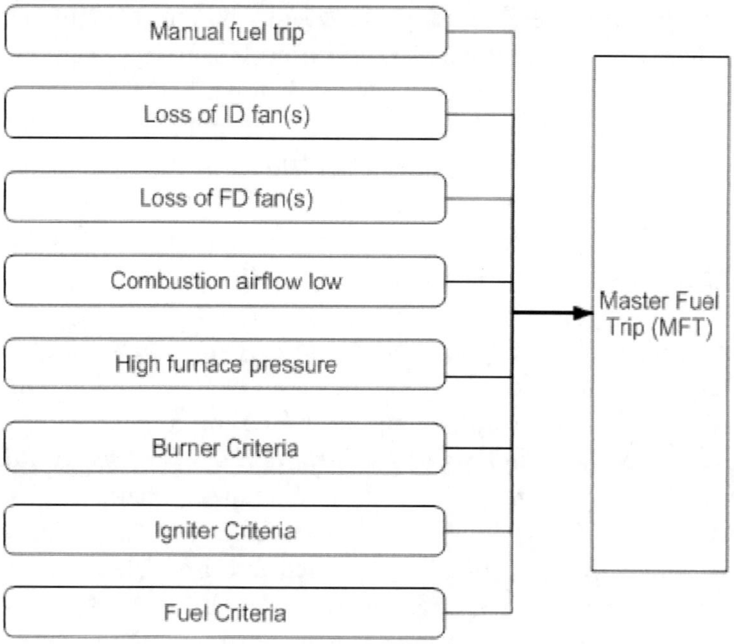

Figure 1. Example trip diagram for a Master Fuel Trip (derived from NFPA 85).

3. Performing PHAs on Boiler Systems

An essential component of the risk management of industrial processes is the use of a qualitative process hazard analysis (PHA) to identify potentially dangerous aspects and to initiate preventative actions. PHAs may take many forms such as Preliminary Hazard Analyses (PrHA), Hazard Identification Studies (HAZID), or Hazard and Operability studies (HAZOP). PHAs can be applied to the physical systems of equipment, controls, personnel, maintenance, and operations. The complexity of the studied industrial process will typically affect the decision on what type(s) of PHAs to perform. The earlier analytical methods are all team-based approaches that rely upon focused brainstorming and systematic evaluation according to a generally defined set of rules. These work well for unique industrial and chemical processes that may have little in the way of prescriptive guidelines for design and operation. Boiler systems share common features of equipment design, maintenance, and controls with other chemical and industrial processes; however, due to the extensive history of boiler operations in industry, a great deal of prescriptive safety information is available as discussed in the Background section above.

In a case where such useful guidelines or corporate experience are available, a PHA may take the form of a Checklist Analysis. A checklist is a written list or table of items that is used to verify the status of a system. Individual checklists may be highly specific to a certain process or company, but they are frequently used to measure compliance with standards and guidelines [4]. A checklist must provide criteria against which a given system is evaluated. Table 1 provides a basic example of a checklist to show some of the sections that may be included. It should be obvious that this checklist is too general to be of much use in a real PHA, but the logical question arises as to how to structure a useful checklist or to improve upon an existing checklist for a boiler system.

Table 1. Example of a basic checklist.

Criteria	Yes	No	N/A	Comments
Mechanical Design	X			Meets ASME
Safety Interlocks and Trips	X			Meets NFPA 85
Procedures		X		Under development
Training		X		No formal program
Energy Conservation Program			X	Ref. Corporate guideline EC-102

To be most useful, checklists must be created by experts in the type of system under study who can be either from in-house or outside sources. The checklist document's scope and utility is dictated by the authors' experience. Hazard analysis guidance documents recommend that checklists be developed by teams of professionals with multi-disciplinary backgrounds and extensive experience in the applicable systems [4]. It is no coincidence that the authors of this article are two Electrical Engineers, a Mechanical Engineer, and a Chemical Engineer. The team constructing, maintaining, or implementing the checklist should possess expertise in boiler operations and maintenance, control systems architecture and implementation, combustion controls, and NFPA and other relevant standards and guidelines.

Further guidance on creation of useful checklists for boiler systems will be discussed below; however, NFPA 85 provides a good starting point for development of a boiler Checklist. From the standard, the Checklist may be expected to cover at least the following areas:

- Fuel Handling
- Mechanical Systems and Equipment Design
- Burner Systems Operation and Safety Management
- Operations – Startup, Shutdown, Normal, Emergency
- Training
- Maintenance
- Commissioning

There are many approaches to take in constructing a useful Checklist Analysis. The checklist may attempt to cover anywhere from a few to all aspects of the person-machine-environment that comprises the operating boiler system in a facility's power plant. The following sections will describe some considerations and provide examples of recommended good practices that we have found useful.

4. The Checklist Process

In this section, we discuss the lifecycle of a checklist for use in a PHA. The steps in the lifecycle include two phases: development and implementation. In the development phase, the important steps are determining the scope of the checklist, gathering sources of guidance documents, evaluating their applicability, melding the references into a useable format, interpreting the requirements of each item, and keeping the list in a usable format. In the implementation phase, the objectives are obtaining data for the list, managing and updating the list, and revisiting the list as appropriate. It is important for team members who are responsible for the checklist lifecycle to understand the requirements of the guidance documents. The inclusion of experienced team members from various disciplines for all phases of the checklist process facilitates successful design and application of the checklist. Each of these aspects presents unique challenges, which we discuss here.

4.1 Checklist Development

4.1.1 Determining the Scope of the Checklist

The first step in the checklist lifecycle is to determine the scope and purpose of the checklist. If the checklist is going to be one of the major documents of the safety plan, it is often more encompassing than it is in those cases where it only reviews selected pieces of equipment or parts of the system. It is reasonable to use a partial system checklist from select portions of guidance documents to augment a larger safety analysis program. While it is may seem admirable to create a checklist that encompasses all systems and is developed from a broad range of sources with every element from every source included, the implementation and management of such a checklist may lead to prohibitively-high costs, excessive time delays, and outright checklist abandonment. Therefore, choosing the appropriate scope for the checklist at the beginning of the process can be an important factor in its ultimate success.

4.1.2 Gathering Sources of Guidance Documents

The next step in the checklist process is gathering a collection of references, guidance documents, standards, regulations, and codes. Choosing standards that directly apply to the process and equipment is a basic requirement during this step, especially where well-established standards exist. It can be helpful during this stage, however, to review and consider references from other fields. For example, the application of good engineering design for control systems is often applicable to most process controls including boiler systems. Some reference sources have been listed earlier, but useful reference documents may be found through NFPA, ABMA,

ASME, ISA, and AIChE publications, standards, and guidance documents as well as the open technical literature. Most companies also have internal standards and guidance documents that should be the starting point for checklist creation.

4.1.3 Reviewing References and Evaluating Applicability

After the various reference documents have been collected, the next step is to review them and determine their applicability and usefulness for the specific process under review and in light of the defined scope of the checklist. In some cases, the applicability of codes may be enforced by an Authority Having Jurisdiction (AHJ), insurance company, or corporate policy. The review process can be divided among the team of specialists, but all team members should have a good grasp over boiler systems operations. Criteria for entries in the checklist may fall into two categories: well-defined requirements and performance-based objectives. An understanding of the difference and how to apply good engineering practices to meet the more subjective performance objectives is essential for the team.

4.1.4 Combining the References into a Useable Format

Once all of the sources have been gathered and their applicability determined, the next step is to combine the references into a useable document. At this point in the checklist development process, it is helpful to consider how the checklist will be used based on the chosen scope. The reference information must be translated into meaningful criteria for the checklist. Many practical considerations are taken into account in this step such as the following list.

- **Accessibility:** It is important to consider who will need to access the checklist document(s) and what tools are required to do so. If the document will be maintained by several people, then choosing a program, such as Microsoft® Excel™ , that the entire team is likely to have available is a good choice.
- **Searchability**: Once the checklist is in use, specific questions often arise and the ability to search the entire document for key words can be helpful. This typically suggests the use of an electronic medium.
- **Entry context**: Many standards are written in such a way that understanding any given block of text requires viewing it in the context of several preceding (or following) entries in order to determine the context. Therefore, it is helpful to organize the checklist in such a way that several entries can be viewed simultaneously.
- **Entry options**: Consistent types of entries should be used in the checklist. One option is to use a drop-down menu with entry options. This will reduce the ambiguity in the meaning of entries. Will the columns have check boxes such as those in Figure 1, or will there be a specific list of entry word choices (i.e., complete, not complete, follow-up)?
- **Responsibility**: Consider adding a column (or columns) to the checklist where specific personnel can be assigned the task of closing punch-list items. Additional information such as due date or follow-up date, approvals, and revision information can also be added.

There are many potential columns that can be added to a checklist, but it is important to keep the checklist simple and workable.

4.2 Checklist Implementation

4.2.1 Obtaining Data for Completing the Checklist

It is good practice to obtain and review applicable engineering documents (P&IDs, electrical drawings, control logic, etc), procedural documents (operating procedures, maintenance documents, etc), and other relevant documentation prior to entering data into the checklist. While it may be tempting to immediately begin populating the checklist entry fields as the materials are being reviewed, it is often helpful to make a first pass through the entire set of documentation (if available) so that consistency among documents can be verified and interactions of various documents will be understood. This process should make populating the list more efficient and help identify potential issues early in the process. Field inspections and personnel interviews are also a vital component of obtaining information for completing the checklist. The operations and maintenance staff can not only provide valuable input, but the checklist process will also act as a useful training aid to inform personnel of the latest safety developments.

4.2.2 Entering Data into the Checklist (Using the List)

Once the checklist has been created and the checklist analysis has been started, the process of populating data entries into the checklist can begin. The formal Checklist Analysis with a limited scope on a simple subsystem may be completed by the team over the course of several hours in a conference room. However, for more comprehensive checklists and complicated systems, the formal Checklist Analysis may take from several days to several weeks. In our experience, "checking" items in the list is best served by including notes that demonstrate that the particular item has been considered or implemented. A balance must be struck between using terse entries that will not be meaningful at a later time when read by another person and using verbose notes that create a list with overwhelming detail. When an item requires large volumes of back-up documentation, we have found it helpful to include a link to such items such as a hyperlink in an electronic document or annotated reference citations.

As the checklist will often be accessed to determine what items are still outstanding, it is helpful to include a mechanism for quickly identifying which items are not yet addressed. We have found that the use of color for this task is helpful, especially when printed copies may be taken to the field. This field can also be used for sorting. For example, a Microsoft® Excel™-based checklist may contain a color-coded column that can be used for quickly determining whether an issue has been addressed, is in progress, or is unresolved. Additionally, for pending or unresolved items, it is helpful to make the pending or unresolved portion of the comment stand out by underlining or highlighting it, as shown in the example checklist portion below.

It can also be helpful to provide a date and/or a document version in the comments, for example, to align the comment with a particular version of an evolving design document if different versions have been reviewed. Presented below is an example portion of a checklist comparing the implementation of a utility boiler to the purge permissive requirements found in NFPA 85, which incorporates the guidelines discussed in this section.

Table 2. Example portion of a detailed checklist for purge permissives.

Row No.	Code Reference	Checklist Requirement[1]	Status	Status Date	Comments	Assigned To
1	NFPA 85 6.4.2.3.4.5 *	Boiler Enclosure Purge Permissives. Boiler enclosure purge permissives shall, at a minimum, include the following:	OK	3/21/10	*Section header, no specific requirement*	
2	Sub (1)	All igniter header and individual igniter shutoff valves proven closed by valve position. Exception: Where the igniter capacity is 1.5 MW (5 million Btu/hr) or less, proof of closure of individual igniter safety shutoff valves by means other than valve position shall be permitted.	OK	3/18/10	In version 2.1 of the functional specification, purge permissives include gas valves closed. Verified at network 10/rung 4 of the BMS ladder logic. Tested on simulator (3/7/10) and during onsite functional testing (3/18/10).	J Martens
3	Sub (2)	If coal is fired on the unit, all pulverizers are stopped and all coal flow to the furnace stopped.	OK	3/22/10	In version 2.1 of the functional specification, coal equipment is verified to be off and coal valves are verified to be closed prior to purge permissive TRUE. Verified at network 10/rung 5 of the BMS ladder logic. Tested on simulator (3/7/10) and during onsite functional testing (3/18/10).	J Martens
4	Sub (3)	If fuel gas is fired on the unit, all main fuel gas header and individual fuel gas burner shutoff valves are proven closed by valve position.	OK	3/21/10	In version 2.1 of the functional specification, gas valves are proven closed by valve position prior to purge permissive TRUE. Verified at network 10/rung 4 of the BMS ladder logic. Tested on simulator (3/7/10) and during onsite functional testing (3/18/10).	J Martens
5	Sub (4)	If fuel oil is fired on the unit, all main fuel oil header and individual fuel oil burner shutoff valves are proven closed by valve position.	OK	3/21/10	In version 2.1 of the functional specification, oil valves are proven closed by valve position prior to purge permissive TRUE. Verified at network 10/rung 4 of the BMS ladder logic. Tested on simulator (3/7/10) and during onsite functional testing (3/19/10).	J Martens

[1] Excerpted from Reference [3].

Table 2. Example portion of a detailed checklist for purge permissives.

Row No.	Code Reference	Checklist Requirement[1]	Status	Status Date	Comments	Assigned To
6	Sub (5)	Any other sources of combustibles that could enter the boiler enclosure or flue gas path proven closed by valve position or other positive means. Exception: Where the capacity of the combustible source is 1.5 MW (5 million Btu/hr) or less, proof of closure of shutoff valves by means other than valve position shall be permitted.	OK	3/21/10	Verified that there are no other sources of combustibles aside from coal, natural gas, and oil.	M Fecke
7	Sub (6)	All required burner air registers are in purge position.	OK	3/21/10	Registers are set and pinned. Position set during commissioning by boiler manufacturer representative. Register position verified to provide purge airflow during onsite functional testing (3/18/10).	M Fecke
8	Sub (7)	At least one FD fan and, if so equipped, one ID fan in service.	Work in Progress		In version 2.1 of the functional specification, must have ID + FD fans running for purge permissive to be TRUE. Verified at network 10/rung 5 of the BMS ladder logic. Tested in simulator (3/7/10). During onsite functional testing (3/18/10), FD fan failed to run. Troubleshooting.	J Cowells
9	Sub (8)	Flue gas recirculation fans shall be operated as recommended by the boiler manufacturer.	Pending		This may be n/a. Boiler believed to not have a flue gas recirculation system. Confirmation required.	J Martens
10	Sub (9)	Total boiler airflow is at purge rate airflow.	OK	4/1/10	In version 2.1 of the functional specification, purge permissives include total airflow between 30% and 40%. Verified at network 10/rung 6 of the BMS ladder logic. Tested on simulator (2/27/10) and during onsite functional testing (3/18/10).	J Cowells

4.2.3 Managing and Updating the List

It is often the case that the design for an upgrade, including the control strategy, is being developed concurrently with the checklist. The checklist itself may provide input to guide portions of the design process. For this reason, the checklist requires continuous updating and management. Changes must be reflected in the checklist as the design evolves. Further, changes may require that previously satisfied requirements be revisited. For this reason, cross-linking of related entries is encouraged. Version control and limiting the personnel who are authorized to make changes to the working checklist may also facilitate successful completion of the checklist.

4.2.4 Revising the List as Appropriate

A checklist is rarely considered complete and as such requires maintenance to continue to reflect current best practices. When changes to equipment are required or updates to processes are made, the checklist should be revisited to understand any impact on previously-satisfied items. Even when changes are not made, it can be a good idea to revisit the checklist at periodic intervals to assure that stated procedures are being followed. The process of revisiting the checklist is another form of ensuring that communication among the stakeholders takes place. Further, updated codes and standards may need to be incorporated into the checklist.

5. Checklist Modules

In this section, we discuss some of the standards and guidance documents for selected subsystems and boiler operations. A full checklist may include each of these principle areas.

5.1 Fuel Handling

The majority of boiler systems are fired on either pulverized coal and/or natural gas; however, systems firing on oil, stoker coal, and biomass also pose distinct hazards. Based on the prevalence of pulverized coal and natural gas, those two fuel types will be discussed here. Controls for natural gas system requirements are delineated in NFPA 85. For example, for multiple-burner boilers, Section 6.6 of NFPA 85 provides requirements for fuel gas systems such as the need for a double block and bleed arrangement in the fuel piping to the burners, fuel line overpressure protection, and provisions for leak testing safety valves. Figure 2 is a piping schematic that depicts a typical natural gas fuel train piping from the supply to the burner. Typical steps in the checklist would be to verify that the drawing meets the code requirements (e.g., 6.6.3.1.2, 6.6.3.1.10, and 6.6.3.1.11 of NFPA 85) and that the as-built equipment in the field matches the drawing. Additional requirements for installation and design of the mechanical components of the fuel gas train can be found in NFPA 54, *The National Fuel Gas Code* [5].

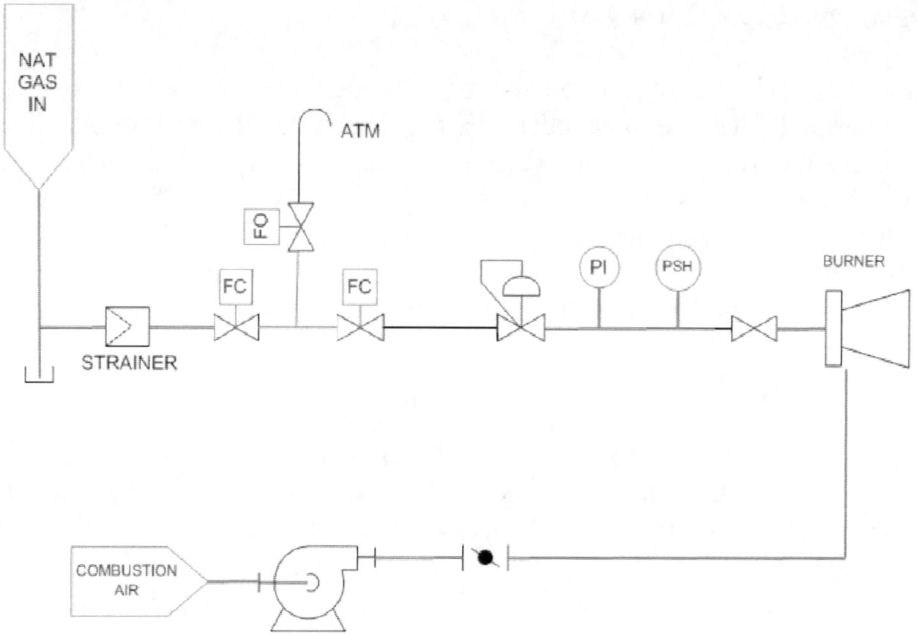

Figure 2. Simplified piping schematic of a typical gas train.

Establishing requirements for coal handling systems are more difficult since there is currently no NFPA or other standard that completely addresses the process. Coal is a solid fuel with much more variable properties than fuel gas. Coal properties can be affected by the type of coal (e.g., lignite versus western coal) and the handling operations prior to use. For pulverized coal boilers, raw coal is supplied as large chunks, which are transported through a coal handling system to storage silos. Coal is transported up through the system to the top of the storage silos via conveyors that may be of the belt type, drag type, or screw type. Coal is then gravity fed down through the silos toward the pulverizers. This large sized coal passes from the silos into pulverizers where it is crushed to form a fine coal powder that is pneumatically conveyed to the burners for combustion. Throughout the coal handling process, the coal may pose significant dust explosion hazards and fire hazards due to self-heating or contact with an ignition source. In addition to the complex safety and design requirements, OSHA has recently given notice that it will be developing a comprehensive combustible dust standard [6]. Potentially useful references for developing solid fuel handling checklists can be found in the following documents among others.

- NFPA 61 *Standard for the Prevention of Fires and Dust Explosions in Agricultural and Food Processing Facilities*
- NFPA 68 *Standard on Explosion Protection by Deflagration Venting*
- NFPA 69 *Standard on Explosion Prevention Systems*
- NFPA 70 *National Electric Code (NEC)*
- NFPA 120 *Standard for Fire Prevention and Control in Coal Mines*
- NFPA 499 *Recommended Practice for the Classification of Combustible Dusts and of Hazardous (Classified) Locations for Electrical Installations in Chemical Process Areas*
- NFPA 654 *Standard for the Prevention of Fire and Dust Explosions from the Manufacturing, Processing, and Handling of Combustible Particulate Solids*

- NFPA 664 *Standard of Fires and Explosions in Wood Processing and Woodworking Facilities*
- NFPA 850 *Recommended Practice for Fire Protection for Electric Generating Plants and High Voltage Direct Current Converter Stations*
- FM Data Sheet 7-76, *Prevention and Mitigation of Combustible Dust Explosion and Fire*

5.2 BMS

According to NFPA 85, the Burner Management System (BMS) of a boiler is, "the control system dedicated to combustion safety and operator assistance in the starting and stopping of fuel preparation and burning equipment and for preventing misoperation of and damage to fuel preparation and burning equipment" [3, Section 3.3.26]. NFPA 85 further states, "The burner management system can include the following functions as specified in this code: interlock system, fuel trip system, master fuel trip system, master fuel trip relay, flame monitoring and tripping systems, ignition subsystem, main burner subsystem, warm-up burner subsystem, bed temperature subsystem, and duct burner system" [3, Section A.3.3.26]. Simply put, the BMS has ultimate authority to stop the fuel supply to the burners in a boiler, regardless of whether it is fired on natural gas, oil, coal, or any other fuel source. Unlike older relay-based BMS systems, today's burner management systems are implemented by some form of controller running control logic software. The BMS monitors inputs (e.g., valve position, damper position, motor status, etc.) and makes decisions about whether or not fuel flow to the burners is allowable based on the state of these inputs. These logic decisions are made in accordance with required safe operation of a boiler. The outputs of a BMS system are typically discrete (i.e., on or off) and have the ability to directly stop fuel.

A checklist covering a utility boiler's BMS should include references to standards on such issues as conditions requiring a master fuel trip (MFT), E-Stop(s), control system hardware and software requirements, required alarms, and sequencing interlocks (e.g. purge before lighting igniters). NFPA 85 provides a prescriptive approach to implementing a BMS.[2] General requirements for the BMS logic can be found in Section 4.6.3 of NFPA 85. The remaining BMS requirements for specific types of boilers (e.g. single burner boilers or multiple burner boilers) may be found in various chapters of the standard.

5.3 CCS

From NFPA 85, the Combustion Control System (CCS) is defined as "the control system that regulates the furnace fuel and air inputs to maintain the air-fuel ratio within the limits that are required for continuous combustion and stable flame throughout the operating range of the boiler in accordance with demand" [3, Section 3.3.44.2]. Unlike the BMS, which has discrete outputs, the CCS typically has analog outputs that can correspond to a range of valve positions, damper positions, and fan speeds. Typical control loops contained in the CCS are air control, fuel control, furnace pressure (or draft) control, and feedwater/drum level control. NFPA 85 provides a prescriptive approach to implementing a CCS. General requirements for the CCS can be found

---

[2] Other standards that can be consulted include ANSI/ISA 84.00.01-2004 (IEC 61511 mod) and IEC 61508 [10, 11].

in Section 4.6.5 [3]. The remaining CCS requirements for specific types of boilers (e.g. single burner boilers or multiple burner boilers) can be found in various chapters of the standard.

Other standards, such as ANSI/ISA 77.41.01 2005 and ABMA (e.g. *Combustions Control Guidelines for Multiple-Burner Boilers*; *Combustion Control Guidelines for Single-Burner Boilers*) pertain to the combustion control system and do not focus on the BMS [7, 8, 9]. As an example, the requirements for air and fuel control and furnace pressure control in the three standards listed above were compared to develop a series of checklist requirements. See the tables with paraphrased requirements below for a comparison of the language for air and fuel requirements in Table 3 and furnace pressure in Table 4.

Table 3. Comparison of air and fuel requirements from guidance documents (paraphrased).

Reference Document:	NFPA 85 [3]	ANSI/ISA-77.41.01-2005 [7]	ABMA Combustion Control Guidelines for Multiple Burner Boilers [8]
Application:	General requirements for all boilers	Boilers with steaming capabilities of >200,000 lb/hr	Multiple burner gas/liquid fuel boilers
Stated Requirements:	4.6.5.2.1 airflow less than purge rate not allowed	4.2.4	...control system design should follow the System Requirements ... of the most current edition of NFPA 85 [p. 16].
	4.6.5.2.3 fuel demand shall not be above available airflow	4.2.5	
	4.6.5.2.4 airflow demand shall not reduce airflow below that required by the actual fuel input(s).	4.2.4	
	4.6.5.3.2 draft control must be in auto for airflow to be in auto	4.2.4	
	4.6.5.3.3.1 airflow must be in auto for fuel to be in auto	4.2.5	

Table 4. Comparison of furnace pressure control requirements from guidance documents (paraphrased).

Reference Document:	NFPA 85 [3]	ANSI/ISA-77.41.01-2005 [7]	ABMA Combustion Control Guidelines for Multiple Burner Boilers [8]
Application:	Multiple burner boilers	Boilers with steaming capabilities of >200,000 lb/hr	Multiple burner gas/liquid fuel boilers
Stated Requirements:	6.5.2.2.2 The furnace pressure control system shall include:	4.2.3 (states "in compliance with the requirements of NFPA 85")	...control system design should follow the System Requirements ... of the most current edition of NFPA 85 [p. 16]. No specific requirements for furnace draft/pressure control beyond general discussion in section 2.2 (8) and VFD description in section 4.2.3.
	(1) Three furnace pressure transmitters, median-select, separate taps	4.1.3.2	
	(2) boiler airflow demand feed-forward signal (not a measured airflow signal)	4.2.3	
	(3) An override action or directional blocking on large furnace draft errors	App. B.7 - points to NFPA. Figure B.2	
	(4) An MFT feed-forward action	App. B.7 - points to NFPA. Figure B.2	

ANSI/ISA 77.41.01 2005 and NFPA 85 are prescriptive standards that are well aligned in the area of combustion controls. While each standard may call out slightly more or less detail, the core requirements are similar. The ABMA standard contains a section entitled, "Control System Recommendations" (Section 3), which provides guidance on a number of control issues, but the scope is at a higher level than some of the specific requirements prescribed by NFPA 85 and ANSI/ISA (such as those listed in the tables above). However, the ABMA standard does address issues not discussed in other standards. For example, Section 3.3 states, "The control system is to send a signal to the BMS anytime there is an internal fault that jeopardizes the ability of the control system to maintain safe operation" [8]. This can be accomplished through the use of a watchdog between the BMS and CCS, which is a recommendation that is not explicitly stated in NFPA 85 or ANSI/ISA 77.41.01 2005.

Only by becoming familiar with the available standards and guidance documents can the checklist designers decide which of these standards or which combination of standards is most applicable to the combustion control system being studied.

5.4 Operations

Boiler operators and maintenance personnel rely on clear, well-defined operating procedures and maintenance procedures to properly and safely accomplish tasks. Operating procedures should be reviewed often and kept up-to-date with modifications to the boiler systems and controls. Likewise, maintenance procedures and calibration schedules need to be updated with changes to the boiler design and/or instrumentation. NFPA 85 contains prescriptive requirements for operating procedures and maintenance. See, for example, Section 6.8.5.2.1.3 for the specific starting sequence required for a multiple burner boiler and Section 4.4.1.1 for a general list of maintenance requirements applicable to all boiler types governed by NFPA 85. Section 6.6 of the ABMA combustion control system standards also provides an abbreviated discussion on operations.

5.5 Training

As mentioned above, boiler operators and maintenance personnel rely on well-defined procedures to properly accomplish tasks safely. NFPA 85 section A.4.3.1 explains that human error is a contributing factor in the majority of explosions. In some cases, human error can be attributed to either insufficient training or no training at all. NFPA 85 lays out general guidelines for both operator and maintenance staff training in Section 4.4.2. It also provides suggestions on what should be included in a training program in Sections A.4.4.2.1.1 and A.4.4.2.2.1. It can be challenging to adequately evaluate the quality of the training program without knowledge and experience gained from seeing several different training programs throughout different industries.

While many companies invest significant time and money into developing training programs, many programs fall short on two accounts. First, companies spend a significant amount of money to develop these programs, but then expect these training programs to be self sustaining. Training programs, like procedures, need to be revisited and updated periodically. This process is made easier if this is incorporating into a company's Management of Change procedures. This may increase the likelihood that regardless of employee and management turnover, the appropriate individual will be responsible for updating or incorporating changes into the training program.

The second way training programs can fall short is by not documenting and tracking both the individual programs and the personnel who have completed the program. Many companies have solved this problem by putting their training program online. This allows companies to easily track personnel training and to document which operators and maintenance staff personnel have been trained. A limitation to some of these online training programs is that it may be difficult to update the training programs with changes in procedures and equipment.

6. Guidance and Conclusions

In our experience, it is difficult to have a 100% comprehensive and 100% completed checklist due to changes in personnel, policies, procedures, requirements, and guidance documents, for example. Nonetheless, the checklist is an excellent tool for managing the requirements from multiple guidance documents and identifying deficiencies for boiler systems. Both the checklist itself and the processes of generating and maintaining the checklist are useful for familiarizing personnel of various levels with the systems and their operation. Success of the checklist process is in part related to the level of management support – management has the ability to give the checklist the appropriate importance and may halt operation of the equipment until they determine that risks have been adequately managed. The finished Checklist Analysis documents provide a snapshot in time. In practice, however, the checklist development and implantation phases may need to be revisited when there are changes in corporate policies, guidance documents, procedures, equipment, among other factors.

7. References

7.1 Cited Documents

[1] Steam, It's Generation and Use, 40th Edition, Steven C. Stultz and John B. Kitto, Editors, The Babcock and Wilcox Company, Barberton, Ohio, 1992.

[2] ASME Boiler and Pressure Vessel Code, American Society of Mechanical Engineers, New York, 2007.

[3] NFPA 85, Boiler and Combustion Systems Hazards Code, National Fire Protection Association, Quincy, Massachusetts, 2007.

[4] Guidelines for Hazard Evaluation Procedures, 3rd Edition, Center for Chemical Process Safety, New York (2008)

[5] NFPA 54, National Fuel Gas Code, National Fire Protection Association, Quincy, Massachusetts, 2009.

[6] News Release Number: 09-1074-NAT, "US Labor Department's OSHA addresses need for combustible dust standard, Advance Notice of Proposed Rulemaking solicits comments from public, October 20, 2009.

[7] ANSI/ISA 77.41.01 2005, Fossil Fuel Power Plant Boiler Combustion Controls, Instrumentation, Systems, and Automation Society, Research Triangle Park, North Carolina, Approved 6 June 2005.

[8] Combustions Control Guidelines for Multiple-Burner Boilers, American Boiler Manufacturers Association, Arlington, Virginia, December 2001.

[9] Combustion Control Guidelines For Single Burner Firetube and Watertube Industrial/Commercial/Institutional Boilers, American Boiler Manufacturers Association, Arlington, Virginia, 1999.

[10] ANSI/ISA 84.00.01-2004 (IEC 61511 mod), Functional Safety: Safety Instrumented Systems for the Process Industry Sector, Instrumentation, Systems, and Automation Society, Research Triangle Park, North Carolina, Approved 2 September 2004.

[11] IEC 61508-1998, Functional safety of electrical/electronic/programmable electronic safety-related systems, Geneva, Switzerland, 1998.

7.2 Useful Sources of Additional Guidance Documents

ABMA – American Boiler Manufacturers Assocition, www.abma.com
AIChE – American Institute of Chemical Engineers, www.aiche.org
ASME – American Society of Mechanical Engineers, www.asme.org
FM – Factory Mutual, www.fmglobal.com
ISA – Instrumentation, Systems, and Automation Society, www.isa.org
NFPA – National Fire Protection Association, www.nfpa.org
IEC – International Electrotechnical Commission, www.iec.ch

Distillation columns risk assessment: when the regular HAZOP evaluation is not enough

Carlos MARENCO
Rhodia Poliamida e Especialidades Ltda
Fazenda São Francisco s/n Paulínia, São Paulo, Brazil
carlos.marenco@br.rhodia.com

Dalva Janine RITA
Ivan MANTOVANI
Fabiana TEDESCHI
Pedro METROPOLO
Nicolau SERRA
Cláudio TAKASE
Rhodia Poliamida e Especialidades Ltda
Fazenda São Francisco s/n Paulínia, São Paulo, Brazil
dalva.rita@br.rhodia.com
ivan.mantovani@br.rhodia.com
fabiana.tedeschi@br.rhodia.com
Pedro.metropolo@br.rhodia.com
nicolau.serra@br.rhodia.com
claudio.takase-External@br.rhodia.com

Prepared for Presentation at
American Institute of Chemical Engineers
2010 Spring Meeting
6th Global Congress on Process Safety
San Antonio, Texas
March 22-24, 2010

UNPUBLISHED

Distillation columns risk assessment: when the regular HAZOP evaluation is not enough

Carlos MARENCO
Rhodia Poliamida e Especialidades Ltda
Fazenda São Francisco s/n Paulínia, São Paulo, Brazil
carlos.marenco@br.rhodia.com

Dalva Janine RITA
Ivan MANTOVANI
Fabiana TEDESCHI
Pedro METROPOLO
Nicolau SERRA
Cláudio TAKASE
Rhodia Poliamida e Especialidades Ltda
Fazenda São Francisco s/n Paulínia, São Paulo, Brazil
dalva.rita@br.rhodia.com
ivan.mantovani@br.rhodia.com
fabiana.tedeschi@br.rhodia.com
Pedro.metropolo@br.rhodia.com
nicolau.serra@br.rhodia.com
claudio.takase-External@br.rhodia.com

Keywords: HAZOP; distillation columns; inertia, VOC emissions

Abstract

The temperature increase is one of the process parameters deviations evaluated during HAZOP analysis in a distillation system. One of the causes of this deviation can be the failure of the cooling system resulting in the reduction of condenser capacity which might cause the emission of volatile organic compounds to the atmosphere. In the presence of an ignition source, a fire with serious consequences on the health of the people might occur, and also important material losses. This risk can be reduced to an acceptable level by installing safety systems (such as steam valve closure). Nevertheless, in some distillation systems, the regular instrumental safety chain is not enough to guarantee a safe unit shut down. After the cooling system failure and the steam valve closure, the volatilization of the most volatile components due to the heat content that remains in the system might occur. This subsequent event might lead to a release of product to the atmosphere. The objective of this work is to alert about this possible scenario and propose an approach for the calculation of the amount of volatiles released to the atmosphere, using simplified heat balances.

1. Introduction

The HAZOP method (HAZard OPerability study) was developed by ICI in the early 70s. In the 1980s risk studies gradually came into use in petrochemicals, oil, chemicals, rail transport, automobiles and other industries. This methodology has been used to assess the safety of new projects or existing units and their modifications. The purpose of this risk analysis is to identify potential accident scenarios that can occur at a facility and to reduce the corresponding risks to acceptable levels. The risk analysis is performed by a multi-disciplinary group.

At Rhodia the HAZOP methodology principles were implemented in the beginning of 80s. The risk analysis is mandatory and performed every 5 years for most of the facilities[1].

2. Assessing and Reducing the Risk in Distillation Columns

In the case of distillation systems, considering column and peripherals, different scenarios are evaluated during the HAZOP study. In steady state operation, normally, the HAZOP group analyses the deviations on temperature, pressure, level, flow rate, etc., and the possible consequences of those failures (such as human and environmental losses). A common cause of the temperature or pressure deviation is the total loss of cooling capacity. In this scenario, the loss of condensation capacity leads to the loss of reflux flow causing, by consequence, the variation of the temperature and pressure profiles of the distillation column. This scenario culminates in the unwanted event, a process accident: the emission of volatile products to the atmosphere via column vent system and its possible ignition. In addition, overpressure in the column above of the Maximum Allowed Working Pressure (MAWP) may occur, and by consequence causing the rupture of the column and/or its peripherals.

Some recent studies presented the use of dynamic simulation for safety analysis in distillation systems [2, 3]. The dynamic simulation was used to simulate the consequences of operational failures including reduction or total loss of cooling capacity and it is possible to observe the dynamic response (such as the pressure increase) and also to evaluate the safety systems installed.

The reduction of the risk associated with the illustrated scenario is normally related to the installation of active safeguards, such as pressure relief devices (PSV) and instrumental safety chain. The Figure (1) shows a classic distillation system, with TISH safety chain type - (high temperature) installed in the column vent actuating by closing the on-off steam valve. The logic of the safety chain is such that an increase of the vent temperature above the stipulated value in steady state conditions, leads to the closure of the steam to the reboiler. Other safety chains might also act on other points of the facility preventing unwanted chain events.

In HAZOP studies carried out at Rhodia [4], the reduction of the scenario risk is associated with the Instrumented Safety Systems reliability, expressed by the SIL (Safety Integrity Level [IEC 61508]).

Figure 1 – Safety barriers in a distillation system

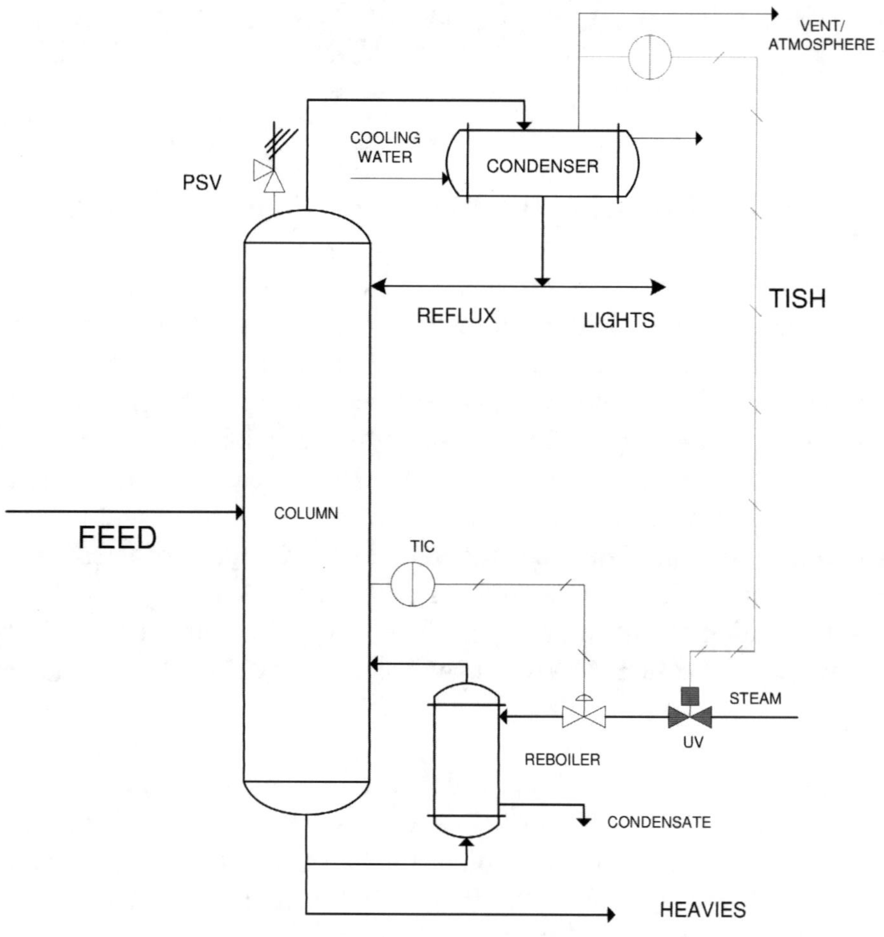

3. Beyond Regular Hazop Analysis

In some distillation systems it is necessary to envisage beyond the horizon of conventional HAZOP analysis and the installed safety barriers.

Let is consider the system showed in the Figure (1). After further analysis, one possible scenario has been identified for distillation systems where the difference of volatility of the compounds is high. In this case the safety barriers must take into account the intrinsic dynamic effect of the heat accumulation in the system.

In this scenario, after losing cooling capacity and the safety barriers taking action (TISH, closing the steam to the column), the most volatile compound might be released to the atmosphere due to great difference of temperatures between the top and the bottom of the column. The necessary energy for the volatilization comes, basically from:

1) the heat content of the metal of the column,
2) the heat content of the metal of the reboiler and other possible peripherals,
3) the amount of heat accumulated in the liquid at the bottom of the column.

In this situation the inertia of the system is observed and this phenomenon might occur from 2 to 30 minutes after the safety barrier actuation. The amount of volatiles released to the atmosphere depends on the column and reboiler dimensions (amount of metal), difference of volatility of the compounds, difference of temperature between the top and the bottom, the holdup of the column, the inertia of the steam valves, and the capacity of the condensers.

4. Methodology Approach

The approach is based on the First Law of Thermodynamics and heat transfer equations and can easily be applied to the distillation columns operating in different conditions. Based on the results it is possible to evaluate the existing safety systems.

The accumulated heat available for the evaporation of the most volatile component of the system takes into account different sources, as shown in the Equations (1) to (4):

1) Heat content due to the steam trapped in the reboiler

$$Q_1 = m_{STEAM\,1}\lambda_{STEAM} \quad \text{[Eq. 1]}$$

2) Heat content due to the steam flow after the security system actuation (the inertia of the steam valves)

$$Q_2 = m_{STEAM\,2}\lambda_{STEAM} \quad \text{[Eq. 2]}$$

3) Heat content due to difference of temperature and composition between top and bottom of the column

$$Q_3 = m_{bottom} C_{P_{bottom}} \Delta T_{(bottom/Tf)} \quad \text{[Eq. 3]}$$

4) Heat content due to the metallic parts (column and reboiler tubes)

$$Q_4 = m_{metal} C_{P_{metal}} \Delta T_{(metal/Tf)} \quad \text{[Eq. 4]}$$

The equilibrium temperature or final temperature of the system (Tf) is the boiling temperature of the most volatile compound at the pressure of the vent system.

Considering the Equations of (1) (4), the amount of product that can be evaporated will be given by:

$$m_{calc} = \frac{\left(\sum_{1}^{4} Q_i\right)}{\lambda_p} \quad \text{[Eq. 5]}$$

m_{calc} is the maximum amount that can be evaporated in the system, but the amount effectively evaporated must take into account the holdup (of volatile product) of the column

If $holdup_{+vol} \geq m_{calc}$, so:

$$m_{evap} = m_{calc} \quad \text{[Eq. 6]}$$

If $holdup_{+vol} < m_{calc}$, so:

$$m_{evap} = holdup_{+vol} \quad \text{[Eq. 7]}$$

The amount of volatiles that effectively will be emitted by the vent system is calculated considering the remaining condensation capacity, as some water remains in the condenser. The reached temperature is Tf.

The equation (8) represents the amount of heat removed due to presence of water and metal in the condenser, with temperature lower than Tf.

$$Q_{cond} = m_{water} C_{P_{water}} \Delta T_{(CW/Tf)} + m_{metal_{cond}} C_{P_{metal}} \Delta T_{(metal/Tf)} \quad \text{[Eq. 8]}$$

The amount of condensed volatiles will be given by:

$$m_{cond} = \frac{(Q_{cond})}{\lambda_p} \quad \text{[Eq. 9]}$$

Finally, the amount of volatiles released to the atmosphere will be:

$$m_{emitted} = m_{evap} - m_{cond} \qquad \text{[Eq. 10]}$$

If the remaining condensation capacity is higher than the amount of evaporated product, there will not be emission of product to the atmosphere. Otherwise, it will be necessary the implementation of extra safety barriers to prevent the emission of product to the atmosphere via the vent system.

The equations (1) to (10) represent a simple method to determine if a product emission will occur in distillation columns operating with very distinct temperatures between the top and the bottom.

It is important to mention that if the system in question have many high volatility components it is necessary to increase the calculation accuracy of the equations (1) to (10) in order to consider these other volatiles present in the mixture, invalidating the simplification indicated in (7).

5. Conclusions And Future Work

In HAZOP studies, the case of distillation columns is evaluated considering that the risk of any unwanted event is reduced by the installation of safety systems, and especially in the case of loss of cooling capacity, the usual safety barrier is the steam valve closure. However, for some distillation systems, where the difference of volatility of the components is high, the simple safety chain suggested might not be enough to guarantee a safe unit shut down. The dynamic effect of accumulated heat in the system might provide energy enough causing the re-vaporization of the most volatile compound. In this case the risk of releasing volatiles to the atmosphere remains.

By using simplified heat balances and construction and operation data of the column and peripherals, it is possible to evaluate the amount of volatile compounds that can be released to the atmosphere.

This simplified approach can be used in different distillation systems and it is an alert to prevent organic emissions to the atmosphere. In addition, it is possible to assess, case by case, the risk level. In continuation to the study, a second approach will be carried out using dynamic simulation to quantify the rate of emission of volatiles to the atmosphere.

6. Nomenclature

λ_{steam}	enthalpy of vaporization (steam)
λ_p	enthalpy of vaporization (most volatile compound)
$\Delta T_{(i/j)}$	difference of temperature between i and j.
$holdup_{+vol}$	*holdup* of the most volatile compound
CW	cooling water
Cp_i	specific heat of the i
m_{water}	amount of water in the condenser
m_{bottom}	amount of product in the bottom of the column
m_{metal}	amount of metal (column and peripherals)
$m_{metalcond}$	amount of metal (condenser tubes)
m_{calc}	possible evaporated amount of volatile compound
m_{evap}	possible evaporated amount of volatile compound, considering the *holdup* of the most volatile compound
$m_{emitted}$	amount of volatile compound released via vent system
m_{steam1}	amount of steam in the reboiler
m_{steam2}	amount of steam released from steam valves after the actuation of the safety system, due to the inertia of the valves
Q_1	heat transferred due to the steam in the reboiler
Q_2	heat transferred due to the steam flow after the security system actuation (the inertia of the steam valves)
Q_3	heat transferred due to the temperature and composition differences between top and bottom of the column
Q_4	heat transferred due to the metallic parts (column and reboiler tubes)
Q_{cond}	amount of heat removed due to presence of water and metal in condenser
T_f	equilibrium (final) temperature

6. References

[1] Rhodia Responsible Care Function, *Procedure-Process Safety Risk Analysis*; Rhodia S.A., 2007.

[2] N.Ramzan, F. Compart, W.Witt, Application of Extended HAZOP and Event-Tree Analysis for investigating Operational Failures and Safety Optimization of Distillation Column Unit; *Process Safety Progress* vol. 26, (3), 2007.

[3] S. Werner, W. Fred and Compart, Assessing Safety in Distillation Column Using Dynamic Simulation and Failure Mode and Effect Analysis (FMEA), *Journal of Applied Sciences,* vol. 7, (15), 2007.

[4] Rhodia Responsible Care Function, *Process Safety Guide-Assessing and Reducing Risks*, Rhodia S.A, 2008.

(26f) Pareto Optimization On Component Inspection Interval for Level Control in An Oil/ Gas Separation System

Xiaole Yang, Artie Mc Ferrin Department of Chemical engineering, Mary Kay O'Connor Process Safety Center, College Station, TX

Carl D. Laird, Artie McFerrin Chemical Engineering Department, Texas A&M University, College Station, TX

Dr M. Sam Mannan, Mary Kay O'Connor Process Safety Center, College Station, TX

This paper deals with component inspection interval optimization of an oil/ gas separation system in offshore plants. In this study, mathematical modeling was developed for assessing the operational risk focusing on overflow scenario in oil/ gas separation system, and component inspection interval were optimized. A numerical pareto optimization technique based on an evolutionary algorism and a technique using a scaling factor to represent the weights of trade-off objectives are used in this study. Pareto optimal solutions were generated to represent the optimal inspection budget and scheduling of pump, control valve, and level transmitter in the system. The results of the optimization process yields an immediate choice of component inspection interval sets for whatever relative weighting is considered as the most appropriate one for the actual design problem by the final decision makers.

See more of this Session: Case Histories of Risk Assessment and Problem Solving

See more of this Group/Topical: Topical 1: Global Congress on Process Safety

(28a) Federal View of IST From the CSB Perspective

John Bresland, United States Chemical Safety Board, Washington, DC

The CSB is an independent federal agency charged with investigating industrial chemical accidents. Headquartered in Washington, DC, the agency's board members are appointed by the President and confirmed by the Senate.

The CSB conducts root cause investigations of chemical accidents at fixed industrial facilities. Root causes are usually deficiencies in safety management systems, but can be any factor that would have prevented the accident if that factor had not occurred. Other accident causes often involve equipment failures, human errors, unforeseen chemical reactions or other hazards. The agency does not issue fines or citations, but does make recommendations to plants, regulatory agencies such as the Occupational Safety and Health Administration (OSHA) and the Environmental Protection Agency (EPA), industry organizations, and labor groups. Congress designed the CSB to be non-regulatory and independent of other agencies so that its investigations might, where appropriate, review the effectiveness of regulations and regulatory enforcement.

The CSB's investigations and mission gives it a unique perspective of IST that crosses legislation, regulations, industry, trade groups, unions, and public boundaries. The CSB Chair will present this perspective.

See more of this Session: Inherently Safer Design/Inherently Safer Technologies (Part II)
See more of this Group/Topical: Topical 1: Global Congress on Process Safety

6TH GLOBAL CONGRESS ON PROCESS SAFETY

Inherently Safer Concepts: The American Chemistry Council Position

Laurie A. Miller
Director, Process Safety
American Chemistry Council
Arlington, Virginia
Laurie_Miller@americanchemistry.com

Peter N. Lodal
Technical Fellow
Plant Protection Technical Services
Eastman Chemical Company
Kingsport, Tennessee
pnlodal@eastman.com

Prepared for Presentation at
American Institute of Chemical Engineers
2010 Spring Meeting
6th Global Congress on Process Safety
San Antonio, Texas
March 22-24, 2010

UNPUBLISHED

Abstract

Inherently Safer (IS) concepts have received increasing attention by Congress, nongovernmental organizations and other public entities in the post-9/11 era. ACC and its member companies believe that IS is one of a number of valuable risk reduction tools within an overall risk management framework. This paper outlines the ACC position.

Introduction

The American Chemistry Council (ACC) represents over 130 leading companies engaged in the business of chemistry. The business of chemistry is an important part of our nation's economy, directly employing nearly 850,000 Americans and producing 19 percent of the world's chemicals. ACC member companies manufacture products critical to everyday items that keep the economy moving and are essential to developing the greener, cleaner, more competitive economy the nation seeks. More than 96 percent of all manufactured goods in the U.S. are directly touched by the business of chemistry. Our members provide the chemistry that is used to produce life-saving medications and medical devices, body armor used by our military and law enforcement officers, lightweight components for vehicles, energy-saving insulation and windows, components for solar panels, wind turbine blades, and so much more.

Because of our critical role in the economy and our responsibility to our communities, both safety and security are priorities for ACC members. ACC is firmly committed to a risk-based approach to both process safety and security in order to effectively reduce potential risks. In 1988, ACC adopted the Process Safety Code, which has been modified twice, most recently in 2008, to better reflect member performance and promote continuous improvement. In 2001, our members adopted an aggressive security program that became the Responsible Care® Security Code (RCSC). It is part of the overall Responsible Care initiative, which is ACC's signature program of ethical principles and management systems designed to continuously improve our members' safety, health, product stewardship, environmental, and security performance. Implementation of Responsible Care is mandatory for all members of the American Chemistry Council, as well as for Responsible Care Partner companies, who represent chemical transporters, distributors, warehouses, logistics planners and others along the supply chain.

Inherently Safer Technology

In the post 9/11 era, this concept has taken on a renewed emphasis. Most recently, the Center for Chemical Process Safety (CCPS) has published its second edition of "Inherently Safer Chemical Processes: A Life Cycle Approach" (ref 1), which contains a wealth of information and numerous pages of references regarding all aspects of inherent safety (IS).

The traditional approach to IS was defined by Trevor Kletz nearly 40 years ago. Kletz defined four main categories for implementation of IS:

1. Substitute
2. Minimize
3. Moderate
4. Simplify

While this framework provides a useful grounding in the fundamental concepts of IS, over time it has significantly limited IS application, due to the "either-or" nature of the alternatives. By definition, IS concepts are consequence-based; that is, IS techniques apply only to the magnitude of the outcome of an event, not to its probability of occurrence. Thus, using IS as an exclusive decision-making criterion precludes the use of some of the best risk reduction methodologies available to industry and government. A true risk-based approach must include consideration of both parts of the risk equation to arrive at an appropriate conclusion. Therefore, ACC believes that IS, as a consequence-only technique, can only be properly used in a holistic risk management framework, and is an important tool in the risk minimization toolkit. The sound basis and usefulness of using IS within a risk framework have been demonstrated in a number of different applications, as shown by the volume of material in the reference section of the CCPS IS book (ref 1). An important aspect of this approach is that it allows evaluation of whether an IS method actually reduces risk or merely shifts risk from one potentially exposed population to another.

Each facility is unique in the chemicals it produces, how these chemicals are produced, handled and stored, and in its community setting. Risk-based evaluation of a chemical process for a potential modification to improve its inherent safety (or an "inherently safer technology (IST)) involves evaluating the modification for its potential effect on the integrity of the chemical product or process and for its own set of potential risks to the public or the environment. The economic feasibility of modification also requires consideration.

Evaluation of one or more of these factors may indicate that the modification would not significantly reduce risk, may shift significant risk elsewhere and/or be technologically or economically infeasible. Even substitution of one chemical for another that may be considered a safer alternative, once thoroughly analyzed in the context of the process in which it will be applied may not adequately address risk or be feasible. Under a risk-based framework, a facility that finds IST alternatives to be infeasible has the flexibility to apply other measures, or "layers of protection" to the process that would not be considered under a consequence-based IS evaluation.

Because of the complexities involved in determining IST alternatives, we are concerned about the efficacy of any government regulatory framework that would evolve out of a mandate to force facilities to implement IST, particularly if it is based on a consequence-only approach. Even if government experts would be included in the regulatory decision-making process, such a mandate could easily drive DHS to simplify the complex by developing a streamlined regulatory approach which out of necessity would be prescriptive rather than address site-specific safety and security risks.

ACC therefore believes that because the current Chemical Facility Antiterrorism Standards (CFATS) already consist of risk-based performance standards, provisions in proposed security legislation that would give DHS authority to mandate process changes are unnecessary. Application of risk-based performance standards not only gives facilities the flexibility to consider all possible risk-reduction options, including "methods to reduce consequences" or "inherently safer" approaches when developing a site security plan, they also provide substantial incentive for facilities in the highest risk tiers - and facing the most stringent requirements and compliance costs - to implement enhancements that could move their facilities to a lower risk tier, or potentially out of the program. While mandating innovation for some IS applications may be straight forward, it is not in the case for all. Consider the following examples:

Example 1

A chlorine user is mandated to switch from the use of one ton (2200 lb) cylinders to 160 lb cylinders, thus reducing the overall ERPG-3 dispersion footprint from 0.5 miles to 200 feet under the most common atmospheric conditions. It has been determined that this change will improve overall health and safety.

In reality, the "improvement" to safety may be illusory, as:

1. There may be no permanent population between 200 feet and 0.5 miles (remote location), so the size of the ERPG-3 plume is irrelevant
2. The number of cylinder changeovers has now increased by a factor of 13. Since the most likely time for a cylinder to leak is when the connections are changing, and since the occupancy at changeover is 100% (there will always be someone present at the cylinder when it is swapped out), the actual risk to the operator has now increased by more than an order of magnitude.

Example 2

A chemical facility is prohibited from using a pneumatic (nitrogen at pressure) test for a 6500 foot in-plant pipeline, being mandated to use the "safer" hydrostatic testing method (water).

In reality, the hydrostatic test method is strongly preferred since it is, in effect, not building a pressurized "bomb". However, the pipeline in question will be handling a strongly water-reactive chemical and should any residual water be left in the pipeline, when chemicals are introduced, a potentially catastrophic chemical reaction with resulting gas generation and potential pipeline burst has a much higher probability of occurrence.

The solution to this problem is to overcome the "inherently unsafe" nature of the pneumatic test by instituting extraordinary safety measures, such as testing in segments, complete evacuation of the test area while under pressure (e.g., weekend testing), and highly detailed leak checking procedures prior to initiating the test.

In Conclusion

The risk-based provisions in CFATS will do more to unleash the ingenuity and resources of the chemical sector in the area of IS than any government mandated IS program. **ACC's position remains that IS is one of a number of valuable risk reduction tools within an overall risk management framework.** Congress should not abandon a strategy to enhance security that employs performance-based security standards to hold facilities accountable but avoid the potential for shifting risk. We agree with Congress that our shared priority is to enhance both safety and security at sites nationwide. The crucial partnership between our industry and the federal government requires each of us to do our part. ACC and its member companies are committed to safeguarding America's chemical facilities, and will continue to work with Congress and all regulatory authorities in that spirit.

Regulating Inherent Safety
Contra Costa County's Industrial Safety Ordinance

Randall L. Sawyer

Contra Costa Health Services

4585 Pacheco Blvd.

Martinez, CA 94553

rsawyer@hsd.cccounty.us

Prepared for Presentation at
American Institute of Chemical Engineers
2010 Spring Meeting
6th Global Congress on Process Safety
San Antonio, Texas
March 22-24, 2010

UNPUBLISHED

Regulating Inherent Safety
Contra Costa County's Industrial Safety Ordinance

Randall L. Sawyer
Contra Costa Health Services
4585 Pacheco Blvd.
Martinez, CA 94553
rsawyer@hsd.cccounty.us

Abstract

There were numerous accidents that occurred in Contra Costa County during the late 1980's and throughout the 1990's. The public and the County's Board of Supervisors were very concerned about the impact these accidents had on the community. The Board of Supervisors decided to expand on any other process accident prevention regulations that existed in the United States to prevent these accidents from occurring in the future. The Industrial Safety Ordinance, which requires facilities to consider Inherently Safer Systems, was approved by the Board of Supervisors and became effective in January 1999. The paper discusses the implementation of applying inherent safety and how Contra Costa Health Services ensures that the requirements are being met.

1. Introduction

The Contra Costa County Board of Supervisors, California, approved the County's Industrial Safety Ordinance in December 1998; it became effective on January 17, 1999. The ordinance was passed to prevent accidents such as the accidents that occurred at Contra Costa County's Petroleum Refineries and Chemical Plants in the late 1980's and 1990's. The Industrial Safety Ordinance expands on the California Accidental Release Prevention Program, which is more stringent than the Federal Risk Management Program. One element of the Industrial Safety Ordinance requires regulated stationary sources to implement inherently safer systems wherever feasible. By requiring stationary sources to consider inherent safety, a reduction of hazards has been accomplished. Since the Industrial Safety Ordinance was implemented, the number of major chemical accidents or releases has decreased.

This paper discusses what has worked and what has not worked by regulating covered facilities. The paper also shows how inherent safety is being considered and implemented by the covered facilities.

2. Background

Contra Costa County is across the San Francisco Bay from San Francisco and is the home to four petroleum refineries and many small to medium sized chemical plants. More than one million people reside in the County. There are nineteen cities in the County including Richmond, Concord, Antioch, Pittsburg, and Walnut Creek.

The California Legislature approved the Risk Management and Prevention Program in 1986, one of the first process safety management regulations in the country. California replaced the Risk Management and Prevention Program in 1997 with the California Accidental Release Prevention Program, which is the Federal Risk Management Program that was made more stringent. Contra Costa Health Services Hazardous Materials Programs has been responsible for implementing the state programs in Contra Costa County.

2.1 Risk Management and Prevention Program

The Risk Management and Prevention Program required facilities that used more than a threshold quantity of an "Acutely Hazardous Material" to submit a Risk Management and Prevention Plan to the local Administering Agency, when requested to do so by the Administering Agency. This program also required facilities to reduce the possibility of accidental releases of toxic chemicals that could impact the public. The statute required facilities to perform Hazard and Operability Studies to assess the risk of accidents and reduce it with additional safeguards when the risk was found to be too high.

Contra Costa County started implementing the Risk Management and Prevention Programs requirements in 1989 and 1990 and the result of requiring facilities to submit a Risk Management and Prevention Plan was a 34 percent decrease in the amount of Acutely Hazardous Materials handled in the County. Figure 1 shows a breakdown of some of the reduction for specific Acutely Hazardous Materials from the largest quantity of the chemical handled in the County between 1990 and 1994 and the quantity in 1994.

The businesses in Contra Costa County started to consider and apply inherently safer changes to their operations as a result of implementation of the Risk Management and Prevention Program. There have been continued and dramatic reductions in the inventories of Acutely Hazardous Materials in the County. All of the wastewater treatment plants have eliminated the use of sulfur dioxide, chlorine, and ammonia through substitution of a different chemical or using ultraviolet light process. All but three of the water treatment plants have eliminated chlorine and ammonia through substitution. All of the refineries have eliminated chlorine use through substitution. Most

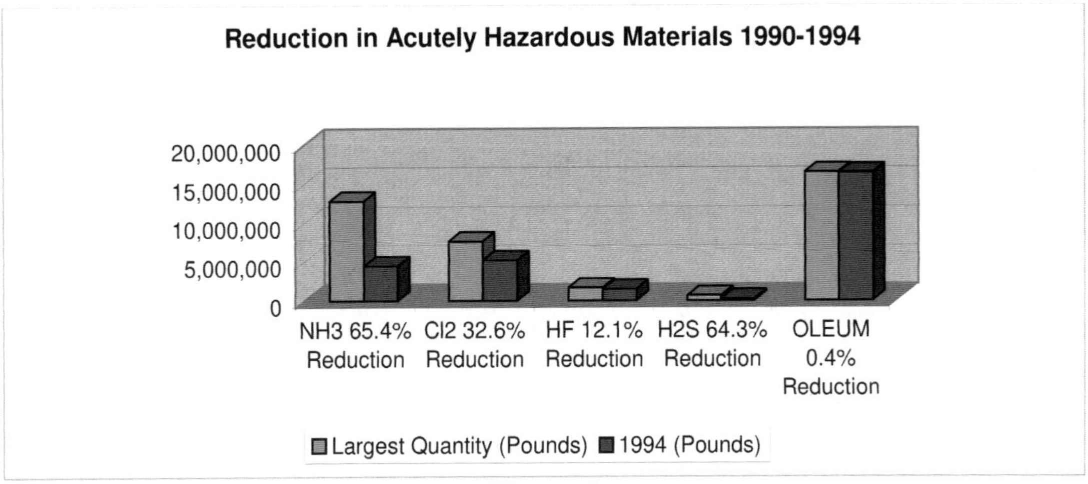

Figure 1

of the selective catalytic reduction process for control of nitrogen oxides has substituted ammonium hydroxide for ammonia. Since 1994, all but one of the large public swimming pools has substituted sodium or calcium hypochlorite or a chlorine generator for chlorine. A spent sulfuric acid regeneration plant reduced the amount of oleum stored by 90% and also reduced the concentration from 35% to 25% by changing the way they produce an electronic grade sulfuric acid. A refinery reduced the amount of ammonia storage from 8.9 million pounds to 1.7 million pounds and also moved the storage location approximately one mile further away from the community then where it was being stored.

2.2 Industrial Safety Ordinance

The Risk Management and Prevention Program was replaced by the California Accidental Release Prevention Program on January1, 1997. The County's Board of Supervisors' concern about accidents that occurred in the County in the late 1980s and 1990s (see Table 1 and Figure 3) led to the passage of an Industrial Safety Ordinance in 1998. The ordinance requires more prevention elements and more public participation for refineries and chemical plants than either the Federal Risk Management or the California Accidental Release Prevention Programs. The ordinance additional requirements include the following:

- Expands the application of human factors beyond the Process Hazard Analysis
- Root Cause Analysis as part of Incident Investigations
- Facilities to perform Management of Organizational changes when there is a change in the number of positions in operations, maintenance, and Health and Safety
- Facilities to implement Inherently Safety Systems for new and existing process
- Perform a Security and Vulnerability Analysis;
- Perform Safety Culture Assessments.

Significantly, the entire facility is covered under the Ordinance, not just the processes with a regulated substance above a threshold quantity.

3. Inherently Safer Systems

Industrial Safety Ordinance §450-8.014(g) defines Inherently Safer System as:

> *"'Inherently safer systems' means "inherently safer design strategies" as discussed in the latest edition of the Center for Chemical Process Safety Publication* <u>*Inherently Safer Chemical Processes*</u>*, and means feasible alternative equipment, processes, materials, lay-outs, and procedures meant to eliminate, minimize, or reduce the risk of a major chemical accident or release by modifying a process rather than adding external layers of protection. Examples include, but are not limited to, substitution of materials with lower vapor pressure, lower flammability, or lower toxicity; isolation of hazardous processes; and use of processes which operate at lower temperatures and/or pressures."*

Industrial Safety Ordinance §450-8.016(d)(3) states when inherently safer systems are to be considered as follows:

> *"For all covered processes, the stationary source shall consider the use of inherently safer systems in the development and analysis of mitigation items resulting from a process hazard analysis and in the design and review of new processes and facilities. The stationary source shall select and implement inherently safer systems to the greatest extent feasible. If a stationary source concludes that an inherently safer system is not feasible, the basis for this conclusion shall be documented in meaningful detail."*

3.1 Initial Implementation of Inherently Safer Systems

Regulated stationary sources are required to implement inherently safer systems to the greatest extent feasible for new processes and facilities and also in the development and analysis of mitigation items from process hazard analysis. Regulated stationary sources are to follow guidance provided by the latest edition of the Center for Chemical Process Safety's (CCPS) Inherently Safer Chemical Processes A Life Cycle Approach book.

Regulated stationary sources are required to implement strategies for reducing risk, including reducing the frequency or the consequences of potential accidents. The strategies can be classified into four categories:

- Inherent – Eliminating the hazard by using materials and process conditions which are non-hazardous; e.g., substituting water for a flammable solvent.
- Passive – Minimizing the hazard by process and equipment design features that reduce either the frequency or consequence of the hazard without the active functioning of any device; e.g., the use of equipment rated for higher pressure.
- Active – Using controls, safety interlocks, and emergency shutdown systems to detect and correct process deviations; e.g., a pump that is shut off by a high-level switch in the downstream tank when the tank is 90% full. These systems are commonly referred to as engineering controls.
- Procedural – Using operating procedures, administrative checks, emergency response, and other management approaches to prevent incidents or to minimize the effects of an incident; e.g., hot-work procedures and permits. These approaches are commonly referred to as administrative controls.

These risk reduction strategies as it applies to inherent safety is illustrated by the diagram shown in Figure 2:

Regulated stationary sources are to apply the different approaches for implementing inherent safety as follows:

- Minimization – Use smaller quantities of hazardous substances (also called *Intensification*)
- Substitute – Replace a material with a less hazardous substance

- Moderate – Use less hazardous conditions, a less hazardous form of a material, or facilities that minimize the impact of release of hazardous material or energy (also called *Attenuation* or *Limitation of Effects*)
- Simplify – Design facilities that eliminate unnecessary complexity and make operating errors less likely, and that are forgiving of errors that are made (also called *Error Tolerance*)

Contra Costa Health Services (CCHS) Hazardous Materials Programs is required to audit and inspect each of the regulated stationary sources at least once every three years. As part of the audit, CCHS ensures that the regulated stationary sources are implementing inherently safer systems to the greatest extent feasible. During the first round of audits, CCHS found that when regulated stationary sources tried to apply inherently safer systems to Process Hazardous Analyses mitigations, the distinction between inherent safety and risk reduction was lost. CCHS found that any risk reduction implemented was considered inherent safety. Regulated stationary sources did use the inherent safety strategies in considering risk reduction, which was an improvement in the consideration of what risk reduction methods to use. Regulated Sources were using risk reduction strategies and not always applying the highest level of these strategies, which include inherent strategy.

TABLE 1

Summaries of major chemical or refinery accidents in Contra Costa County in 1990's

Date Accident Occurred	Accident Description	Onsite Impact	Offsite Impact
March 25, 1999	Petroleum fire that required the sounding of sirens. A valve stem blew out causing a release of hydrocarbons and hydrogen sulfide that ignited, resulting in a major fire.	Shutdown a major part of the refinery. Two people slightly injured fighting the fire.	Smoke was visible over much of the North Bay Area. Disruption in traffic, sirens sounded, people were asked to shelter-in-place.
February 23, 1999	Flash fire at a crude unit killing four people and seriously injuring another.	Shutdown of the crude unit and later the whole refinery. Four people died with another seriously injured.	None.
April 23, 1997	An upset in the distillation unit sent hydrocarbons to the sulfur recovery units. The sulfur recovery units released hydrogen sulfide.	Shutdown parts of the refinery until the problem was found and the different units were stabilized.	People were asked to shelter-in-place.
January 27, 1997	A hydrocracker had a runaway reaction that raised the temperature and pressure of the outlet gases. An outlet pipe failed and caused an explosion and fire. One worker was	One worker was killed and 46 other sought medical attention. Many of the people seeking medical attention were injured while running away from the explosion and fire. The outlet piping and	There was little or no offsite impact from this explosion and fire. People were not asked to take any protective action, such as sheltering place.

Summaries of major chemical or refinery accidents in Contra Costa County in 1990's

Date Accident Occurred	Accident Description	Onsite Impact	Offsite Impact
	killed and 46 people sought medical attention.	some of the equipment and instrumentation was damaged. The hydrocracker unit was shutdown until the unit could be repaired.	
May 1996	A hot coke drum was dumped and a fire was initiated.	Fire damage to the coker unit. The coker unit was run at partial capacity until repairs could be made.	People were asked to shelter-in-place. A few people sought health care assistance complaining of the smoke and odors.
April 1, 1996	Fire at their cracked gas unit. The fire lasted more than three hours.	Fire damage was extensive onsite. The unit was shutdown until repairs could be made.	People were asked to shelter-in-place downwind of the refinery. The smoke from the fire impacted the area downwind.
February 1, 1996	Piping failed because of corrosion and a release of hydrogen occurred with an explosion. The fire lasted approximately 10 minutes.	Shutdown of the hydrogen unit.	Noise of the explosion was heard offsite for well over a mile. Some minor damage offsite.
June 16,	Petroleum tank fire that continued	Response by the facility along with local	Smoke impacted the surrounding

Summaries of major chemical or refinery accidents in Contra Costa County in 1990's

Date Accident Occurred	Accident Description	Onsite Impact	Offsite Impact
1995	to burn off and on for approximately one week.	fire departments.	communities. The communities were asked to shelter-in-place. Odors lingered in the communities for a week after the fire. The refinery put up approximately 100 people in motels until the incident was 100% secure.
August 22 - September 6, 1994	A column used to strip carbon dioxide developed a leak. The column continued to operate for a 16-day period. The hole in the column became larger where a solution of Catacarb® was atomized and released. Estimates of between 80-225 tons of Catacarb® was released.	The Catacarb® affected the health of many of the workers downwind of the column. At the end, the hydrocracker processing unit was shutdown.	The release impacted the surrounding communities. Many homes and cars were cleaned to remove the Catacarb® solution. A health clinic setup by the refinery that treated more than 1,200 people over the next year.
July 26, 1993	Overheating of a tank car of 35% oleum led to a rupture disk relieving that released between 4 and 8 tons of sulfur trioxide over a	The loss of the contents of a tank car of 35% oleum and the shutting down of their processing.	Over 22,000 people sought medical assistance, with 15-22 people staying in the hospital overnight. People were asked to

Summaries of major chemical or refinery accidents in Contra Costa County in 1990's

Date Accident Occurred	Accident Description	Onsite Impact	Offsite Impact
	three hour and forty-five minute period. The sulfur trioxide formed a sulfuric acid cloud.		shelter-in-place downwind of the release.
May, 1992	Sulfuric acid mixed with hydrocarbon spilled. The hydrocarbons ignited. The fire was difficult to put out because the sulfuric acid would break down the foam used by the fire fighters. One employee was killed and another was seriously injured.	Fire damaged some of the equipment onsite. One person died and another was seriously injured	Smoke went offsite and impacted the community. The smoke also included the degradation products of the sulfuric acid (e.g., sulfur trioxide, sulfur dioxide). Shelter-in-place was called for the community downwind of the fire.

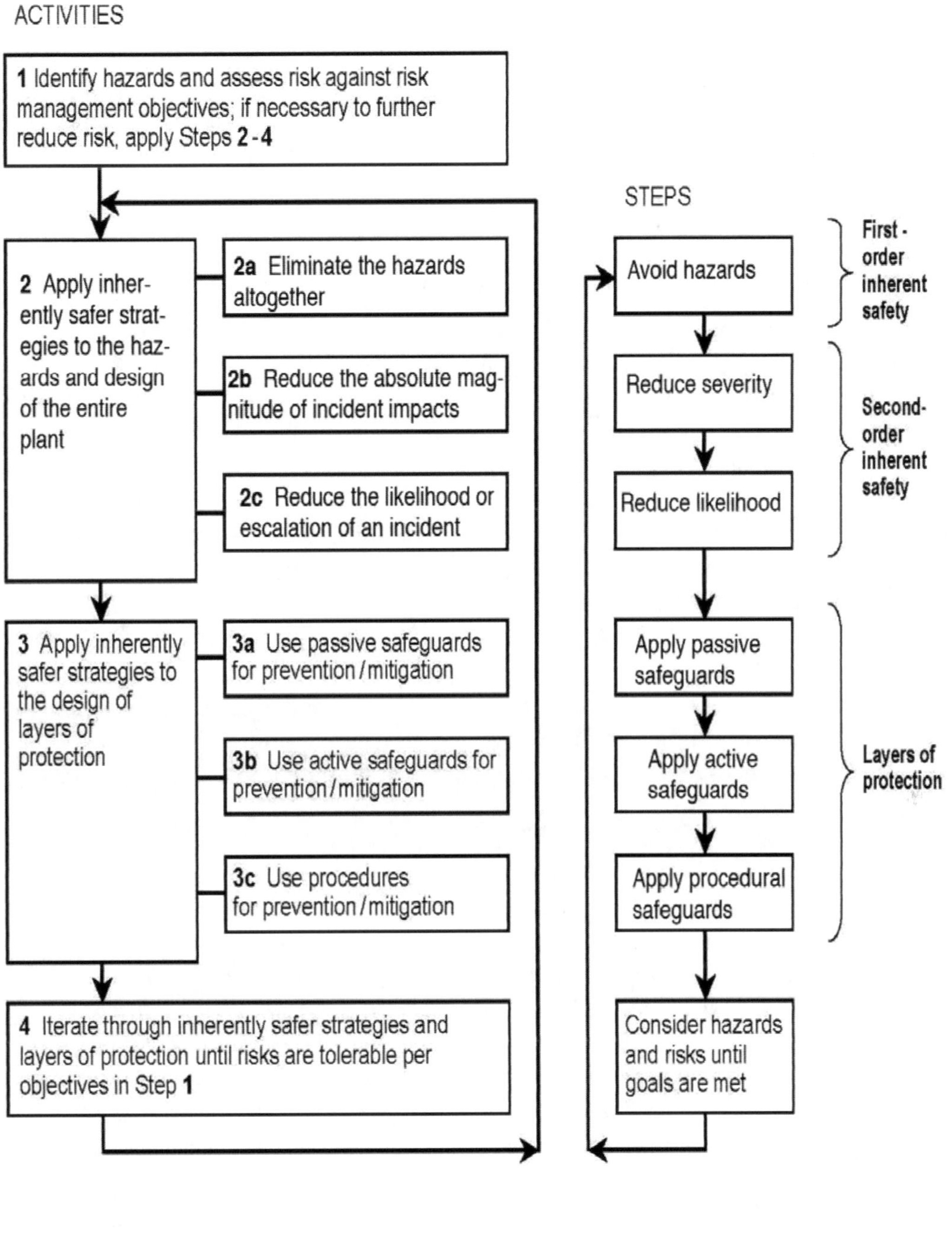

Figure 2

Inherent Safety Considerations in Process Risk Management (adapted from Amyotte, et al., 2006)[1]

---

[1] Center for Chemical Process Safety, Inherently Safer Chemical Processes, A Life Cycle Approach, 2nd Edition, 2009, page 16

3.2 Inherently Safer Systems Guidance

It became clear to CCHS that additional guidance was needed on when and how to consider inherent safety. CCHS worked with the regulated stationary sources and developed this guidance, which was broken down into seven sections. The first section addresses new covered processes; the second addresses existing processes; the third addresses mitigations resulting from Process Hazard Analysis (PHA); the fourth defines feasibility; the fifth addresses recommendations from process hazard analyses; the sixth addresses Inherently Safer System Reports; and the seventh contains definitions. The first four sections are discussed below.

3.2.1 New Process or Facilities

Inherent safety must to be considered for new processes at different phases in the development of a project. Inherently Safer Systems should be reviewed early in the development phase of a new covered process and then reviewed throughout the different project design phases. The objectives for an inherent safety review are to employ synergistic teams. Composition of the review teams will vary at different phases of the development cycle of the project and with the nature of the process. In product development and design scope, the team may comprise of chemists, process design engineers, industrial hygienists, safety engineers, environmental, and control engineers etc. In the hazard and operations review phase, the team may include operations and maintenance personnel as well. A stationary source may choose not to use an Inherently Safer System review team at a particular phase. If this does occur, the stationary source should document the reasons for not using a team for performing an Inherently Safer System review during this particular phase. A multi-disciplined group should be used throughout the various phases of development, implementation, and operation. The team's purpose is to:

- Understand the hazards
- Find ways to reduce or eliminate the hazards

> *"The first major objective for the inherent safety review is the development of a good understanding of the hazards involved in the process. Reducing and eliminating hazards and their associated risks is the second major objective. Applying inherent safety principles early in the product/process development effort provides the greatest opportunity to achieve the objectives of the inherent safety review process for the project at hand."*[2]

The stationary source should use a review process for new processes that includes an Inherently Safer Systems Review at different phases of the design process, such as the following phases when applicable:

- During the chemistry forming (synthesis) phase for product/process research and development to focus on the chemistry and process

[2] Center for Chemical Process Safety, Inherently Safer Chemical Processes, A Life Cycle Approach, 1996

- During the facilities design scoping and development prior to completion of the design basis to focus on equipment and configuration
- During the basic design phase of the project

3.2.2 Existing Processes

The Industrial Safety Ordinance requires that stationary sources consider hazards as part of the process hazard analyses.

> *"The process hazard analysis shall be appropriate to the complexity of the Covered Process and shall identify, evaluate, and control the hazards involved in the Covered Process. The process hazard analysis shall address: the hazards of the process; the identification of any previous incident which had a likely potential for catastrophic consequences; engineering and administrative controls applicable to the hazards and their interrelationships such as appropriate application of detection methodologies to provide early warning of releases."*[3]

Stationary sources should use one of the following methods to ensure that inherently safer systems (inherent and passive categories) are considered and documented for the covered processes:

- An independent inherently safer system analysis that is done in addition to a Process Hazard Analysis (PHA)[4]. These analyses should review the covered processes for ways to eliminate or reduce hazards that are present in the covered process. This may be achieved by using a checklist or guideword analysis that incorporates inherently safer systems. If the stationary source decides to use some other inherently safer systems checklist or other methods to evaluate inherently safer systems, Contra Costa Health Services must approve these prior to their use.

- An inherently safer system analysis that is incorporated into the existing process hazard analysis review process. This would require that each covered process in its entirety has an initial inherently safer systems analysis conducted. (Incorporating inherently safer systems into a revalidated process hazard analysis may not be sufficient to satisfy the initial inherently safer system review if the whole process is not evaluated.) This may be achieved using a checklist or guideword that incorporates inherently safer systems considerations into a Hazard and Operability Study where a Major Chemical Accident or Release could reasonably occur. These analyses would review the covered processes for ways to eliminate or reduce hazards as well as risks that are present in the covered process.

---

[3] Contra Costa County's Industrial Safety Ordinance, Chapter 450-8 §450-8.016(d) (1)

[4] If the stationary source decides to do an independent inherently safer systems analysis, CCHS suggests that this be done in conjunction with the process hazard analysis, but it may be appropriate for a stationary source to perform an inherently safer systems analysis that is done at a different time than the process hazard analysis. Either approach meets the guidance from this document, as long as the inherently safer systems analysis is revalidated at least once every five years.

3.2.3 Process Hazard Analysis Recommendations and Mitigations

The concepts as addressed in the CCPS book Inherently Safer Chemical Processes A Life Cycle Approach, Second Edition for looking at all four categories of risk reduction (Inherent, Passive, Active, and Procedural) should be used in the development and mitigation of recommendations from process hazard analysis as well as for considering Human Factors. Chapter 6 of the CCPS book Inherently Safer Chemical Processes A Life Cycle Approach, Second Edition discusses many inherently safer system strategies that can be incorporated in the development of mitigations to address the recommendations from process hazard analysis.

The stationary sources should provide guidance to personnel responsible for developing and analyzing recommendations and mitigation items resulting from the unit PHA. The guidance should include the concepts of inherently safer systems including:

- The different categories of risk reductions
- Moving up the different levels from procedural to active to passive to inherent
- Approaches to apply inherently safer systems including minimization, substitution, moderation, and simplification

3.2.4 Feasibility

One of the areas of greatest concerns to the regulated stationary sources is what would be considered feasible. The fourth section of the guidance defines feasibility. To assist in the determination of feasibility, Contra Costa Health Services is using a modification of guidance from OSHA and the U. S. EPA. (Federal OSHA provided guidance for justifiably declining recommendations from incident investigations in the September 1994, OSHA Instruction CPL 2-2.45A CH-1. These criteria have since been applied to recommendations formulated during Process Hazard Analysis.) The version used by Contra Costa Health Services uses the following criteria for infeasibility**:**

- The analysis upon which the recommendations are based contains factual errors.
- The recommendation is not necessary, i.e., the safeguards may be inadequate, but the consequences are operational or the consequence or severity of the scenario would not result in a Major Chemical Accident or Release.
- An alternative inherently safer system would provide a sufficient level of hazard reduction (NOTE: Implementing only one option to address identified hazards may not be adequate to address the greatest hazard reduction or elimination. However, it is not necessary to implement more than one inherently safer system if the implementation of a second inherently safer system does not add any significant hazard reduction or has been documented as infeasible.)
- The recommendation is in conflict with existing federal, state, or local laws.
- The recommendation is in conflict with Recognized and Generally Accepted Good Engineering Practices (RAGAGEP).
- The recommendation is economically impractical, such that the process unit can no longer be financially operated. This determination can include the following factors:
    - Capital investment
    - Product quality

  - Total direct manufacturing costs
  - Operability of the plant
  - Demolition and future cleanup and disposal cost
- The recommendation would have a negative social impact such that the project should not be implemented. Some examples of social impact are a visual or noise impact on the community that is not acceptable or increased traffic congestion.
- The recommendation may violate a license agreement and the license agreement cannot be modified and must remain in effect.
- The recommendation may decrease the hazard, but would increase the overall risk.
- An alternative measure would provide more risk reduction than the inherently safer system.

If the inherently safer system recommended is determined not to be implemented because it will create more risk, or if other modifications that are not inherently safer system are made such that the overall risk is less than if the inherently safer system were implemented, the stationary source will need to document how this determination was made. A qualitative risk assessment/analysis could be used as a basis for the determination. The facility needs to document how they determined the qualitative severity and likelihood for the existing or modified conditions and for the conditions if the Inherently Safer System is implemented. If the qualitative risk analysis shows the same level of risk, then a quantitative risk assessment/analysis should be performed to compare the risk of the existing or modified conditions to the risk if the Inherently Safer System is implemented. The documentation should include the background information that was used to do the comparison of the existing or modified conditions to the conditions if the Inherently Safer System is implemented. Another method may be used by the stationary source, such as a weighted scoring decision matrix as shown on page 23 of CCPS book Inherently Safer Chemical Process A Life Cycle Approach, First Edition if the method is approved by Contra Costa Health Services prior to the use of the method.

County Ordinance Code Chapter 450-8, §450-8.016(d)(3) requires the following: "This documentation shall include: (1) sufficient evidence to demonstrate to the County's satisfaction that implementing this inherently safer system is impractical, and (2) the reason for this conclusion." The documentation should include the applicable background information, calculations, and the reasons why an inherently safer system was not implemented. If there is any concern that the reasons for not implementing an inherently safer system would not satisfy Contra Costa Health Services, the stationary source should consult with Contra Costa Health Services to determine if the justification is satisfactory. The stationary source should then receive in writing from Contra Costa Health Services their decision and how they came to their decision, including background information, calculations, and alternatives considered.

3.3 Implementing Inherently Safer Systems

The regulated stationary sources have applied inherently safer systems considerations to their existing and new processes since the guidance has been completed; include applying independent Inherently Safer System analyses and incorporating the Inherently Safer

System evaluation as part of their existing process hazard analyses. Also, the regulated stationary sources have applied the guidance for any new processes or facilities, and in the mitigations of recommendations from process hazard analyses. Applying inherently safer systems to process hazard analyses mitigations allows a regulated stationary source to use the thought process for implementing risk reduction measures, including the first and second orders of inherent safety. Table 2 shows some of the Inherently Safer Systems that have been implemented by the regulated stationary sources.

4. Effect of the Industrial Safety Ordinance including Inherently Safer Systems

Figure 3 shows the incidents that occurred during the 1990's prior to the passage of the County's Industrial Safety Ordinance where a fatality, serious injuries, or major onsite and/or offsite damage occurred, which is the definition of a Severity Level III Major Chemical Accident or Release as defined below.

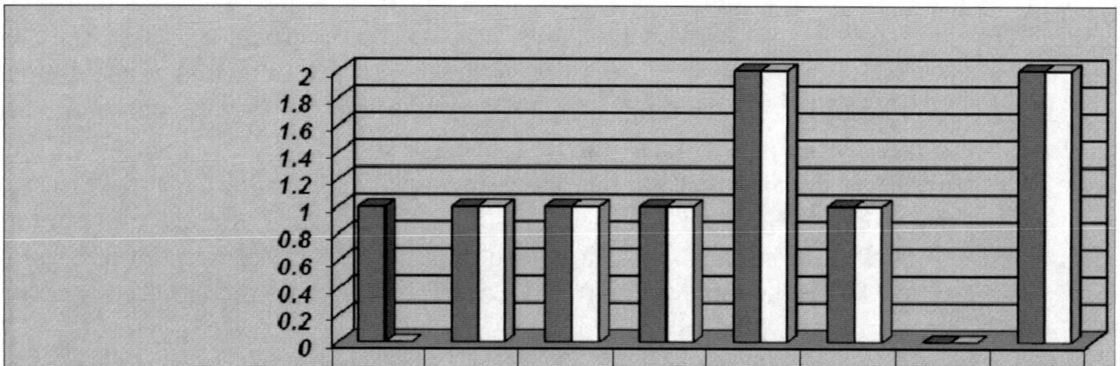

	1992	1993	1994	1995	1996	1997	1998	1999
■ All facility	1	1	1	1	2	1	0	2
□ County & Richmond ISO	0	1	1	1	2	1	0	2

Figure 3
Major Chemical Accident or Release Severity Level III

Figure 4 indicates that since the Industrial Safety Ordinance was implemented, the number of major chemical accidents or releases has decreased. A major portion of the County's ordinance is inherently safer systems. Further analysis and audits over a longer period of time will be needed to determine how much of the reduction may be attributed to the requirement to implement the above described Inherently Safer System process.

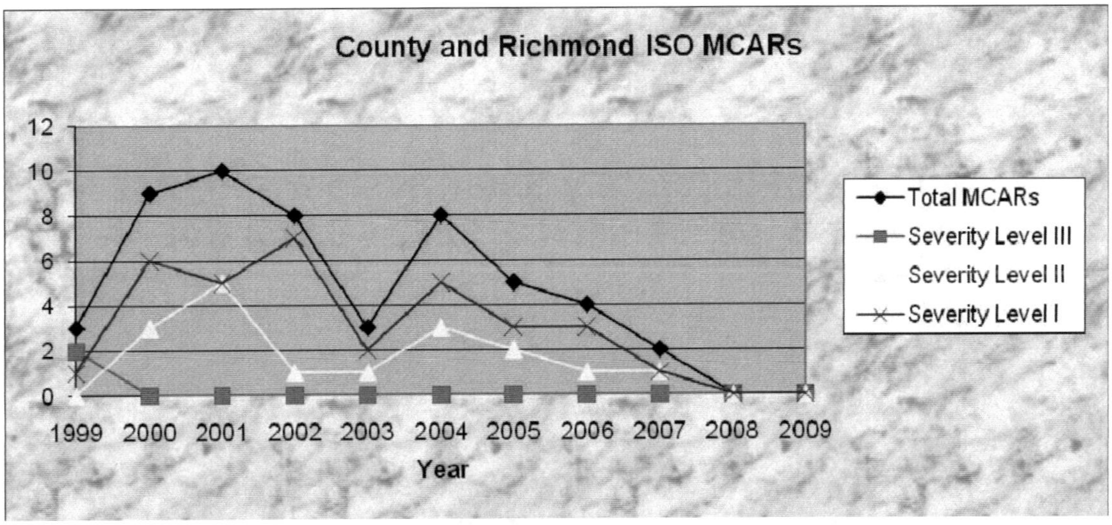

Figure 4
Major Chemical Accident or Releases

Major Chemical Accidents and Releases are defined in the Industrial Safety Ordinance. The types of Major Chemical Accidents can be broken down further as shown in Figure 4 and defined as follows:

- Severity Level III – A fatality, serious injuries, or major onsite and/or offsite damage occurred
- Severity Level II – An impact to the community occurred, or if the situation was slightly different the accident may have been considered major, or there is a recurring type of incident at that facility
- Severity Level I – A release where there were no injuries or minor injuries, the release had no impact or slight impact to the community, or there was no damage or minor onsite damage

TABLE 2

Summary of Inherently Safer System and Risk Reduction Implemented in Contra Costa County

2003 - 2009

Risk Management Strategies	Design Strategies	Actions	Frequency
Inherent:	Minimization	Reduction of inventory	51
		Eliminated Process Unit	1
		Eliminated Hazard	2
		Reduced equipment and piping	50
	Substitute	Used less hazardous chemicals	2
	Moderate	Reduced the hazard by changing hazard conditions form	7
		Decreased concentration	1
	Simplify	Simplified process	11
Passive:	Minimization	Revised equipment design features	1
	Substitute	Change of energy source	1
	Moderate	Design and material of construction changes	90
		Relocated chemicals to a safer location	9

5. Conclusions

Contra Costa Health Services has found that risk management and accident prevention regulations alone have reduced the amount of acutely hazardous materials that are handled in Contra Costa County by 34 percent within the first four years of requiring such a program. CCHS has also found that it was necessary to provide detailed guidance to ensure consistent and thorough implementation of inherently safer concepts by the regulated stationary sources. By requiring stationary sources to consider inherent safety, a reduction of hazards has been accomplished. Is it necessary for regulations to be developed to ensure application of inherently safer systems analyses and a reduction of hazards? Requiring stationary sources to consider inherent safety has been effective in Contra Costa County in reducing the hazards that exist at regulated stationary sources. Industry and the process safety community have the responsibility and obligation to share knowledge and encourage the use of inherent safety beyond what is required.

6TH GLOBAL CONGRESS ON PROCESS SAFETY

Modeling Small Releases from Tanks

Fredrick H. (Rick) Knack, PE, CSP
Knack Engineering, PC
13016 Eastfield Rd., Ste 200-334, Huntersville, NC 28078
rick.knack@knackengineerng.com

Michelle L. Brown
FMC Lithium Division
PO Box 795, Bessemer City, NC 28016
michelle.brown@fmc.com

Fredrick W. (Rick) Knack
Virginia Polytechnic Institute and State University
106 Lucas Drive, Apt 2, Blacksburg, VA 24060
rknack@vt.edu

Prepared for Presentation at
American Institute of Chemical Engineers
2010 Spring Meeting
6th Global Congress on Process Safety
San Antonio, Texas
March 22-24, 2010

UNPUBLISHED

Modeling Small Releases from Tanks

Fredrick H. (Rick) Knack, PE, CSP
Knack Engineering, PC
13016 Eastfield Rd., Ste 200-334, Huntersville, NC 28078
rick.knack@knackengineerng.com

Michelle L. Brown
FMC Lithium Division
PO Box 795, Bessemer City, NC 28016
michelle.brown@fmc.com

Fredrick W. (Rick) Knack
Virginia Polytechnic Institute and State University
106 Lucas Drive, Apt 2, Blacksburg, VA 24060
rknack@vt.edu

Keywords: Release modeling, Water-ethanol solution; Carbon dioxide

Abstract

The methodologies in the chapter on "Dispersion Modeling" in the CCPS book *Guidelines for Consequence Analysis of Chemical Releases*, provide means for modeling outdoors releases at relatively long distances from a large source. At times it is necessary to model releases at relatively short distances from a small source. Making the assumption that the model as developed is valid at the near range, it was used to demonstrate that very low levels of ethanol are emitted from indoor mixing tanks and that accidental operator exposure to hydrogen bromide fumes in a lithium bromide manufacturing operation is acceptably safe. This paper presents the development, implementation, and results of using the model for small releases from tanks.

1. Introduction

From time to time, we have wanted to be able to predict the concentration of emissions from tanks or reactors that are indoors in order to estimate whether more action was required to ensure safety of personnel and facilities. There are two specific cases which we will discuss here.

The first case was a fire code issue. A manufacturer of alcoholic beverages had grown from a small distilling and blending facility in the early 1900s to a major supplier of liqueurs and other products. Over the course of time, they had introduced more and more tankage to their warehouse-like facility. These were storage vessels containing material with flashpoints ranging from approximately 23°C (120 proof) to 33°C (52 proof) but the material still classified as flammable liquids and so NFPA 30 and OSHA required that they be vented to the outside. For the purposes of discussion with the authorities, the company wanted an estimate of the flammable envelope around the tank vents.

The second case did not involve regulatory or code issues. A company was conducting a reaction in an open vessel that produced carbon dioxide. As with the storage tanks mentioned above, this was a legacy process that did not have all the controls that it would have if built today. A question came up in discussions as to whether, in upset conditions, carbon dioxide could be released to the surrounding area in concentrations sufficient to cause harm to personnel.

As we considered the issues and reviewed the available references on modeling, we proposed that, as a first estimate, we could use a model that had been developed for outdoor releases. We recognized that this required making strong assumptions about the validity of such a model, but felt that for discussion purposes, it was a place to start.

Our results showed that for the ethanol tanks, a flammable concentration could be expected up to approximately 3 meters from the tank at floor level or approximately 4.5 meters at the release level. For carbon dioxide releases, we estimated that there would be a small area of about 18 cubic meters (1 meter high by 24 meters wide by 4.5 meters long) where an operator could be exposed to approximately one-fourth the IDLH concentration of carbon dioxide.

2. Model development

The CCPS book *Guidelines for Use of Vapor Cloud Dispersion Models* provides useful background on vapor cloud dispersion models, however, we found the information in *Guidelines for Consequence Analysis of Chemical Releases* to be more complete and therefore, that volume became our primary reference. Chapter 2 discusses source models including discharge rate models, flash and evaporation models, and dispersion models. We determined that the plume model described in section 2.3.1.2 and shown below in Figure 1 was the most suitable for the cases we wanted to examine.

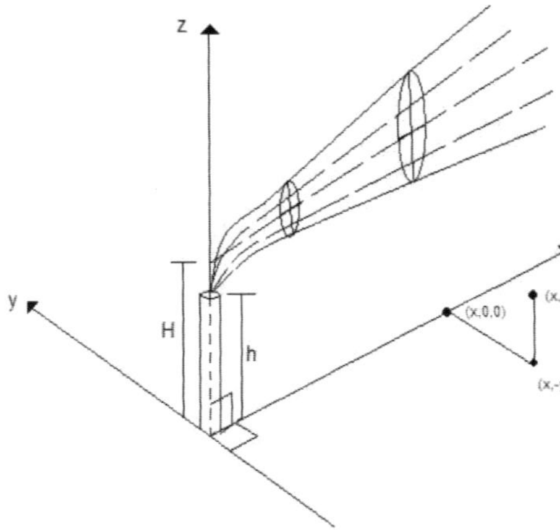

Figure 1. Plume Release

In this model, the average concentration at a point x, y, z downstream from a release is calculated using the equation:

$$\langle C\rangle(x,y,z) = \frac{G}{2\pi\sigma_y\sigma_z u}\exp\left[-\frac{1}{2}\left(\frac{y}{\sigma_z}\right)^2\right]\left\{\exp\left[-\frac{1}{2}\left(\frac{z-H}{\sigma_z}\right)^2\right] + \exp\left[-\frac{1}{2}\left(\frac{z+H}{\sigma_z}\right)^2\right]\right\} \quad \text{[Eq. 1]}$$ [1]

where

$\langle C\rangle(x,y,z)$ is the average concentration (mass/volume);
G is the continuous release rate (mass/time);
σ_x, σ_y, and σ_z are the dispersion coefficients in the x, y, and z directions (length);
u is the wind speed (length/time);
x is the downwind distance (length);
y is the crosswind direction (length);
z is the distance above ground (length); and
H is the height of the source above ground level plus the plume rise (length).

Values for σ_y, and σ_z are found from formulas in *Consequence Analysis of Chemical Releases* Table 2.12. For Pasquill-Gifford stability class F (the most stable), these are:

$$\sigma_y = 0.04x(1 + 0.0001x)^{-1/2} \quad \text{[Eq. 2]}$$

and

$$\sigma_z = 0.016x(1 + 0.0003x)^{-1} \quad \text{[Eq. 3]}$$

for rural terrain.

Similarly:

$$\sigma_y = 0.11x(1 + 0.0004x)^{-1/2} \quad \text{[Eq. 4]}$$

and

$$\sigma_z = 0.08x(1 + 0.0015x)^{-1/2} \quad \text{[Eq. 5]}$$

for urban terrain.

Then for any distance, x, equation 1 may be solved for concentration.

For releases above ground level, the maximum ground concentration occurs downwind along the centerline and the location of the maximum is found using the value for the vertical dispersion coefficient at that location:

$$\sigma_z = \frac{H}{\sqrt{2}} \quad \text{[Eq. 6]}$$

---

[1] The equations are from Chapter 2 *Consequence Analysis of Chemical Releases*, This is Equation 2.61. The other equations are not referenced directly.

The ground concentration at that point is found from:

$$(C)_{max} = \frac{2G}{e\pi u H^2}\left(\frac{\sigma_z}{\sigma_y}\right) \quad \text{Eq. 7]}$$

where e is the natural logarithm base.

The procedure for finding the maximum concentration and its downwind distance is:

1. Use equation 2 to determine the dispersion coefficient, σ_z, at the maximum downwind distance.
2. Use the equation for σ_z taken from Table 2.12 (examples above) to determine the location of the maximum ground concentration.
3. Use equation 3 to determine the maximum concentration.

Examining equations 2 through 5, it is seen that σ_y and σ_z become zero at the point of release (x = 0). Considering this and examining equation 1, it is seen that for small distances from the source, the predicted value for concentration becomes large and at the source itself becomes infinite. This is not an issue for situations when an approximation of the concentration is desired a long distance from the source. For concentrations near the source, however, a correction must be applied. Reference 2 suggests that the appropriate correction is to define a "virtual source", (x_v) some distance upwind from the true source which gives the actual concentration at the source. Assuming that at the virtual source $x_v = y_v = z_v$. Then from equation 1

$$\sigma_y(x_v)\bullet\sigma_z(x_v) = \left(\frac{G}{\pi u\langle C\rangle^*}\right) \quad \text{[Eq. 8a]}$$

or, rearranging

$$\langle C\rangle^* = \frac{G}{\sigma_y\sigma_z\pi u} \quad \text{[Eq. 8b]}$$

evaluated at x_v, where $\langle C\rangle^*$ is the concentration at the source.

Equation (8b) is solved by trial and error, guessing values for x_v, calculating σ_y and σ_z, calculating $\langle C\rangle^*$ and using that to refine the next assumption of x_v. Having identified the virtual source location, concentrations at locations of interest or alternatively distances to concentrations of interest can be readily found.

Using the model to calculate the concentrations of carbon dioxide, the material being released from the reactor can be considered to be pure carbon dioxide. The ethanol concentrations required one more step, specifically estimating the concentration at the release point and the liquid in the tank is an ethanol water-solution

In applying this to our situation, the concentrations of interest are the concentrations above the liquid in a tank of a given proof value and temperature and, if that concentration is above the LFL, the distance from a tank vent that the concentration will be reduced to below the LFL.

Calculation of the concentration of ethanol in the vapor space above the liquid in a tank at various temperatures and liquid concentrations is done using the common equation for vapor-liquid equilibrium ratios

$$K_i = y_i / x_i \qquad \text{[Eq. 9]}$$

where:

K_i is the equilibrium coefficient for substance i

y_i is the mole fraction of i in the vapor phase

x_i is the mole fraction of i in the liquid phase

K_i is calculated from the equation

$$K_i = \gamma_i \Phi_i^L / \Phi_i^V \qquad \text{[Eq. 10]}$$

where

γ_i is the activity coefficient for substance I and

ϕ_i^L and ϕ_i^V are the mixture fugacity coefficients of substance i in the liquid and vapor phases respectively.

Using the Magules representation, the activity coefficients are calculated from the equations:

$$\begin{aligned} \gamma_1 &= \exp\{[A_{12} + 2(A_{21} - A_{12})x_1]x_2^2\} \\ \gamma_2 &= \exp\{[A_{21} + 2(A_{12} - A_{21})x_2]x_1^2\} \end{aligned} \qquad \text{[Eq. 11]}$$

The A_{ij} coefficients have been determined empirically and are tablulated. For ethanol (1), water (2) solutions, $A_{12} = 1.6022$ and $A_{21} = 0.7947$.

To verify that the equations were valid and the concentration of ethanol in the vapor was accurate, they were tested against the tabulated values for flashpoint of various concentrations as shown below.

Table 1. Calculated versus Experimental Flash Points

Mole Fraction EtOH (x_1)	Proof	Flash Point (°C)	Flashpoint (Calculated) (°C)	Error Percent
50	153	20.6	21.1	2.4
40	137	21.7	22.5	3.7
32	120	22.6[2]	23.6	4.4
30	112	22.7	23.9	5.3
20	90	25.0	26.2	4.8
17	80	27.3	27.4	0.4
10	52	32.8	32.4	-1.2

While the agreement is not exact, it is well within the level of precision needed for our model.

The release rate for the materials must be known or estimated. For carbon dioxide, the rate was known to be 0.15 kg/s based on normal operations. For ethanol, the rate was estimated with filling being the worst case. During filling, the air and ethanol/water vapor over the top of the fluid in the tank is expelled at the same rate as the tank is being filled, approximately 30 – 40 gallons per minute (gpm). In order to be conservative, the 40 gpm rate was used in this analysis.

Finally, the wind speed must be estimated. A visit to the facility where ethanol was the material of concern in March while the building was essentially closed showed no strong odor of ethanol or other substances, reasonable comfort from a temperature and humidity standpoint, and no disagreeable human order. Based on this, it was estimated that the building experienced about 6 changes of air per hour[3] which would be equivalent to approximately 0.03 m/sec for air movement. The operation where carbon dioxide is produced is conducted with large doors open and a value of 0.2 m/s was used for the carbon dioxide modeling.

3. Results

Using the information and assumptions above, the following results shown in Tables 2 through 4 were calculated for the ethanol releases.

[2] NFPA values for 120 and 80 proof were interpolated from tabulated values.

[3] Two to 6 air changes per hour are recommended for warehouses. In this case, because of the potential for significant odors and the lack of their presence, the higher change rate is used.

Table 2. Distance to LFL (Ground Level Release)

Proof	Temperature of Interest (°C)	Initial Concentration (% EtOH)	Distance to LFL[4] (m)
80	32	4.74	2.1
90	27	3.96	0.7
90	32	5.37	3.2
120	27	4.54	1.8
120	32	6.20	4.5

In the case in question, the release would take place at the top of a tank and the distance of interest is not the direct horizontal distance, but rather the distance to a height where an ignition source might be found. While the tanks were of different heights, 2.4 meters is a reasonably representative height. Looking at this, we see that the distances to the LFL is approximately 70% of that found previously.

Table 3. Distance to LFL (2.4 meters Release Height, Horizontal)

Proof	Temperature of Interest (°C)	Initial Concentration (% EtOH)	Distance to LFL (m)
80	32	4.74	1.5
90	27	3.96	0.5
90	32	5.37	2.3
120	27	4.54	1.2
120	32	5.84	3.1

The following table shows the distance from the release point at 2.4 meters above floor level to a point 1.8 meters above floor level, approximately an operator's head level.

Table 4. Distance To LFL (2.4 meters Release Height, 1.8 meters Above Floor)

Proof	Temperature of Interest (°C)	Initial Concentration (% EtOH)	Distance to LFL (m)
80	32	4.74	0
90	27	3.96	0
90	32	5.37	0.8
120	27	4.54	0
120	32	5.84	1.8

Thus, we see that, based on this model, only in the case of 90 or 120 proof material at 32°C or above is there any potential for a flammable mixture to occur at a level

[4] The lower flammable limit (LFL) for ethanol in air is 3.3 percent by volume from NFPA 325, *Fire Hazard Properties of Flammable Liquids, Gases, and Volatile Solid*, 1994.

0.6 meters below the release point and in those cases, the distance is still relatively low. Figure 2 shows an example of the results for 90 proof material at 32°C.

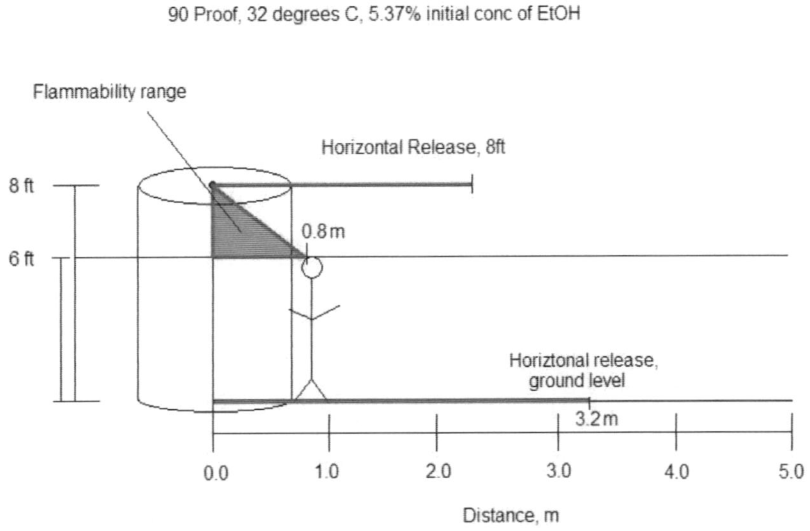

Figure 2. Example of Flammable Region

The results for the carbon dioxide modeling are shown in Table 5, below. For a release rate of 0.15 kg/s (based on normal operations), assuming a release height of 3 meters and a wind speed of 0.2 m/s, concentrations were calculated at "nose height", 1.7 meters as shown in Table 5. Note that the IDLH for carbon dioxide is 40,000 ppm.

Table 5. Calculated CO_2 Concentrations – Downwind

	Rural Terrain	Urban Terrain
Distance (m)	Concentration (ppm)	Concentration (ppm)
1	0	0
2	0	7
4	0	20,000
8	0	53,000
10	4	45,000
15	1700	27,000
20	11,000	17,000
Maximum Ground Concentration	6000 (112 m)	11,000 (22.5 m)

If the values are calculated at one-half meter and one meter off the wind, the values shown in Tables 6 and 7 are found.

Table 6. Calculated CO_2 Concentrations – 1/2 Meter Off Wind

	Rural Terrain	Urban Terrain
Distance (m)	Concentration (ppm)	Concentration (ppm)
1	0	0
2	0	0.5
4	0	1100
8	0	45,000
10	0.3	41,000
15	420	25,000
20	5000	17,000

Table 7. Calculated CO_2 Concentrations – 1 Meter Off Wind

	Rural Terrain	Urban Terrain
Distance (m)	Concentration (ppm)	Concentration (ppm)
1	0	0
2	0	0
4	0	1500
8	0	28,000
10	0.3	30,000
15	420	22,000
20	5000	15,000

Table 8 shows values for urban terrain for an approximate cube from 1 meter off the floor to 2 meters off the floor, two meters off the wind, and 4.5 to 9 meters from the source (approximately 18 cubic meters).

Table 8. Calculated CO_2 Concentrations – 2 Meters Off Wind

	Urban Terrain	
	1 m Off Floor	2 m Off Floor
Distance (m)	Concentration (ppm)	Concentration (ppm)
4.5	165	39
9	13,000	6,100

Recognizing that IDLH values are defined as the concentration of a hazardous agent that will produce irreversible health effects or death following a 30-minute exposure, and that the maximum concentration outside the cube is approximately one-third the IDLH for carbon dioxide, the cube represents a conservative estimate of the danger area for an operator. The area layout for carbon dioxide producing operations is approximately 5600 square meters equaling a volume of 5600 cubic meters from 1 to 2 meters off the floor. The potential hazard area for an operator is approximately 0.003 of this volume. Figure 3 shows the potential hazard area.

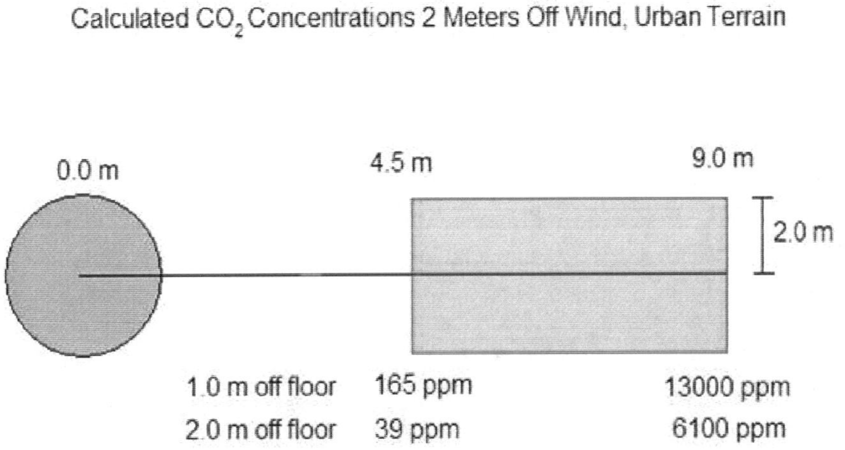

Figure 3. Potential Carbon Dioxide Hazard Area

4. Summary

We were able to use a plume model as described in *Guidelines for Consequence Analysis of Chemical Releases* to estimate the concentrations of ethanol and carbon dioxide at near range from storage tanks and reaction vessels. The results showed that the extent of flammable concentrations of ethanol and hazardous concentrations of carbon dioxide are small with the assumptions used. We recognize that the assumptions are, of necessity, strong, and therefore do not propose this methodology for making critical safety decisions without additional work. None the less, barring refutation of the assumption that outdoor, long-distance models can be adapted for indoor, small-distance use, we believe that this methodology provides a good rough-order-of-magnitude means for estimating hazards from such releases.

5. References

[1] *Guidelines for Use of Vapor Cloud Dispersion Models Second Edition,* Center for Chemical Process Safety of the American Institute of Chemical Engineers (AIChE), New York, 1996

[2] *Guidelines for Consequence Analysis of Chemical Releases,* Center for Chemical Process Safety of the American Institute of Chemical Engineers (AIChE), New York, 1999

[3] *Perry's Chemical Engineers Handbook*, 6th edition, McGraw-Hill, New York, 1984, Chapter 13

[4] NFPA 325, *Fire Hazard Properties of Flammable Liquids, Gases, and Volatile Solid*, 1994.

[5] NFPA values for 120 and 80 proof were interpolated from tabulated values.

[6] *Mechanical Engineering Reference Manual Ninth Edition*, Michael R. Lindeburg, P.E. Professional Publications, Inc. Belmont, CA 1994, Chapter 11.

[7] Documentation for Immediately Dangerous to Life or Health Concentrations (IDLH): NIOSH Chemical Listing and Documentation of Revised IDLH Values (as of 3/1/95)

[8] J. W. Vincoli, CSP, *Basic Guide to Industrial Hygiene*, Van Nostrand Reinhold, New York, 1995

Atmospheric storage tank explosion modeling

Jérôme TAVEAU
Institute for Radiological Protection and Nuclear Safety
Plants, Laboratories, Transports and Waste Safety Division
Industrial Risks, Fire and Containment Assessment and Study Department
Fire and Explosion Study and Assessment Section
31 avenue de la Division Leclerc
92 260 FONTENAY AUX ROSES cedex, FRANCE
jerome.taveau@irsn.fr

Jérôme RICHARD
Institute for Radiological Protection and Nuclear Safety
Plants, Laboratories, Transports and Waste Safety Division
Industrial Risks, Fire and Containment Assessment and Study Department
Fire and Explosion Study and Assessment Section
31 avenue de la Division Leclerc
92 260 FONTENAY AUX ROSES cedex, FRANCE
jerome.richard@irsn.fr

Prepared for Presentation at
American Institute of Chemical Engineers
2010 Spring Meeting
6th Global Congress on Process Safety
San Antonio, Texas
March 22-24, 2010

UNPUBLISHED

Atmospheric storage tank explosion modeling

Jérôme TAVEAU
Institute for Radiological Protection and Nuclear Safety
Plants, Laboratories, Transports and Waste Safety Division
Industrial Risks, Fire and Containment Assessment and Study Department
Fire and Explosion Study and Assessment Section
31 avenue de la Division Leclerc
92 260 FONTENAY AUX ROSES cedex, FRANCE
jerome.taveau@irsn.fr

Jérôme RICHARD
Institute for Radiological Protection and Nuclear Safety
Plants, Laboratories, Transports and Waste Safety Division
Industrial Risks, Fire and Containment Assessment and Study Department
Fire and Explosion Study and Assessment Section
31 avenue de la Division Leclerc
92 260 FONTENAY AUX ROSES cedex, FRANCE
jerome.richard@irsn.fr

Keywords: atmospheric storage tank, explosion modeling, FLACS, land-use planning

Abstract

Many flammable products are usually stored in large vessels at atmospheric pressure. Ignition of an hydrocarbon-air mixture in such tanks can lead to an explosion and cause lethal casualties or damage the surrounding facilities and buildings. To prevent that, safety distances shall be defined.

Several simple methodologies have been published allowing the calculation of safety distances. The study compares results from simple methodologies with more sophisticated calculations. Three-dimensional simulations of gas explosions using the CFD code FLACS have been performed to define a typical pressure load profile to apply on the inner side of the vessel's walls. The structural response of the tank (deformation and displacement), under the loading condition previously obtained, has been computed with LS-DYNA.

Consequently, we have been able to justify that some of the assumptions used in simple methodologies are not relevant. We also have established specific input data that should be used in order to obtain a reasonably conservative estimation of the safety distances.

1. Introduction

Many flammable products e.g. liquid hydrocarbons at normal temperatures and pressures such as gasoline, naphtha, heavy oil, kerosene, diesel oil, etc. are usually stored in large vessels at atmospheric pressure or under low pressure in smaller tanks (up to 0.5 bar) **[1, 2, 3]**. In those tanks, a more or less important amount of flammable gaseous hydrocarbon-air mixture may be present above the liquid phase (accidentally for floating roof tanks or continuously for fixed roof tanks). Then, assuming that a sufficient amount of energy is supplied to the mixture, it may ignite and lead to either or both an explosion and a fire (depending on the volume, the chemical and physical properties of the flammable mixture involved in the combustion process and also on the structural properties of the vessel).

When a tank explosion occurs, overpressure waves that are generated might cause devastating effects as in 2006 at Partridge-Raleigh oilfield, USA **[4]**. To prevent that, safety distances shall be defined (this is also a legal requirement in the EU with the Seveso II directive). Therefore, several simple methodologies (that we will call 'simple models') allowing the calculation of safety distances corresponding to overpressure thresholds, have been published (see for instance **[5, 6, 7, 8]**).

The study proposes guidelines for engineers in order to improve their safety distance assessment's methodology, based on a comparison of 'simple models' with more sophisticated calculations (LS-DYNA, FLACS).

2. Industrial accidents

J. I. Chang and C. C. Lin **[9]** have listed 242 accidents during the last forty years involving atmospheric storage tanks containing petroleum or chemicals:

- 18 accidents concerned fixed roof tanks;
- 6 accidents concerned fixed roof tanks with internal floating roof.

H. Persson & A. Lönnermark **[10]** have also listed 480 fires involving hydrocarbon storage tanks between 1951 et 2003 (table 1).

Table 1. Number of fires involving hydrocarbon storage tanks

Years	50's	60's	70's	80's	90's	2000-2003
Number of fires	13	28	80	135	161	62

Several atmospheric storage tank explosions have also occurred during the last thirty years; on the basis of the SP report **[11]**, Hailwood et al. identified 21 tank explosions followed by a fire. More recently, GEXCON has investigated a tank explosion **[12]** which happened on 24th May 2007 at Sløvåg in Norway.

So it seems the risk of explosion of a tank is not unlikely and needs to be correctly assessed.

3. Safety distance assessment methodologies

In France, the law n°2003-699 of July 30th 2003 adopted after the AZF disaster asks for the investigation of all representative scenarios.

Because of the potential consequences in terms of land-use planning, there is an increasing need to have relevant methodologies to evaluate safety distances for each kind of dangerous phenomenon, and in our case for tank explosions.

Numerous methodologies are used in France to estimate safety distances (table 2).

Table 2. Overview of methodologies' assumptions

Methodology	ΔP		F	Pressure decay curves
A	P_W ([1])		0.2	TNT
B	$H/D > 1$	1 bar	0.6	TNT
	$H/D \leq 1$	0,5 bar		
C	0.5 x P_W		0.6	TNT
D	1.5 bar		0.6	BAKER
E	-		-	TNT
F	2.5 bar		-	BK WAVE
G	-		-	TNO Multi-energy
H	2 to 9 bar		-	BAKER

Methodologies A, B, C & D: the French Ministry of the Environment (A, B), the French petroleum operators' association (C) and a French petroleum operator (D) use [Eq. 1] to calculate the energy available for pressure waves:

$$E = \underbrace{[\Delta P \times V / (\gamma-1)]}_{(1)} \times \underbrace{[1-F]}_{(2)} \times \underbrace{[1/(1+\varepsilon)^{3\gamma-3}]}_{(3)} \quad \text{[Eq. 1]}$$

E : energy [J]
ΔP : bursting pressure [Pa]
V : volume of the tank [m^3]
γ : specific heat ratio [dimensionless]
F : fracture distribution energy factor [dimensionless]
ε : carbon steel deformation at fracture factor [dimensionless]

The first factor ΔP x V / (γ-1) represents the Brode energy, which is an estimation of the energy released by the explosion.

[1] **tank working pressure**

The second factor (1-F) represents the energy loss due to ejection of fragments. According to Cox & Saville **[13]**, F equals to 0.2 for a brittle failure and 0.6 for a ductile rupture.

The third factor $1 / (1+\varepsilon)^{3\gamma-3}$ used in methodologies (C, D) represents the energy loss due to tank deformation. ε is assumed to be equal to 0.38 for carbon steel.

Some methodologies (A, D, E) also introduce the concept of TNT equivalence to calculate the energy available for pressure waves.

Then safety distances corresponding to a given overpressure threshold are estimated thanks to [Eq. 2]:

$$d_i = \lambda_i \times E^{1/3} \quad \text{[Eq. 2]}$$

E = energy [J]

λ_i = scaled distance regarding to energy $[m.J^{-1/3}]$

The scaled distance λ_i is a function of a given threshold and can be read using:

- TNT curves **[14]** for (A, B, C, E) methodologies;
- Baker curves **[15]** for (D, H) methodologies;
- TNO multi-energy curves **[7]** for (G) methodology.

Methodologies E & F: Electricité de France (EDF) has used a TNT equivalence method (E), but has also developed its own method called BKWAVE (F), which uses specific curves with different bursting pressures defined thanks to the BKWAVE software.

Methodology G: INERIS has also developed its own method called PROJEX, which is quite different from other methodologies. Near the tank, the peak overpressure is estimated using shock tubes theory and is supposed to be constant, whereas far away from the tank, the PROJEX methodology uses the TNO multi-energy curve n°10.

Methodology H: TNO has developed a method using the Brode energy and based on Baker curves. The energy is multiplied by a factor 2 to take into account the effect of the ground.

We can notice that these methodologies were originally developed to evaluate safety distances in case of pressure vessel bursting.

It can be argued that even though several types of tanks exist (fig. 1), same methodologies are used to estimate safety distances. But it is well-known that mechanical strength of tanks leads a predominant role in the assessment of safety distances. Only methodology B, developed by the working group "Flammable liquids depots" created by the French Ministry of the Environment in 2006, tries to take into account the influence of the H / D ratio by applying different bursting pressures (table 2).

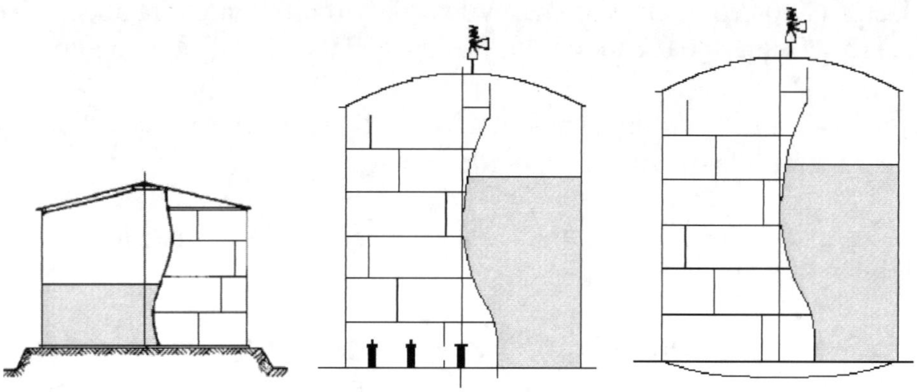

Figure 1. From the left to the right:
fixed roof tank, vertical tank with flat bottom, vertical tank with spherical bottom

4. Application of current methodologies to the Herne chemical site accident

In 1984, lightning provoked the bursting of a 10,000 m^3 isopropyl tank at the Herne chemical site in Germany. The roof was ejected and the overpressure broke windows of the surrounding houses up to 1 kilometer from the centre of the explosion.

Figure 2. Tank fire, Herne

In France, an overpressure of 20 mbar corresponds to the indirect effects threshold by broken windows. The corresponding safety distance has been calculated using data available from this accident, using current safety distance assessment methodologies.

Table 3: Application of current methods to the Herne chemical site explosion considering a 50% filled tank

Method	Safety distance (20 mbar[2])	Overpressure (500 m)	Calculated bursting pressure[3]
A	380 m	TNT chart overtaken	4.6 bar
B	240 m		9.2 bar
C	200 m		16.5 bar
D	230 m		41.3 bar
E	460 m		3.5 bar[4]
F	600 m	19 mbar	3.5 bar[5]
G	370 m	11 mbar	5.1 bar
H	360 m	11 mbar	5.3 bar

This example shows the huge differences between current methodologies' results. In this case, most of them underestimate safety distances. Only one method (F), proposed by A. LANNOY (EDF), gives reasonable results. Bursting pressure is largely overestimated and unrealistic in many cases (C, D). Charts employed by some methods (particularly TNT chart) do not allow estimating the safety distance corresponding to a 20 mbar overpressure with sufficient precision.

Nevertheless, these results need to be considered with caution, as breakage of windows also largely depends on the impulse, the type of window, pressure wave's reflection, etc.

5. Gas explosion and structural response simulations conducted by IRSN

5.1 Gas explosion simulations

Gas explosion simulations were conducted using the FLACS software in order to identify key parameters for a tank explosion assessment. The following parameters have been studied:

- roof-to-shell connection failure pressure;
- stoechiometric ratio;
- location of the ignition source;
- roof specific mass;
- H / D ratio;
- tank volume.

[2] When using TNT chart, the safety distance corresponding to a 20 mbar overpressure threshold in the table is obtained by multiplying by two the safety distance corresponding to a 50 mbar overpressure threshold
[3] Corresponding bursting pressure to obtain 20 mbar at 500 m
[4] Bursting pressure assumed in the methodology
[5] Bursting pressure assumed in the methodology

It has been assumed that the tank is full of vapors (no liquid) and is not anchored into the ground. It has also been supposed that only the roof is ejected because most of the tanks have weak (frangible) roof-to-shell attachment nowadays.

Tank roof has been modeled as a wall with an evolutive porosity. The roof is supposed to be non deformable and does not fracture, which is a conservative assumption because it maximizes containment of combustion products inside the tank. According to M. R. Baum **[16]**, the effect of the roof is neglected once it is distant from half of the tank radius.

The main conclusions of the simulations are:

- key parameters for the assessment of a tank explosion are, by order of importance: volume of gases, failure pressure, roof specific mass;
- bursting pressure can be superior from 16% to 51% to the failure pressure;
- fraction of burned gas depends on bursting pressure and stoechiometric ratio; it reaches only few percents for the simulated cases.

5.2 Structural response simulations

Structural response simulations were conducted with LS-DYNA software (fig. 3).

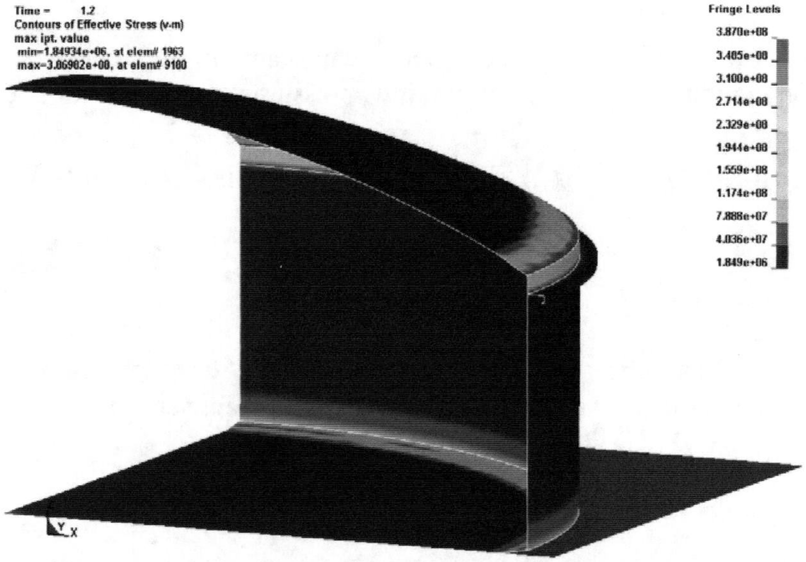

Figure 3. LS DYNA simulation for a compact tank showing the distribution of stress

LS-DYNA simulations mainly showed that:

- atmospheric tanks experience very limited plasticity, mainly at the junction between the tank's shell and the roof (figure 3) and only when the roof failure pressure is high;
- in this case, tank's base can lift up by half a meter;
- the tank's shell does not deform significantly (figure 3).

5.3 Energy conversion

When a tank explodes, the energy liberated by the combustion process is converted into different forms:

- pressure wave;
- kinetic energy, gravity forces and drag forces for the roof;
- tank's deformation.

Simulations were conducted using FLACS and LS-DYNA software in order to evaluate the fraction of energy devoted to each form.

Results showed in our study that only a small fraction of the energy is lost by:

- ejection of the roof (from 1% to 9%);
- tank's deformation (1%);
- gravity forces (1%) and drag forces (less than 1%).

These simulations also showed that the Brode energy is below the energy released by the combustion process.

5.4 Pressure wave propagation

FLACS simulations were performed to assess safety distances for two types of tanks:

- elongated tank (H / D > 1): D = 9,5 m; H = 14 m; P_F[6] = 1 bar (fig. 4abc) ;
- compact tank (H / D < 1): D = 60 m; H = 14 m; P_F = 0.5 bar (fig. 5abc).

[6] **roof failure pressure**

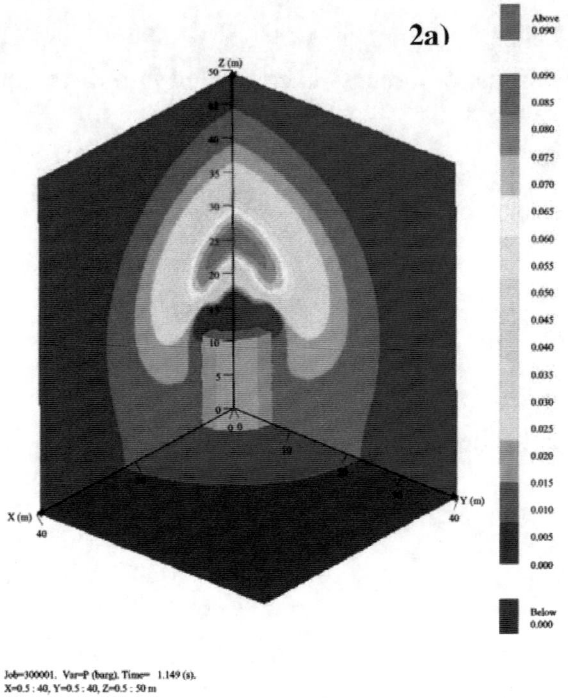

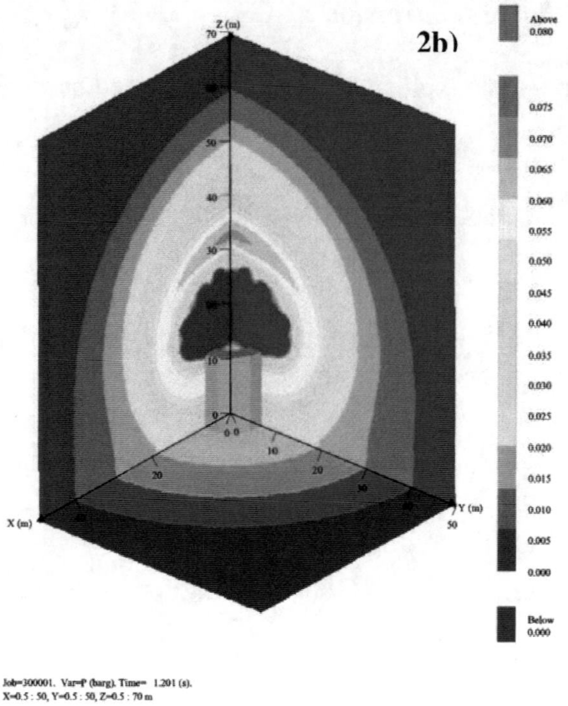

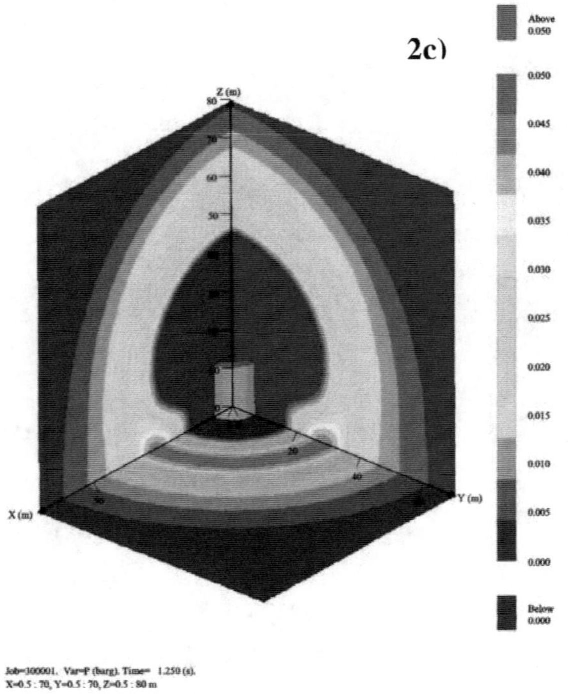

2a) After the ejection of the roof, the pressure wave begins to propagate above the tank (1.149 s after ignition)

2b) Then the pressure wave is expanding: after diffraction on the tank's shell, the pressure wave begins to propagate to the ground (1.201 s)

2c) Later, the pressure wave becomes hemispherical (1.25 s)

Figures 4abc. FLACS simulation of a tank explosion with H / D > 1

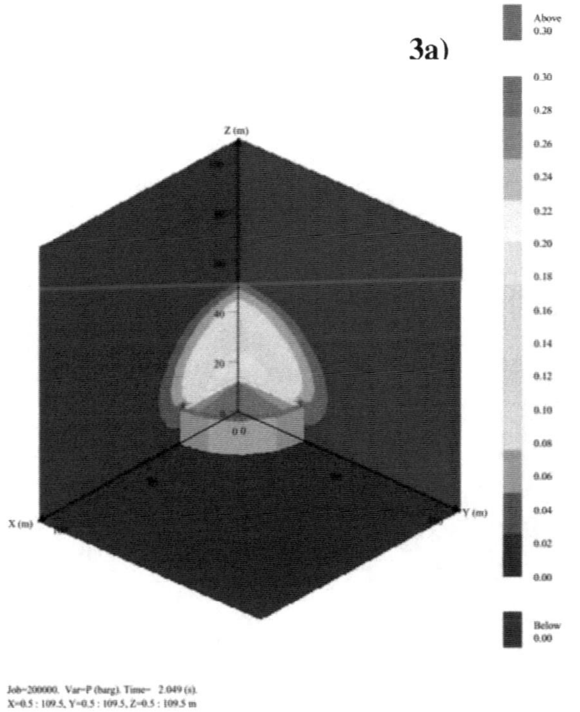

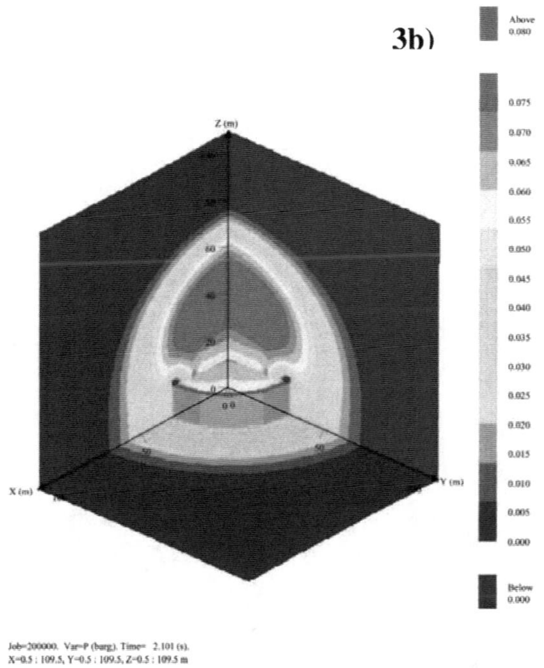

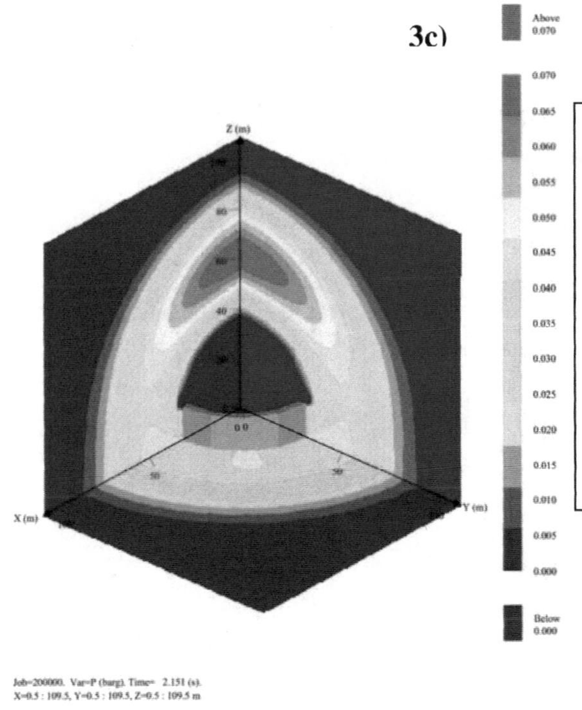

3a) The pressure wave begins to propagate outside (2.049 s after ignition)

3b) Diffraction of the pressure wave is less clear than in the previous case, because of the tank geometry (2.101 s)

3c) Later, we can see that the pressure wave is stronger along the vertical axis and weaker at the ground level (2.151 s)

Figures 5abc. FLACS simulation of a tank explosion with H / D < 1

From these figures, we can see that directional effects are very important for pressure effects: the 50 mbar overpressure threshold can be reached on the vertical axis whereas the 20 mbar overpressure threshold is not reached at the ground level. Also pressure wave propagation is very sensitive to the geometry of the tank. These results obviously depend on the assumptions used for the calculations (in our case, only roof is ejected).

Comparisons were made between all methodologies and FLACS simulations. An example (for a compact tank) is given at figure 6.

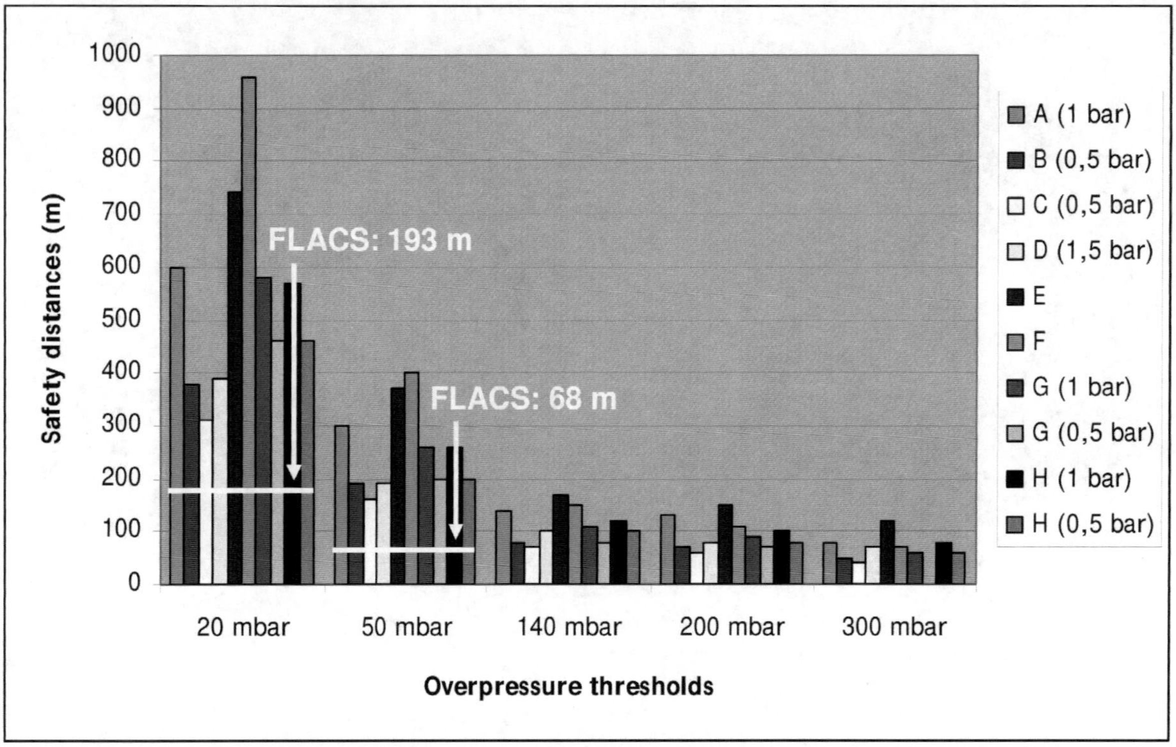

Figure 6. Comparison of methodologies for a tank with H / D < 1 (bursting pressure indicated into brackets)

6. Conclusion

Safety engineers need accurate methodologies to assess safety distances around their plants. After reviewing current safety distance assessment methodologies, this paper has presented the work done by IRSN in order to compare the assumptions made by 'simple models' with CFD modelling and evaluate the influence of several parameters for a tank explosion assessment.

Simulations of the combustion process with FLACS have showed that only a small fraction of gases effectively burns during a tank explosion; we have also noticed that a conservative approximation of the bursting pressure could be 1.5 times the failure pressure.

We have as well been able to justify that some of the assumptions of 'simple models' are debatable, and particularly:

- LS-DYNA simulations have showed that atmospheric tanks should not experience plasticity, or only in a limited proportion;
- FLACS and LS-DYNA simulations have demonstrated that only a small fraction of the energy is lost by the ejection of the roof (1 to 9%) and the deformation of the tank (less than 1%); these simulations have also showed that the Brode energy is below the energy released by the explosion.

These results are based on the assumptions used for our calculations.

IRSN is currently conducting some further studies (theoretical and numerical) in order to develop a new methodology for the assessment of tank explosion safety distances. Future goals of numerical simulations could be the study of the influence of the presence of an anchoring system or a large amount of liquid inside the tank at the moment of the explosion.

7. References

[1] R. Chamayou. *Réservoirs métalliques: stockage des liquides - Généralités.* BM 6 590, 1997, Éditions Techniques de l'ingénieur, Paris

[2] R. Chamayou. *Réservoirs métalliques: stockage des liquides à température ambiante.* BM 6 591, 2004, Éditions Techniques de l'ingénieur, Paris

[3] R. Chamayou. *Réservoirs métalliques pour stockage des liquides.* BM 6 593, 1997, Éditions Techniques de l'ingénieur, Paris

[4] *Case study. Hot work control and safe work practices at oil and gas production wells. Partridge-Rayleigh Smith County oilfield.* U.S. Chemical Safety and Hazard Investigation Board, report n°2006-07-I-MS, June 2007

[5] *Guide méthodologique pour la réalisation des études de dangers en raffineries, stockages et dépôts de produits liquides et liquéfiés.* Union Française des Industries Pétrolières, 2002

[6] *Modélisation des effets de surpression dus à une explosion de bac atmosphérique.* Groupe de Travail Dépôts Liquides Inflammables, 2006

[7] *Methods for calculation of physical effects. Chapter 7: Rupture of vessels.* 3rd Edition, revised print 2005, TNO, The Hague, The Netherlands

[8] L. Heudier. *Les éclatements de réservoirs, phénoménologie et modélisation des effets.* Rapport Ω-15, INERIS, octobre 2004

[9] J. I. Chang & C-C. Lin, *A study of storage tank accidents.* Journal of Loss prevention in the process Industries, Volume 19, pp.51-59, 2006

[10] H. Persson & A. Lönnermark. *Tank fires : Review of fire incidents 1951-2003.* SP Swedish National Testing and Research Institute, SP REPORT 2004:14

[11] M. Hailwood, M. Gawlowski, B. Schalau & A. Schönbucher. *Conclusions drawn from the Buncefield and Naples incidents regarding the utilization of consequence models.* Chemical Engineering Technology, Volume 32, Issue 2, pp. 207-231, 2009

[12] T. Skjold, K. van Wingerden, R. Abiven & Ø. Larsen. *Accident investigation following the Vest Tank explosion at Sløvåg.* GEXCON, 3rd revision, 26th September 2008

[13] B. G. Cox & G. Saville. *High Pressure Safety Code.* London, Imperial College, High Pressure Technology Assessment, 1975

[14] UFC 3-340-02. *Structures to resist the effects of accidental explosions.* December 2008

[15] W. E. Baker, P. A. Cox, P. S. Westine, J. J. Kulesz & R. A. Strehlow. *Explosion hazards and evaluation.* Elsevier, Amsterdam, 1983

[16] M. R. Baum. *Rupture of a gas-pressurized cylindrical vessel: the velocity of a detached end-cap*. Journal of Loss Prevention in the Process Industries, Volume 8, Issue 3, pp. 149-161, 1995

6TH GLOBAL CONGRESS ON PROCESS SAFETY

Update of "Guidelines to Vapor Cloud Explosion, Pressure Vessel Burst, BLEVE and Flash Fire Hazards"

Quentin A. Baker
Baker Engineering and Risk Consultants, Inc.
3330 Oakwell Court, Suite 100
San Antonio, TX 78218
(210) 824-5960
QBaker@BakerRisk.com

Adrian J. Pierorazio
BakerRisk Canada
5515 North Service Road, Suite 304
Burlington, Ontario L7L 6G4
Canada
(289) 288-0100
AdrianP@BakerRisk.com

John L. Woodward, Ph.D.
Baker Engineering and Risk Consultants, Inc.
3330 Oakwell Court, Suite 100
San Antonio, TX 78218
(210) 824-5960
JWoodward@BakerRisk.com

Ming Jun Tang, Ph.D.
Baker Engineering and Risk Consultants, Inc.
3330 Oakwell Court, Suite 100
San Antonio, TX 78218
(210) 824-5960
MTang@BakerRisk.com

Prepared for Presentation at
American Institute of Chemical Engineers
2010 Spring Meeting
6th Global Congress on Process Safety
San Antonio, Texas
March 22-24, 2010

UNPUBLISHED

Update of "Guidelines to Vapor Cloud Explosion, Pressure Vessel Burst, BLEVE and Flash Fire Hazards"

Quentin A. Baker
Baker Engineering and Risk Consultants, Inc.
3330 Oakwell Court, Suite 100
San Antonio, TX 78218
210-824-5960
QBaker@BakerRisk.com

Adrian J. Pierorazio
BakerRisk Canada

John L. Woodward, Ph.D.
Baker Engineering and Risk Consultants, Inc.

Ming Jun Tang, Ph.D.
Baker Engineering and Risk Consultants, Inc.

Keywords: Vapor cloud explosion, VCE, pressure vessel burst, PVB, boiling liquid expanding vapor explosion, BLEVE, flash fire

Abstract

The first edition of the CCPS book *Guidelines to Evaluating Vapor Cloud Explosions, Flash Fires and BLEVEs* was published in 1994. Since that time there have been considerable advances in research for these types of explosions and fires, and hazard prediction methods have improved. The second edition was completed in October 2010. It is a major revision of the *Guideline* with contents completely reorganized, recent research results added, revised and new prediction methods inserted, and sampled problems worked for each prediction method. The treatment of pressure vessel burst was expanded and placed in a dedicated chapter, resulting in a new name for the book as shown in the paper title.

1. Introduction

The second edition is intended to provide methods for practicing engineers to estimate the characteristics of a flash fire, vapor cloud explosion (VCE), pressure vessel burst (PVB), or boiling liquid expanding vapor explosion (BLEVEs). Knowledge of the consequences of flash fires, VCEs, PVBs, and BLEVEs has grown significantly in recent years as a result of international study and research efforts, and incident investigations. Insights gained regarding the generation of overpressure, thermal radiation, and fragmentation has resulted in the development of reasonably descriptive models for calculating the effects of these phenomena.

The book summarizes and evaluates these methods, identifies areas in which information is lacking, and provides an overview of ongoing work in the field. The arrangement of this book is considerably different from previous editions, including separating pressure vessel bursts into its own chapter. The book remains the most in-depth technical material produced in a CCPS project.

The phenomena of each type of hazard (i.e., VCEs, PVBs, BLEVEs, and flash fires) are addressed in individual chapters. These chapters include a description of the relevant phenomena, an overview of the related past and present experimental work and theoretical research, and selected consequence estimation methodologies. Each chapter includes sample problems to illustrate the application of the methodologies presented. The material presented is supported by approximately 400 references, with all references provided in a separate chapter..

1.1 Scope of the Book

The goal of this book is to provide the reader with an adequate understanding of the basic physical principles of flash fires and explosions and the current state of the art in hazard estimation methodologies. It is not the goal of this book to provide a comprehensive discussion of all of the experimental work and theoretical research that has been performed in the field of flash fire and explosion evaluation.

The book does not address subjects such as toxic effects, confined explosions (e.g., an explosion within a building), dust explosions, runaway reactions, condensed-phase explosions, pool fires, jet flames, or structural responses of buildings. Furthermore, no attempt is made to address the frequency or likelihood of accident scenarios.

1.2 Book Audience

The book was written for two target audiences: safety professionals involved in consequence analysis and managers responsible for chemical processing facilities. For a person new to the field of explosion and flash fire hazard evaluation this book provides a starting point for understanding the phenomena covered and presents methods for calculating the possible consequences of incidents. It provides an overview of research in the field and numerous references for readers with more experience.

Managers will be able to utilize this book to develop a basic understanding of the governing phenomena, the calculation methods to estimate consequences, and the limitations of each method.

2. Management Overview

The Management Overview chapter was written specifically for managers of chemical processing companies, and it contains an overview of the hazards associated with flash fires, VCEs, PVBs, and BLEVEs. The potential consequences of these types of incidents are illustrated by brief incident accounts, including data on number of fatalities and injuries and amount of property loss for each event. The chapter attempts to educate managers on the availability of models for each of the hazards and the benefits of understanding the threat posed by each hazard in managing the risks associated with a chemical processing facility. The chapter is purposefully non-technical for an audience that is not familiar with modeling of explosion and fire hazards.

3. Case Histories

The Case Histories chapter provides a review of incidents involving flash fire, VCE, PVB, and BLEVE hazards. These case histories illustrate the conditions present at the time of the event, highlighting the serious consequences of such events and the need for evaluation of the hazards.

The selection of incidents was based on the availability of information, the kind and amount of material involved, and the severity of damage. A total of 17 case histories are included. The incidents described in this chapter cover a range of factors:

- Materials: hydrogen, propylene, propane, cyclohexane, ethylene oxide, and natural gas liquids
- Period of time: between the years 1964 and 2007
- Quantity released: 90 kg (200 lb) to 40,000 kg (85,000 lb)
- Site characteristics: settings ranging from rural to very congested industrial areas
- Severity: injuries and damage vary widely in cases presented

4. Flash Fires

A flash fire is the combustion of a flammable gas/air mixture that produces relatively short term thermal hazards with negligible overpressure (blast wave). A flash fire burns through a pre-mixed vapor cloud, consuming the fuel as it burns. If a flash fire burns back to the source, it can become a jet fire or a pool fire.

Flash fires are the only non-explosion hazards treated in the book. Flash fires were included in the first edition because of similarities with some aspects of VCE and BLEVE phenomena including:

- The flammable material discharge and dispersion for a flash fire is similar to that of a VCE; however, the flammable cloud is not in congestion and confinement and lacks turbulence to result in flame acceleration like a VCE.
- A flash fire is a transient event, albeit with a duration that is typically longer than a BLEVE or VCE.
- The radiant energy intensity or surface emissive power (SEP) of a flash fire is determined by the same mechanisms as a BLEVE, including absorption of radiant energy by carbon particles, although the flash fire SEP is weaker than that of a BLEVE.

Similar phenomena need to be described in an explosion or a flash fire model:

- The dimensions of an unignited vapor cloud of flammable material should first be modeled.
- Potential overlap of the flammable cloud with congested zones should be calculated to ascertain whether a VCE or flash fire is likely.

- The flash fire model must predict flame speed, both into and with the wind.
- Flame speed is lowest at the flammable limits and highest near the stoichiometric concentration, and depends on the turbulence within the vapor cloud. Thus, the concentration gradients in a flammable cloud translate to complexity such as a flame that envelopes the cloud before burning into the cloud center. An example of this was captured in experiments by the Health and Safety Laboratory (HSL) in Figure 1.
- As with BLEVEs, accurate radiation impact from a flash fire requires accounting for atmospheric transmissivity and view factors.

The second edition of the book updates the experimental data base for flash fire tests and correlations for atmospheric transmissivity.

The book retains an example calculation of radiant energy reaching a target. This calculation requires integration of the transient radiation from a passing flame front. The difficulties evident in this example illustrate the need for using a flash fire model.

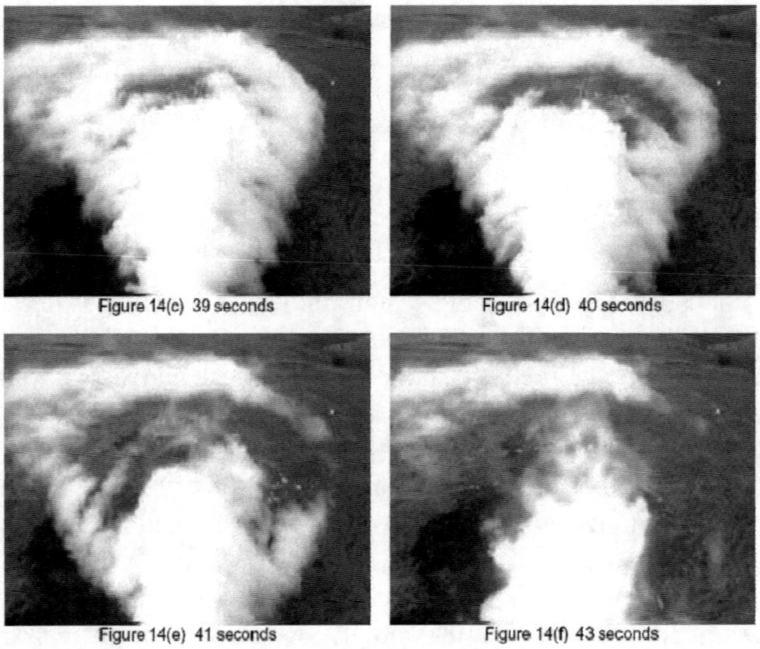

Figure 1. Flame front progression in LPG vapor cloud (2.0 m/s wind, 2.6 kg/s discharge for 51 s, ignition 25 m from source, [HSL, 2001).

Reproduced by permission of the HSL.

The flash fire model of Raj and Emmons (1975) is still commonly used. This model incorporates emissive power correlations from data observed in the China Lake tests, in the Maplin Sands experiments, and in the Montoir LNG fire test (Nedelka, 1990). It also applies flame speed correlations and flame width to length ratios calibrated to experiments by Steward (1964). The model describes some of the complex behavior of flash fires. For example, if the unburned fuel concentration is lean (between the LFL and the stoichiometric) the flame does not require further diffusive input of air, so the flame front is roughly as high as the plume. Otherwise, since a rich fuel mixture requires diffusive input of air, the burning front is higher than the original plume because the adiabatic expansion of hot combustion products pushes unburned fuel upward.

A new CFD gas dispersion and flash fire model is reviewed, the Kameleon FireEx model. This is a field model fire simulator developed at Norwegian University of Science and Technology (NTNU) and SINTEF Energy, Department of Applied Thermodynamics and Fluid Dynamics in Trondheim, Norway. This model has a pre-processor to set up the geometry and specify boundary and initial conditions and sub-models. The sub-models include the k-ε turbulence model, the Eddy Dissipation Concept of turbulent combustion, the Eddy Dissipation Soot Model and the Discrete Transfer Model of radiative heat transfer.

5. Vapor Cloud Explosions

A vapor cloud explosion is the result of a release of flammable material in the atmosphere, dispersion of the flammable material in air, and delayed ignition of the flammable portion of the vapor cloud. An accidental release of gaseous or liquid fuel may not necessarily lead to an explosion. If the released material ignites promptly, a fire is the outcome. Delayed ignition of a vapor cloud in an uncongested area leads to a flash fire. Some releases do not ignite, averting any fire or explosion consequences. Certain conditions must be met in order for a VCE to occur. First, there must be a release of flammable material into a confined/congested area. Second, ignition must be delayed long enough to allow the formation of the ignitable mixture, with the fuel-air concentration lying between the flammable limits. Third, there must be an ignition source of sufficient energy to ignite the fuel-air mixture.

An introduction of vapor cloud explosion (VCE) phenomena and terminology is given in the VCE chapter, followed by an overview of experimental research on vapor cloud deflagrations and detonations. Mechanisms of flame acceleration, which are essential for a VCE to occur, are presented. Three important factors affecting flame acceleration, namely confinement, congestion, and fuel reactivity, are discussed in order to better understand the large amount of experimental results on VCE deflagrations. Experimental measurements and empirical equations to predict the effects of VCE detonations are also presented.

VCEs are complex explosions. In a vapor cloud deflagration, the combustion and pressure wave are decoupled. The overpressure generated in a VCE deflagration varies with combustion rate: minimal overpressure is produced at low flame speed, and high overpressures are produced at higher flame speeds. Consequently, the damage to the surroundings caused by VCE deflagration ranges from minimal to more severe. VCE detonations are typically more severe than deflagrations due to the high overpressure generated by a supersonic wave. The situation for VCE deflagrations is complex because the flame speed and pressure buildup in the deflagration are not unique for a given cloud composition, but vary in a wide spectrum depending on many factors. Moreover, the composition of fuel and combustion products at the flame front, within the cloud, which supports the deflagration, changes continuously. The vast majority of accidental VCEs are deflagrations.

The physical layout of a process area within a flammable cloud has a direct bearing on the outcome of a VCE. The layout is described by two parameters: confinement and congestion. Confinement refers to solid surfaces that prevent movement of unburned gases and a flame front in one or more dimensions. Congestion refers to obstacles in the path of the flame that generate turbulence. Turbulence produces flame folding and wrinkling, increasing flame surface area and combustion rate. Repeated, closely spaced obstacles may create conditions favorable to creating a VCE. Experimental data are organized in the book according to the type of confinement and degree of congestion to illustrate the effect of confinement and congestion on VCE blast pressure.

VCE blast prediction methods are discussed and arranged into three categories: TNT equivalency methods, blast curve methods, and numerical simulations. The TNT method is included for historical perspective and due its continued use by the insurance companies and some government agencies, but this method is no longer recommended. Three blast curve methods are presented: the TNO Multi-Energy Method, the Baker-Strehlow-Tang (BST) Method, and the Shell Congestion Assessment Method (CAM). Sample problems are provided to illustrate each methodology. Numerical simulations using computational fluid dynamics models are also described, but due to the complexity of this subject, the book content was limited to an overview and illustration.

6. Pressure Vessel Bursts

A major change in the book was separation of PVBs into its own chapter. The first edition included PVB treatment as part of BLEVE predictions. However, pressure vessel accidents that do not involve flashing liquid are a hazard that warrants separate consideration.

PVB, as the name implies, is a type of explosion that involves burst of a pressure vessel containing a gas at elevated pressure. The term “pressure vessel” in PVB is not necessarily synonymous with the definition of pressure vessel used in the ASME pressure vessel code; rather, any vessel or enclosure that can build significant pressure before bursting can generate a PVB. Upon burst, the sudden expansion of a compressed gas generates a blast wave that propagates outward from the source. Pieces of the vessel shell along with any external appurtenances are thrown, creating a fragment hazard. It is not necessary for the vessel contents to be flammable or contain reactive chemicals. PVBs occur with flammable gases, inert gases, reactive gases, or mixtures of these materials.

Among the types of explosions addressed in the book, PVBs can potentially generate the highest blast pressures at the explosion source due to the internal pressure achieved in a vessel at the time it bursts. However, the explosion energy of a PVB is usually small compared to VCEs and BLEVEs because the energy is limited by volume of compressed gas in the vessel. In comparison to VCEs, PVBs can produce high damage, but the damage zone is relatively small due to the relatively low energy involved.

In spite of the significant design safety factors and pressure relief devices, accidents involving pressure vessels can still occur for a variety of reasons. Accident scenarios can involve one or more of the following causes of failure:

- Loss of mechanical integrity – the pressure vessel’s attached connections are compromised in some fashion; as a consequence, the stresses in a pressure-carrying component can exceed the component’s ultimate capacity. Failures due to a loss of mechanical integrity may occur at normal working pressure.

- Pressurization beyond ultimate failure pressure – pressure goes beyond the vessel's ultimate capacity with scenarios considering the rate of pressurization versus the ability of the system to relieve pressure through safety devices.
- External agencies – an external source compromises mechanical integrity, such as impact, cutting, and fire.
- Fabrication, transportation and installation – Flaws, weaknesses and deficiencies induced prior to usage can cause the installed vessel differ from the design intent.

Damage to the surroundings from a PVB may be caused by blast pressure and fragments. Blast effects provide the primary source of damage from a PVB. Fragmentation typically causes damage relatively close to the vessel. Unlike BLEVEs, the thermal effects of PVBs are usually minor relative to the potential blast pressure and fragment damage. Experimental and analytical results are presented in the book for blast pressure and fragmentation. Blast curves for the prediction of airblast parameters are included. The simplified correlation of directional blast effects from cylindrical pressure vessels has been replaced with figures that provide adjustment factors to calculate pressure and impulse at various angles from a horizontal vessel. Lastly, fragment velocity and throw prediction methods are provided.

7. Boiling Liquid Expanding Vapor Explosions

A boiling liquid, expanding vapor explosion (BLEVE) is defined as a sudden loss of containment of a pressure-liquefied gas existing above its normal atmospheric boiling point at the moment of its failure. Such a failure will result in rapidly expanding vapor and flashing liquid. BLEVEs are typically similar to pressure vessel bursts in terms of their blast effects, with an increase in the energy released due to the involvement of flashing liquid. The flashing liquid can lead to hazards that are not typically of primary concern for pressure vessel bursts; these include rocketing fragments and fireballs (where the liquid contents are flammable).

The initiating mechanisms and failure pressures are similar to that of the pressure vessel bursts discussed in the pressure vessel burst chapter, but the involvement of flashing liquid can increase the energy released quite substantially and thereby have farther ranging effects. The BLEVE chapter builds on the material presented in the pressure vessel burst chapter, focusing on the changes due to the flashing liquid. The chapter opens with an introduction to the thermodynamics of boiling and flashing liquids. This discussion includes a discussion of nucleate boiling, homogeneous boiling, and the superheat limit temperature. The chapter then proceeds to discuss the modes of failure. In BLEVEs, a crack can release vapor, causing the pressure in the vessel to reduce. If the vessel contains only vapor, this crack may then be stable and the result would be a jet or vapor flowing from the vessel. When a liquid above its normal atmospheric boiling point is also present, the decrease in pressure can cause the liquid to boil and this can lead to a re-pressurization of the vessel that can drive the initial failure to grow. These "two-step" BLEVEs have been observed in experimental studies and are distinctly different than the failure modes presented for pressure vessel bursts.

After presenting a narrative of a "typical" BLEVE scenario (i.e., what most readers will think of when they consider BLEVEs), the book presents experimental and analytical methodologies that are currently accepted for assessing the consequences of BLEVEs. These include airblast, thermal hazards (i.e., fireball effects, drawing on the material presented under flash fires), and fragment and debris throw models, including correlations to predict rocketing fragments.

8. Book Publication

The book was printed during the first quarter of 2010 and is available through AICHE Publications department. It can be purchase online at http://www.aiche.org/publications/.

A Resilience Assessment Framework for Infrastructure and Economic Systems: Quantitative and Qualitative Resilience Analysis of Petrochemical Supply Chains to a Hurricane

Eric D. Vugrin
Sandia National Laboratories
Albuquerque, NM
edvugri@sandia.gov

Drake E. Warren
Sandia National Laboratories
Albuquerque, NM
dewarre@sandia.gov

Mark A. Ehlen
Sandia National Laboratories
Albuquerque, NM
maehlen@sandia.gov

Prepared for Presentation at
American Institute of Chemical Engineers
2010 Spring Meeting
6th Global Congress on Process Safety
San Antonio, Texas
March 22 – 24, 2010

UNPUBLISHED

A Resilience Assessment Framework for Infrastructure and Economic Systems: Quantitative and Qualitative Resilience Analysis of Petrochemical Supply Chains to a Hurricane

Eric D. Vugrin
Sandia National Laboratories
Albuquerque, NM
edvugri@sandia.gov

Drake E. Warren
Sandia National Laboratories
Albuquerque, NM
dewarre@sandia.gov

Mark A. Ehlen
Sandia National Laboratories
Albuquerque, NM
maehlen@sandia.gov

Keywords: resilience, petrochemical supply chain, hurricane, infrastructure protection, disruptions, critical infrastructure, key resource

Abstract

In recent years, the nation has recognized that critical infrastructure protection should consider not only the prevention of disruptive events, but also the processes that infrastructure systems undergo to maintain functionality following disruptions. This more comprehensive approach has been termed critical infrastructure resilience. Given the occurrence of a particular disruptive event, the resilience of a system to that event is the system's ability to efficiently reduce both the magnitude and duration of the deviation from targeted system performance levels. Under the direction of the U. S. Department of Homeland Security's Science and Technology Directorate, Sandia National Laboratories has developed a comprehensive resilience assessment framework for evaluating the resilience of infrastructure and economic systems. The framework includes a quantitative methodology that measures resilience costs that result from a disruption to infrastructure function. The framework also includes a qualitative analysis methodology that assesses system characteristics affecting resilience to provide insight and direction for potential improvements. This paper describes the resilience assessment framework and demonstrates the utility of the assessment framework through application to two hypothetical scenarios involving the disruption of a petrochemical supply chain by hurricanes.

1. Introduction

Historically, U.S. government policy toward critical infrastructure protection (CIP) has focused on physical protection and asset hardening (for examples, see [1], [2], [3], and [4]). Recently, the federal government has realized "protection, in isolation, is a brittle strategy" [5] and not all disruptive events, natural or manmade, can be prevented. Hence, national CIP policies must prepare the nation for unavoidable disruptive events.

With the formation of the Department of Homeland Security's (DHS's) Critical Infrastructure Task Force in 2005, this shift in policy become a national priority as the task force made critical infrastructure resilience (CIR) its top-level strategic objective. CIR is the concept concerned with how critical infrastructures absorb, adapt, and recover from the effects of a disruptive event to ensure delivery of critical infrastructure services. As a result of this shift, the federal government has started a coordinated set of resilience initiatives to understand what features create resilience in critical infrastructures/key resources (CIKRs) and has initiated calls to agencies to start measuring the resilience of their infrastructure systems.

Private industry, including the chemical sector, would benefit through the inclusion of resilience concepts in business plans. By understanding how their businesses (and those that they depend upon) are affected by and recover from various types of disruptions, private industry can develop comprehensive emergency plans and evaluate the costs and benefits associated with various recovery strategies. Hence, the DHS Science and Technology Directorate has tasked Sandia to develop a resilience assessment methodology that can be applied to the chemical sector and other CIKR systems.

A uniform, methodical approach for assessing resilience of infrastructure systems is required to successfully incorporate resilience into CIP policies and business planning practices. This approach needs to be general enough to apply to all types of infrastructure systems to account for dependencies between different infrastructure types and establish standards across all infrastructure types. Furthermore, resilience assessment approaches should explicitly account for the costs of recovery processes in comprehensive disruption cost evaluations.

With these two requirements in mind, we have developed a novel framework for evaluating the resilience of infrastructure and economic systems [6]. The framework includes a new definition of resilience, a mathematical resilience cost measurement approach, and a qualitative analysis methodology that assesses system characteristics that affect resilience. This framework can be applied to studies of natural and manmade disruptions. This paper describes the three components of the resilience assessment framework in detail. Furthermore, we demonstrate the application of the framework through analysis of the resilience of the national petrochemical supply chain during two different hurricane disruption scenarios.

2. A Framework for Resilience Assessment

To define resilience, we identified the factors that determine resilience and need to be quantified when measuring resilience costs. We then described how these factors need to be combined to measure resilience costs and developed a mathematical algorithm for that purpose. We also developed a qualitative analysis approach to explain the system's resilience costs. Additionally, the qualitative approach can also be used to determine improvements needed to enhance the system's resilience. This section describes the individual framework components in greater detail.

2.1 A Definition of System Resilience

We define system resilience as follows:

> Given the occurrence of a particular disruptive event (or set of events), the resilience of a system to that event (or events) is the ability to efficiently reduce both the magnitude and duration of the deviation from targeted system performance levels.

Elements of this definition can be further defined as follows:

- Disruptive event - This definition considers resilience of a system to a specific disruption. Different disruptions may affect a system in different ways and, thus, necessitate different recovery processes. Hence, a system may have different levels of resilience to different disruptions.
- System performance - Given the flexibility of many systems to adjust and reconfigure to a disruptive event, maintaining system structure is not as important as maintaining system performance. Hence, measurement of resilience should evaluate how a disruption affects system performance and causes productivity to decrease relative to targeted system performance levels; i.e., how the system should behave during and after disruptive events.
- Efficiency - The term "efficiency" means using the lowest possible amount of resources during recovery processes; depending on the domain, these resources could be dollars, repair man-hours, infrastructure replacement assets, or time.

2.2 Calculation of Resilience Costs

We have developed a mathematical resilience costs measurement approach that can be used to objectively determine the impacts of disruptions on a system and the resilience costs associated with disruptions. The resilience cost measurement approach requires quantification of two key components of the definition of system resilience: systemic impact (*SI*) and total recovery effort (*TRE*). *SI* is the impact that a disruption has on system productivity and is measured by evaluating the difference between a targeted system performance (*TSP*) level and the actual system performance (*SP*) following the disruption. *TRE* refers to the efficiency with which the system recovers from a disruption and is measured by analyzing the amount of resources expended during the recovery process. The measurement of system resilience costs requires the quantification of both *SI* and *TRE*.

Figure 1 graphically represents systemic impact for a hypothetical system that has been disrupted. In this example, system performance decreases immediately following the disruption shock. With the onset of recovery actions, performance levels eventually increase and ultimately attain targeted system performance levels. At this point, recovery is considered complete. *SI* is quantified by calculating the area between the *TSP* and the actual *SP* curves in Figure 1. This area is calculated using the formula in Eq. 1.

Figure 1: Systemic Impact

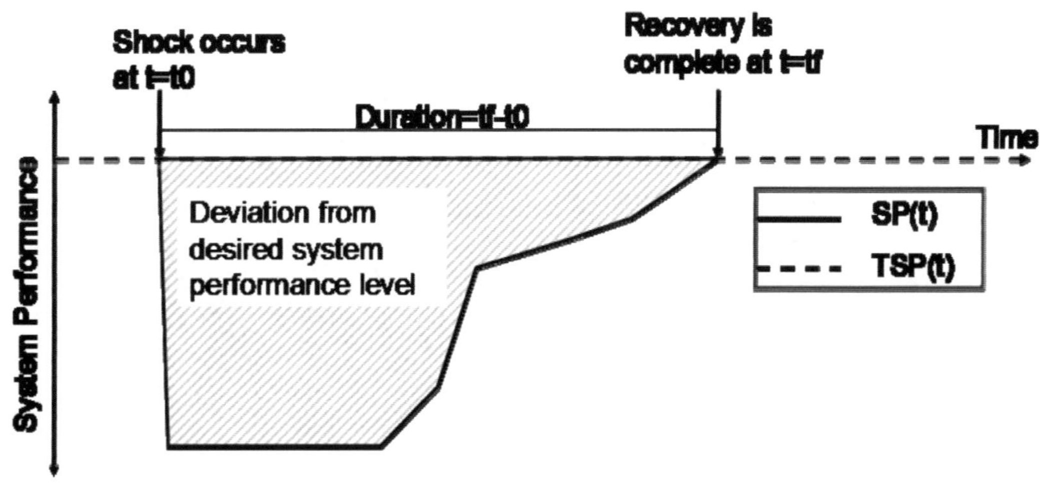

$$SI = \int_{t0}^{tf} \left[TSP(t) - SP(t) \right] dt \qquad \text{[Eq. 1]}$$

Figure 2 illustrates the recovery response for the system shown in Figure 1. After the disruption initiates, the recovery response begins and resources are expended in this effort. The *TRE* is the cumulative amount of resources expended during the recovery period and is represented by the area under the recovery effort (*RE*) curve in Figure 2. This area is calculated by Eq. 2.

Figure 2: Total Recovery Effort

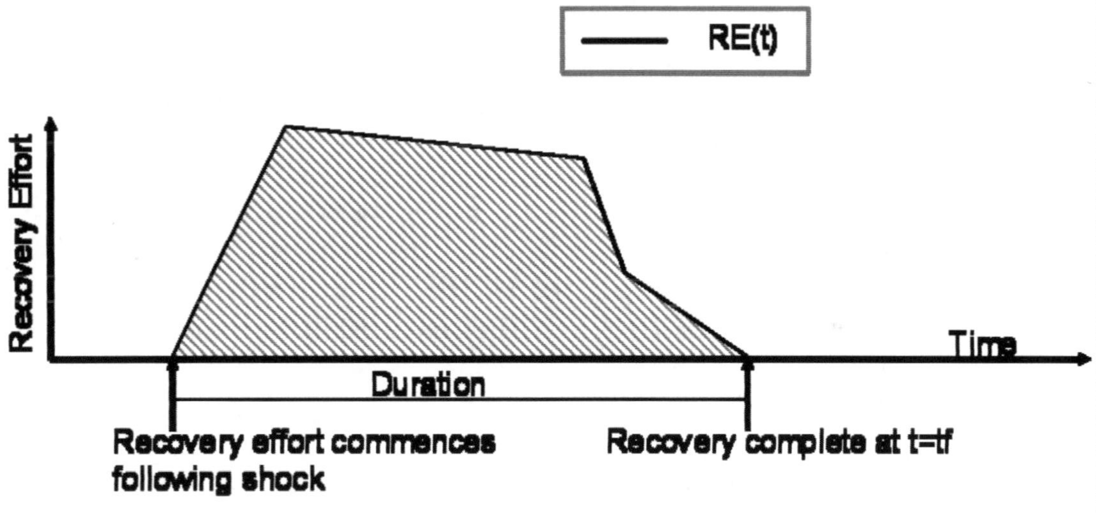

$$TRE = \int_{t0}^{tf} \left[RE(t) \right] dt \qquad \text{[Eq. 2]}$$

The system performance is determined by the *RE*. That is, different *RE*s lead to different system performances. For example, if no *RE* is made following the disruption, the loss of system performance may be great. In contrast, if recovery resources are deployed shortly after the system shock, system performance may not be significantly affected, and SI may be small. The recognition that *SI* is implicitly determined by the selected recovery strategy leads to the development of two types of resilience cost measurements: optimal resilience (*OR*) costs and recovery-dependent resilience (*RDR*) costs. *OR* costs are the resilience costs for a system when the optimal recovery strategy, minimizing the combined *SI* and *TRE* costs, is employed. This calculation is beyond the scope of this paper. *RDR* costs are the resilience costs of a system under a particular recovery strategy and are calculated with Eq. 3.

$$RDR(RE) = \frac{\int_{t0}^{tf} \left[TSP(t) - SP(t) \right] dt + \alpha \times \int_{t0}^{tf} \left[RE(t) \right] dt}{\int_{t0}^{tf} \left| TSP(t) \right| dt} = \frac{SI + \alpha \times TRE}{\int_{t0}^{tf} \left| TSP(t) \right| dt} \qquad \text{[Eq. 3]}$$

RDR costs are linear combinations of *SI* and *TRE*. The denominators in Equation (3) are normalization factors that permit the comparison of the resilience of systems of different magnitudes. Because resilience represents a balancing of *SI* and *TRE* costs, the calculation of *RDR* costs includes the parameter α, which is a weighting factor that allows an analyst to assign the relative importance of the systemic impact and total recovery effort terms. Assigning a small positive value to α weighs the systemic impact more heavily; a large positive value for α weighs the cost of recovery more heavily. To equally weigh *SI* and *TRE*, α is set to 1.

Several things about this resilience measurement approach should be considered. First, smaller *RDR* costs indicate increasing resilience, with zero being the minimum possible value. Additionally, no finite *RDR* costs correspond with the concept of a minimally "resilient" system. Also, the approach for measuring system resilience is neither model- nor domain-specific. It only requires time series data (either historical or from a model) that represent system output and recovery efforts. However, the summation of the *SI* and $\alpha \times TRE$ terms in Eq. 3 requires either that the *SI* and *TRE* be measured in the same units and α be a dimensionless constant or that α be assigned units that are appropriate for converting *TRE* to the same units as *SI*.

Because the *RDR* costs are dimensionless quantities, they are most informative when used in a comparative manner. For example, they can be used to compare the resilience of different systems to the same disruption. The system that has lower resilience costs will be the more resilient system. *RDR* values can also be used to compare the resilience of the same system to different types of disruptions. The system is more resilient to the disruption that results in smaller *RDR* values. Moreover, they can be used to compare the resilience of a system to a disruption under different recovery strategies. Each different recovery strategy will result in different *SI* and *TRE* values. The recovery strategy that results in the smallest *RDR* values will provide maximal resilience for the system.

2.3 Qualitative Resilience Analysis

Joseph Fiksel [7] suggests that "it is important to assess not only performance outcomes but also the intrinsic characteristics that contribute to system resilience." Consequently, the framework described in this paper features a qualitative analysis component that can be used to explain the results of quantitative measurements or can take the place of quantitative results when no data are available. This analysis is done through consideration of system structures, characteristics, and features.

This portion of the framework uses three fundamental system capacities (*absorptive capacity, adaptive capacity, and restorative capacity*) to formulate how properties of a system can determine system resilience, specifically by reducing *SI* and *TRE*.[1]

These capacities are affected by resilience enhancement features; that is, the features of the system that are in place before a disruption and that affect one or more of the system's capacities. Identifying resilience enhancement features enables a better understanding of fundamental characteristics that contribute to resilience. Most importantly, pre-disruption preparatory actions can target these resilience enhancement features to increase the resilience of the system.

The following subsections describe the characteristics of resilience capacities and related resilience enhancement features and provide examples of each. Figure 3 summarizes the distinguishing characteristics of the capacities.

Figure 3: Resilience Capacities

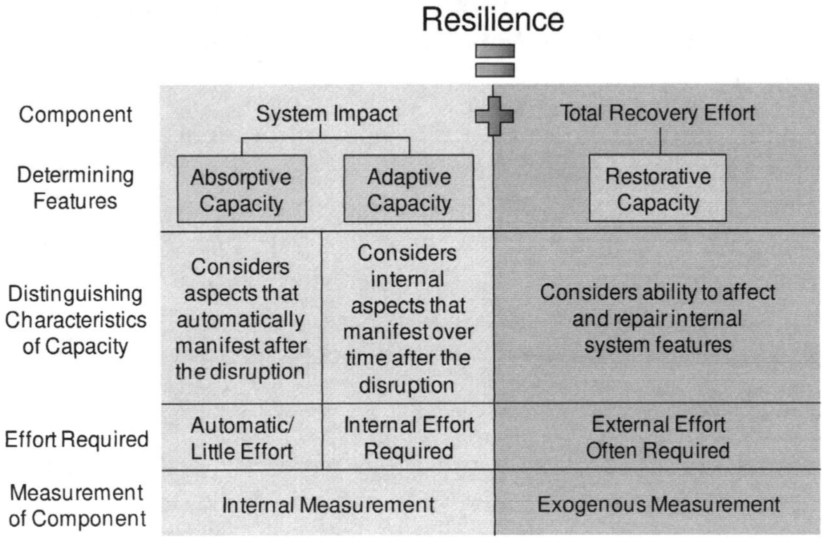

[1] These capacities are similar to the abilities to "absorb, recover from, or successfully adapt to adversity or a change in conditions" in the official Department of Homeland Security definition of resilience (U.S. Department of Homeland Security Risk Steering Committee 2008), but do not include that definition's ability to resist.

2.3.1 Absorptive Capacity

Absorptive capacity is the degree to which a system can automatically absorb the impacts of system perturbations and minimize consequences with little effort. The absorptive capacity is an endogenous feature of the system.

For example, storage can enhance the absorptive capacity; if a chemical plant is disabled but a large amount of collocated storage of its product is undamaged, customers can continue to be supplied by the stored quantities, with little cost to the producer or customer, while the plant is repaired.

Other examples of absorptive capacity resilience enhancement features include the system robustness and redundancy. System robustness decreases *SI* through the strength of individual connections in the system. Levees that prevent hurricane storm surges from damaging a chemical facility are an example of system robustness within the chemical sector. System redundancy decreases *SI* through providing alternate pathways for the system mechanics to operate. A specific example of system redundancy in the chemical sector is purchasing inputs from multiple, geographically-dispersed suppliers.

2.3.2 Adaptive Capacity

Adaptive capacity is the degree to which the system is capable of self-organization for recovery of system performance levels. It is a set of properties that reflect actions that result from ingenuity or extra effort over time, often in response to a crisis situation. It reflects the ability of the system to change endogenously during the recovery period.

Consider the scenario in which a hurricane destroys many high voltage power lines, leaving many customers without electricity. Customers with emergency generators enhance system adaptive capacity because the system can be changed (customers adapt to the disruptive event by generating power from a fuel source such as gasoline rather than connecting to the electric grid) so that some portion of system performance is regained at a relatively low amount of effort.

Substitutability, the ability to replace one system component or input with another, is a resilience enhancement feature that can affect adaptive capacities. Chemical production processes can sometimes substitute inputs or use alternate technologies. Economic systems with a high adaptive capacity can easily adjust to shocks through ordinary means such as input substitution, which can occur in situations where an input is scarce. Other resilience enhancement features that increase adaptive capacity tend to be more difficult to identify because they often rely upon the ingenuity of people faced with adversity.

2.3.3 Restorative Capacity

Restorative capacity is the ability of a system to be repaired easily, and these repairs are considered to be dynamic. In the case of massive catastrophic events, systems may not be able to repair themselves or they may not be able to do so rapidly enough to prevent unacceptably large consequences. In these circumstances, repairs may be performed by entities external to the system. In the context of infrastructure policies, the government can be an external, enabling or repairing entity. The government agencies may not directly perform the repairs, but may serve as a lead restoration planner or restoration planning coordinator. These repairs usually restore the system to near its original pre-event state, but can also restore the system to a completely new state or regime that anticipates future system requirements. Therefore, the repairs are a form of investment.

An infrastructure owner, such as a corporation, may also take action, often by drawing on external resources such as other units of the corporation or by outsourcing. Businesses may be able to perform repairs using their own local resources (such as custodial staff), but in most cases these repairs could be better described as maintenance. However, a business with a high degree of restorative capacity may be able to eliminate the need for repairs by an external entity.

Restorative capacity directly affects the *TRE*, although repairs to the system enabled by the system's restorative capacity also increase the system performance and may reduce recovery duration, thereby reducing *SI*. Whereas adaptive capacity reflects the ability of a system to be changed endogenously, restorative capacity reflects the ability to be repaired. Most importantly, adaptive capacity involves changes that can radically alter the structure of the system to restore system performance, while restorative capacity most often involves repairs that are implemented with the goal of returning a system to something near its original structure. For example, the electric power grid has monitoring systems that can automatically detect when and where a break in the grid emerges. Such technologies enhance the restorative capacity of the power grid because repair crews can be sent to the location of the break. These technologies result in a shorter disruption that is easier to repair (in terms of cost and time) than it would be if crews had to search large portions of the grid to find the break before repairing it.

Another differentiation between adaptive capacity and restorative capacity is that restorative capacity may affect a system's ability to be permanently changed (an investment decision) while adaptive capacity is primarily concerned with features that are temporary. The restorative period provides an opportunity to remake the system. For example, a chemical plant using an inefficient technology may decide to invest in a superior, more efficient technology during restoration. The restoration period is also an opportunity to modify the system to increase resilience to future events. Therefore, a system with a greater restorative capacity may indirectly have higher absorptive and adaptive capacity. Systems with a greater ability to be upgraded during the restoration period will have a greater restorative capacity.

3. Application to the National Petrochemical Sector

This section describes an application of the framework to a petrochemical supply chain model to demonstrate the utility of the resilience assessment. Specifically, the framework was used to compare the resilience of the national petrochemical supply chain to two hurricane scenarios. To perform this analysis, we used an agent-based simulation to simulate the effects of a hurricane disruption on the national petrochemical supply chain. The model reported economic data representing system performance and recovery processes, and these data were then processed, using the resilience assessment framework

3.1 Chemical and Economic Models

To perform the comparative analysis, we used the National Infrastructure and Simulation Analysis Center (NISAC) Petrochemical Supply Chain Model (version #1, 2007) that Sandia developed as a part of NISAC. This NISAC petrochemical supply chain model consists of two primary components. The first component, the chemical data model (CDM), is a database of domestic and foreign chemical plants, chemical productions, commodity flows, and chemical infrastructure (for example, pipelines, rail networks, and water-transport networks). The version of the CDM used for this analysis contains data for almost 4,000 domestic and foreign consumers and producers of 63 commodity petrochemicals. Each of the firms in the CDM makes a primary feedstock petrochemical (benzene, toluene, ethylene, propylene, xylene, o-xylene, or p-xylene), converts these chemicals into other petrochemicals, or produces other chemical and non-chemical products based on these chemicals. Using the stoichiometric (chemistry-based) and other production "recipes" for each chemical, the CDM can identify the basic relationships among these chemicals.

Sandia uses the petrochemical CDM in conjunction with the NISAC Agent-Based Laboratory for Economics (N-ABLE™) microeconomics simulation tool [8] to simulate disruptions to the petrochemical sector from various types of disturbances. N-ABLE™ is a collection of tools that Sandia has developed to perform supply-chain analysis: the analysis of ways individual firms within multi-tiered, multi-product economic systems purchase input goods, produce products, sell them in markets, and ship them via different modes of transportation.

In N-ABLE™, each agent-based enterprise firm is composed of buyers, production workers, supervisors, sellers, and strategic planners who conduct their real-world analog tasks within the enterprise and among enterprises. Firms representing the range of economic activity in a supply chain (such as manufacturing, transportation, and consumption) are modeled by specifying such things as particular production functions, buying and selling behaviors, inventory capacities, and long-term strategic planning. Production decisions are made by a production manager, independent of input costs. The production manager adapts production based on the inventory of finished goods and the presence of market signals (orders). Every day, the plant orders enough to raise its inventory position to a predetermined level, taking into account expected usage patterns and historic averages of delivery times. Entire supply chains are constructed from collections of firms, based on this enterprise design, with each participating firm interacting with others through markets and physical infrastructure.

When applied to the petrochemical CDM, N-ABLE™ simulations can provide quantitative and qualitative information necessary for resilience analyses. For example, if a hurricane temporarily shuts down a set of chemical production facilities, N-ABLE™ can estimate economic impacts resulting from a decreased chemical supply to downstream facilities (e.g., customers of the closed facilities, the customers of the customers, etc.). N-ABLE™ can also predict losses resulting from decreased demand of input chemicals used by the closed production plants to upstream facilities (e.g., suppliers to the closed plants, suppliers of the suppliers, etc.). These economic impact and loss estimates can be used to measure the *SI*s to the petrochemical supply chain from a hurricane.

In addition, N-ABLE™ can predict how the petrochemical sector will adapt to and recover from a disruption. The tool has the capability to estimate production curtailments by the customers of the closed plants who cannot find new suppliers, the higher transportation costs associated with new suppliers, the use of chemical substitutes, and the implementation of different production technologies and recipes to adapt to a disruption. The cost estimates associated with the recovery and adaptation processes are crucial to calculating the *TRE* in a resilience analysis.

3.2 Analysis Methodology

In this analysis, we consider the resilience of the national petrochemical sector to two different hurricane scenarios. In the first scenario, a Category 2 hurricane makes landfall in the Houston, Texas, area (Figure 4). It is common practice for Gulf Coast petrochemical production facilities in the projected path of a hurricane to shut down operations 48 hours prior to hurricane landfall. On average, the petrochemical facilities within the electric power outage contours will be without power for a few weeks. Production at these facilities will not likely be restored immediately following restoration of power. Following a plant shutdown, petrochemical facilities often require additional startup time to perform system checks, such as purging lines and vessels with inert gases such as nitrogen, to ensure the unit's operability. Hence, to simulate the cumulative effects of the electric power outage on production levels at the affected petrochemical facilities, it is also assumed that all petrochemical facilities within the outage contours are shut down for 25 days.

Figure 4: Electric Power Outages: Houston Scenario

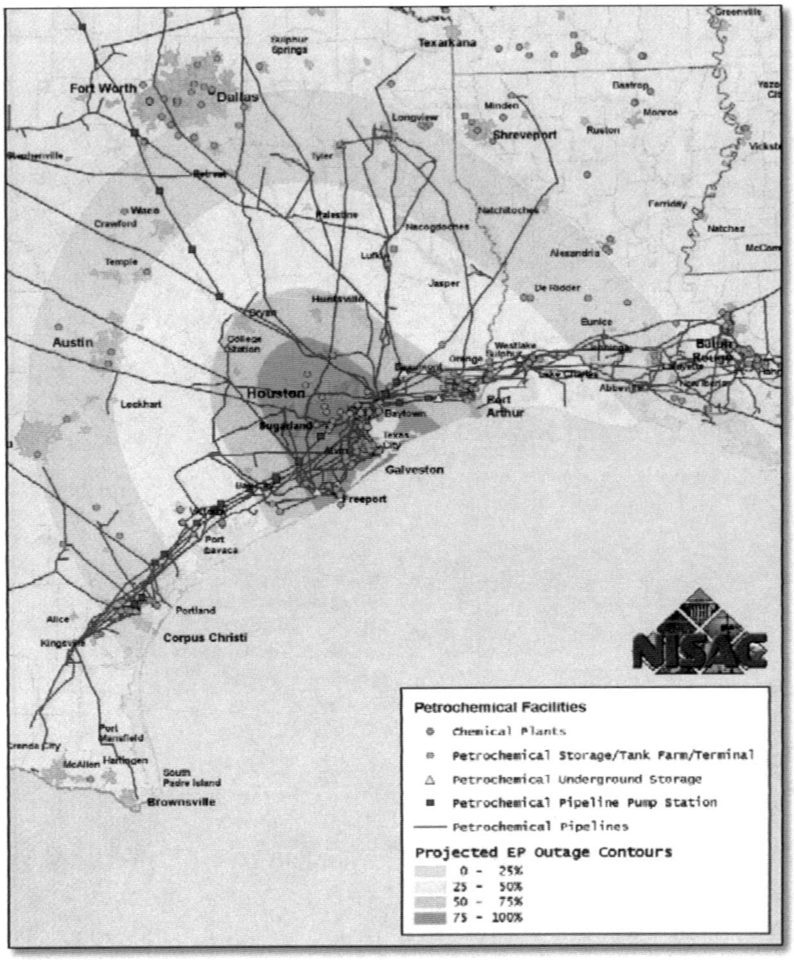

For the Houston hurricane scenario, 1,390 chemical firms (36 percent of all firms in the model) will be affected. These firms represent 570,000 daily short tons of product supply (81 percent of production capacity for the entire petrochemical sector) and 349,000 daily short tons of product demand (53 percent of the sum of all demand for the particular chemicals if and when these firms are running at full capacity).

In the second scenario, a Category 2 hurricane makes landfall near New Orleans, Louisiana (Figure 5). It is assumed that all petrochemical facilities that lie within the outage contours will be shut down for 25 days.

Figure 5: Electric Power Outages: New Orleans Scenario

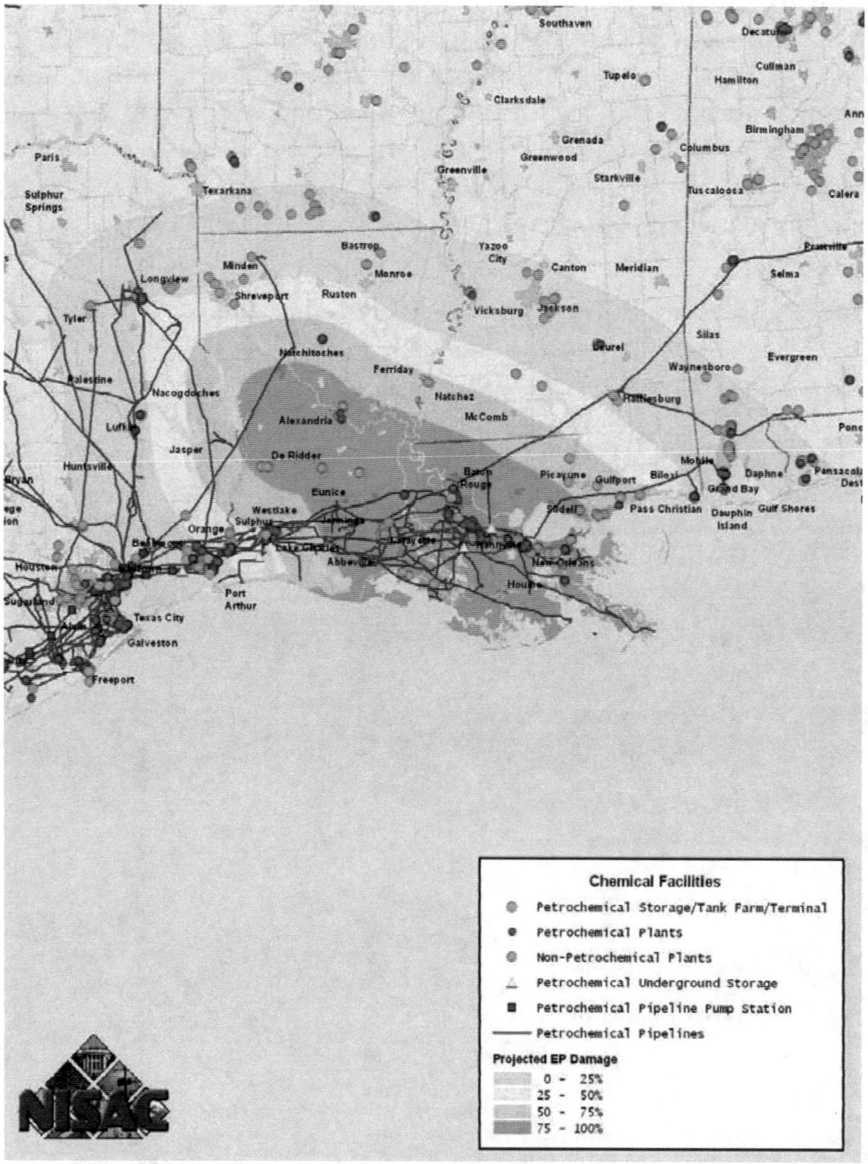

For the New Orleans scenario, 886 chemical firms (23 percent of all firms in the model) will be affected. These firms represent 475,000 daily short tons of product supply (68 percent of production capacity for the entire petrochemical sector) and 493,000 daily short tons of product demand (75 percent of the sum of all demand for the particular chemicals if and when these firms are running at full capacity). Some of the same petrochemical facilities are affected in both the Houston and New Orleans scenarios.

It is projected that the petrochemical sector will be less resilient to the Houston hurricane scenario for two reasons. First, a greater fraction of production capacity is shut down in the Houston hurricane scenario than the New Orleans scenario (81 percent versus 68 percent). Second, the product demand affected in the New Orleans scenario is larger than that of the Houston scenario (75 percent versus 53 percent). That is, more product consumers and, thus, a greater fraction of product demand will still be in operation during the Houston scenario. Thus, unmet demand is expected to be higher in this scenario, and more consumers will be expending more resources to receive their necessary products, thereby driving up recovery costs in an attempt to limit system impacts.

To quantitatively evaluate the resilience of the petrochemical supply chain, we ran three sets of N-ABLE™ simulations. In the baseline scenario, we assumed no disruptions. In the Houston disruption scenario, we assumed that a hurricane is projected to make landfall on day 202 of the simulation and the electric power outage shown in Figure 4 is expected to occur. On day 200, all petrochemical facilities within the contours shut down in anticipation of the storm. Normal production capabilities are assumed to return on day 225 of the simulation. The New Orleans disruption is identical to the Houston disruption scenario with the exception that different petrochemical facilities are affected.

The market value of production (*MVP*) is the metric used to measure system performance. *MVP* captures total "street value" of every step of production. It is similar to the sale value of end products, but it counts production at every stage in the production process, whereas the sale value only counts chemicals that are sold on the merchant market. *MVP* equals sale value of end products if there is absolutely no vertical integration; i.e., outputs of every stage of the production process are sold on the merchant market

For this analysis, two factors are considered in the recovery effort variable: additional aggregate transportation costs (*TC*) and additional aggregate market costs (*MC*). When a disruption decreases the supply of available chemicals, consumers of those chemicals will seek new suppliers. These suppliers will likely be farther from the consumers than the original suppliers, so the cost of transporting chemicals from the new suppliers will likely be greater due to the increased transportation distances. Marketing costs are the costs associated with the process that a buyer uses to find a supplier. These costs are expected to increase in times of shortages. Unmet demand will generally be low during baseline conditions, when long-term, supply-demand ratios and relationships ensure that most buyers can find product most of the time, but generally higher during the disruption, when these buyers must try to find product from alternative sources. These market search costs will increase if buyers are unable to find product, either due to lack of supply or lack of sufficient time to find suppliers.

We recognize that, in the event of an actual hurricane, recovery of individual chemical plants and the entire supply chain could involve additional expenses, including repair of damage caused by winds and/or flooding, startup costs, etc. However, for the sake of simplicity, we only consider the *TC*s and *MC*s when calculating the *TRE* for this example. To calculate *RDR* costs, we set α to 1 in Eq. 3 and we approximate the integral with 1-day time-step intervals because N-ABLE^{TM} reports data on a daily basis.

3.3 Simulation Results

Figures 6 and 7 show the *MVPs* and recovery costs for all scenarios. *MVP* is immediately affected by the disruptions in each scenario. The initial *MVP* decrease is less than the fraction of shut-down production capacity because the system uses stored inventories of chemicals and seeks new suppliers from further distances. However, inventories start to deplete near the end of the shutdown period, and *MVP* decreases by more than 80 percent for the Houston scenario. Maximum *MVP* reductions for the New Orleans scenario are more modest, representing an approximate 30-percent reduction. When full production capacity is restored, *MVP* levels for the disruption scenarios actually exceed baseline levels, initially (days 240 to 275), because the plants have ramped up production levels to meet not only previously unmet demand but also to restore inventory levels. After day 275, no consistent differences are observed between the baseline and disruption *MVP* levels; differences between the *MVP* levels are attributed to the stochastic nature of the supply chain model.

Figure 6: Simulated Market Values of Production

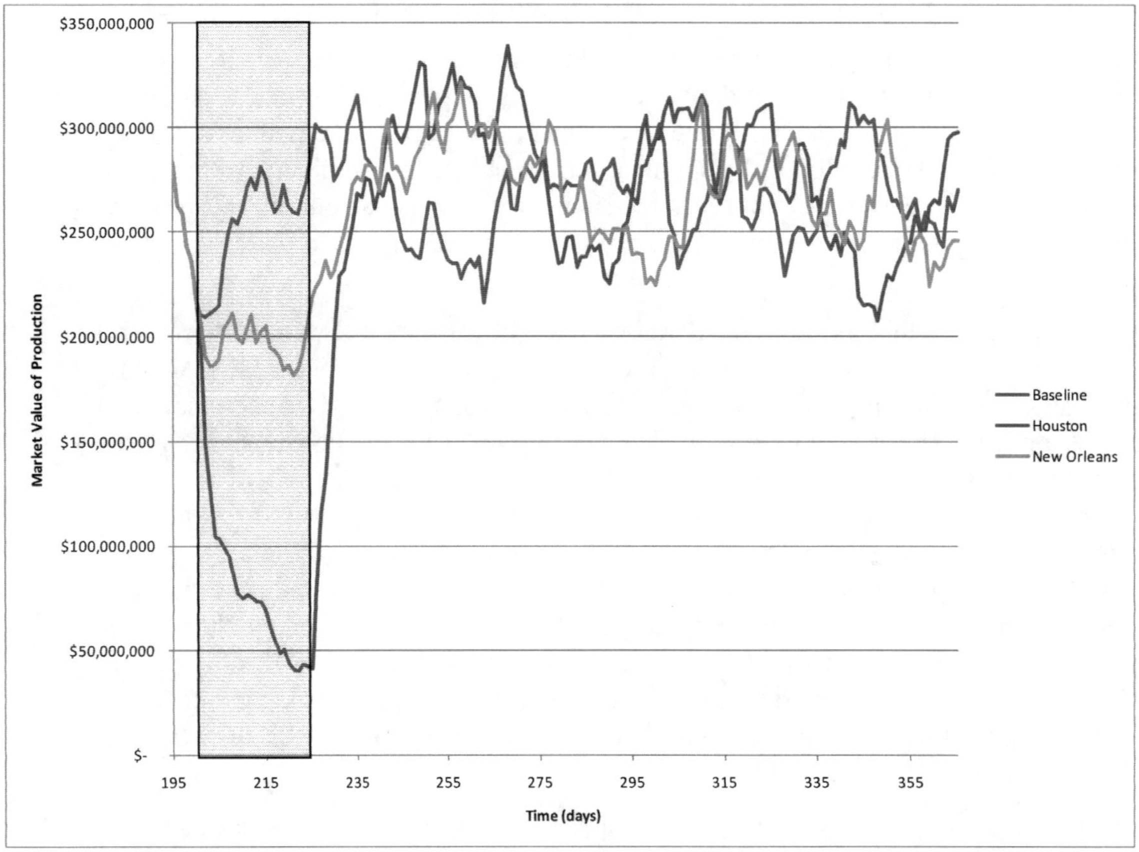

Figure 7: Recovery Costs

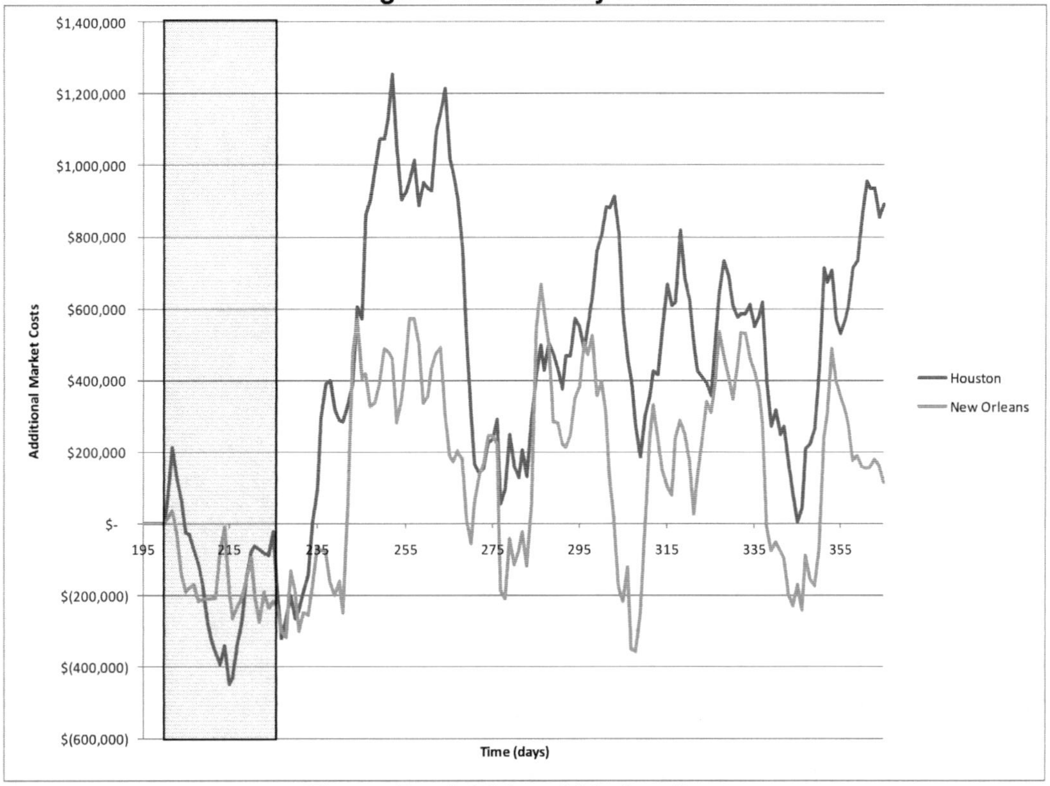

Figure 7a: Additional Market Costs

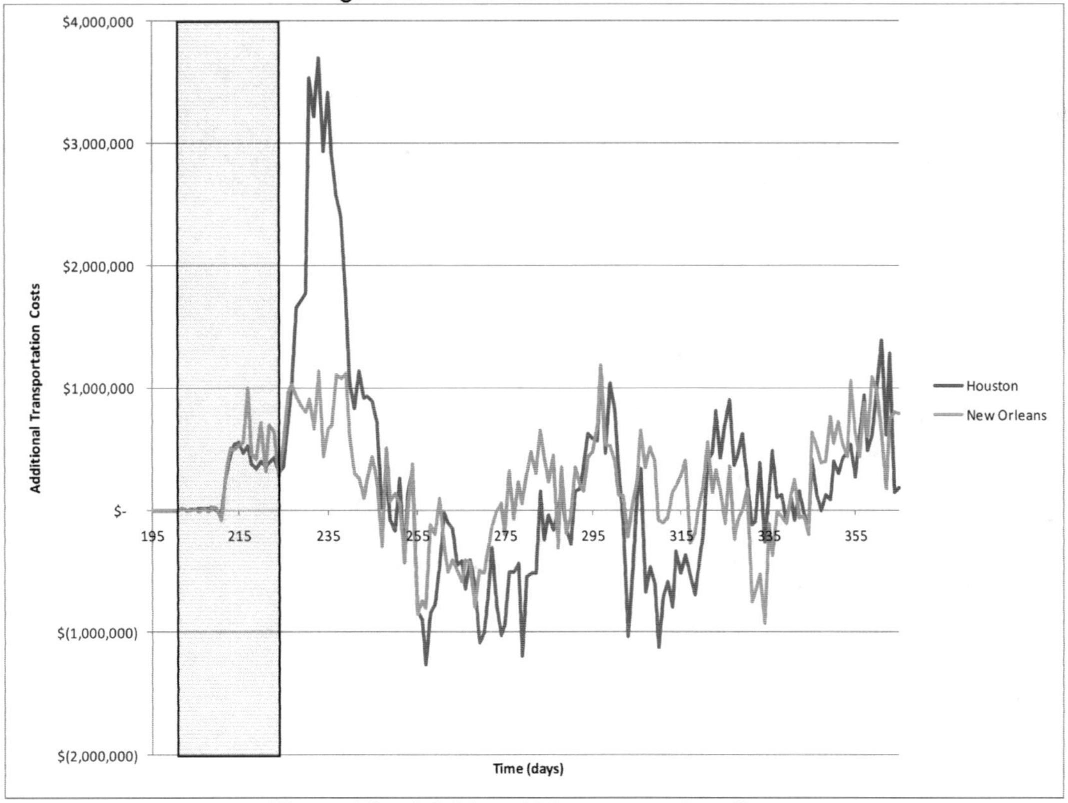

Figure 7b: Additional Transportation Costs

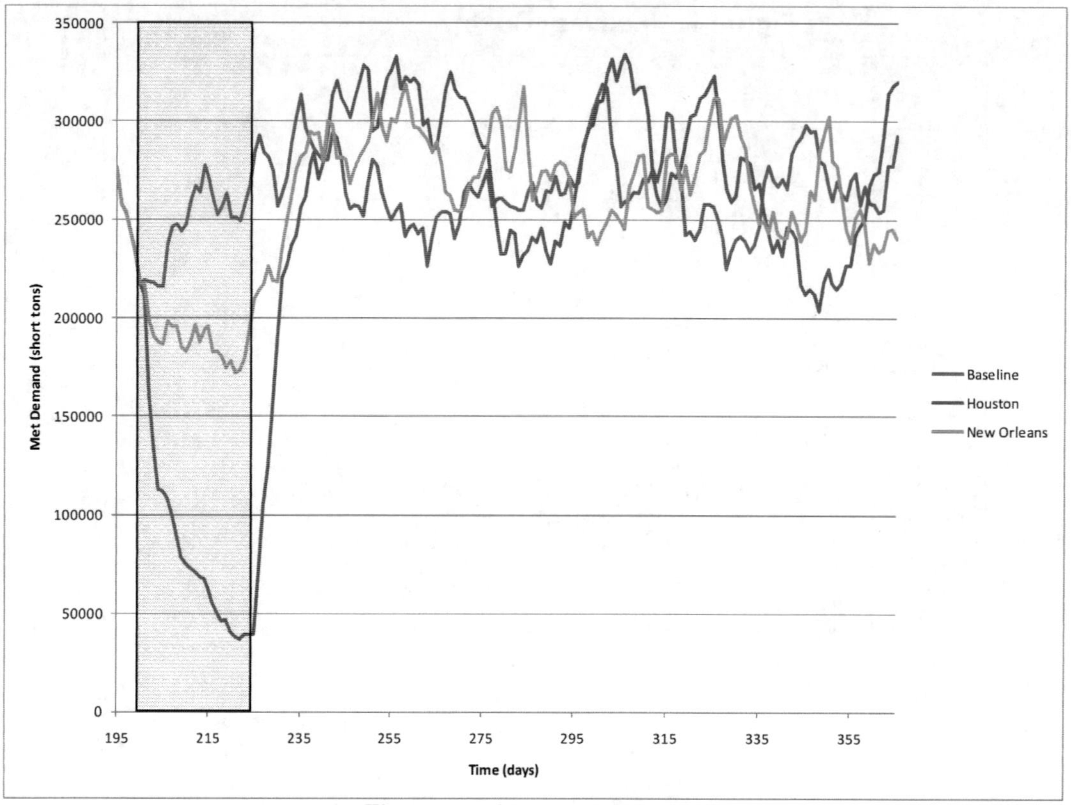

Figure 7c: Unmet Demand

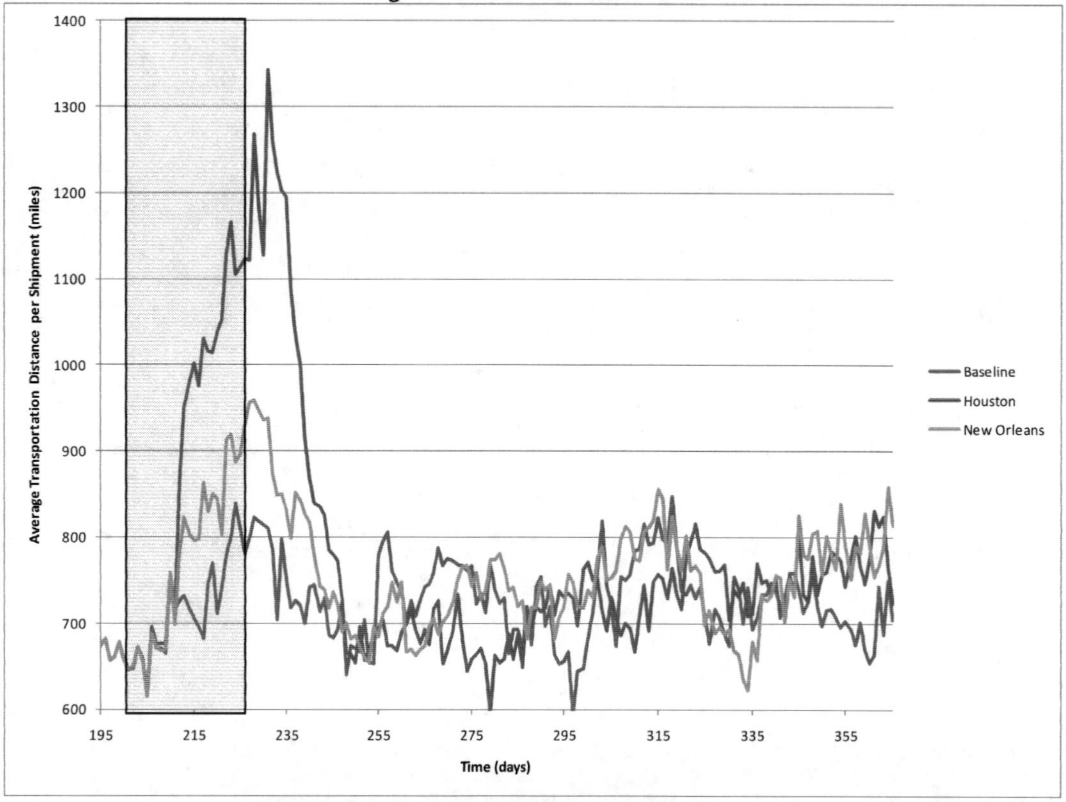

Figure 7d: Average Transportation Distance

Additional transportation costs are determined by two factors: average distance travelled by a chemical shipment and met demand. Travel distances for the disruption scenarios start to exceed baseline distances around day 210 (Figure 7d). Travel distances are highest for the Houston scenario, with average travel distances peaking above 1,300 miles per shipment (a 70-percent increase over baseline distances) around day 230. Average distances for the New Orleans scenario peak around 950 miles per shipment (approximately a 20-percent increase) during the same period of time. However, met demand decreases dramatically during the shutdown period (Figure 7c). This decrease in met demand and, consequently, the number of chemical shipments offsets the increase in transportation distances, keeping increases in transportation costs relatively moderate. However, after production capacity is restored, met demand increases at the same time that travel distances are highest, and additional transportation costs peak during this period (near day 230). After this period, disruption scenario travel distances decrease and become similar to baseline levels. Differences between the disruption scenarios and baseline scenario are attributed primarily to the stochastic nature of the supply chain model.

In this analysis, differences between disruption and baseline *MVP* levels and transportation costs after day 275 appear to be primarily due to the stochastic nature of the supply chain model; they do not appear to be caused by production shutdown. Hence, recovery is considered complete at this time.

3.4 Resilience Assessment

Table 1 lists the systemic impact, recovery efforts, and resilience costs for both disruption scenarios. The resilience costs for the Houston scenario are higher than the New Orleans scenario, so the petrochemical sector is more resilient to the New Orleans hurricane scenario than the Houston hurricane scenario. Furthermore, resilience costs are primarily determined by systemic impact as recovery costs are two orders of magnitude less than systemic impacts.

Table 1. Resilience Costs (unitless)

Measure	Houston Hurricane	New Orleans Hurricane
Target Market Value of Production ($M)	20,000	20,000
Systemic Impact ($M)	3,600	600
Recovery Effort: Market ($M)	23	3.9
Recovery Effort: Transportation ($M)	32	14
Resilience Cost	0.19	0.03

Resilience cost is calculated according to Eq. 3; i.e.,

$$\text{Resilience Cost} = \frac{\text{Systemic Impact} + \text{Market Recovery Effort} + \text{Transportation Recovery Effort}}{\text{Target Market Value of Production}}$$

The resilience framework allows qualitative assessment of attributes that enhance the supply chain's absorptive, adaptive, and restorative capacities. In this analysis, the supply chain is more resilient to the New Orleans hurricane because, when compared with the Houston hurricane, a larger fraction of the overall supply chain is unaffected, giving the entire supply chain greater ability to keep chemical production running. Thus, the diversity in location increases the system's absorptive capacity and *SI*s are lower for the New Orleans scenario.

Recovery costs are relatively small when compared with *SI*. This result suggests two resilience enhancement strategies. First, companies or policymakers could seek ways of enhancing the robustness of petrochemical supply chains to decrease *SI*. Alternatively, they could try more aggressive recovery strategies that may be more costly than current recovery processes but are still relatively low cost when compared to *SI*.

This example illustrates some general and important considerations when computing *TRE*. The calculation of *TRE* ideally should account for all costs associated with the supply chain's recovery to targeted system performance levels. While this example only considered transportation and market costs in the estimation of *TRE*, we recognize that recovery costs in actuality would be more extensive and include additional expenses such as repair of damage caused by winds and/or flooding, startup costs, etc. More work is needed to identify and estimate a complete set of variables that contribute to recovery costs (in addition to those listed in Table 1) so that better estimates of *TRE* can be calculated.

4. Summary

The resilience assessment framework presented herein consists of three primary components. First, our definition of system resilience indicates what factors need to be considered when assessing the resilience of a system. Second, we quantitatively evaluate these factors using our resilience cost measurement methodology. Third, we use the qualitative analysis component to examine resilience capacities and resilience enhancement features of the system, to explain or replace quantitative results.

The chemical sector (and other industries) must make cost-benefit decisions on a regular basis, and the resilience cost measurement approach described in this paper provides a structured, quantitative means for conducting those cost-benefit studies. For example, the resilience framework separates resilience costs into two categories: costs resulting from decreased system productivity (*SI*) and costs attributed to recovery activities (*TRE*). This information provides firms with a strategy for decreasing resilience costs by measuring tradeoffs between *SI* and *TRE*. If the *SI* far outweighs total recovery effort, a firm may want to focus on approaches for decreasing *SIs* (perhaps by increasing inventories, adding redundancy, etc.) without significantly increasing *TRE*. If *TRE* outweighs *SI*, a firm may want to focus on developing cheaper, more efficient strategies for recovery that do not drastically increase *SI*. Additionally, if the resilience costs for a sector or supply chain are calculated before and after resilience enhancements are made, the framework can be used to compare benefits of the decreased resilience costs with costs of making the system modifications. The framework can also be used to choose recovery strategies by comparing the resilience costs under different recovery strategies. Thorough application of this resilience assessment framework can result in a comprehensive evaluation of a system's resilience and provide information about how to further enhance system resilience.

5. Acknowledgements

The authors would like to thank Adam Rose and Tony Barrett from the National Center for Risk and Economic Analysis of Terrorism Events (CREATE) for the thoughtful insights that they provided to improve and clarify the text.

The authors performed this work with funding from the DHS Science and Technology Directorate.

Sandia is a multiprogram laboratory operated by Sandia Corporation for the Department of Energy's National Nuclear Security Administration under Contract DE-AC04-94AL85000.

6. References

[1] Reagan, Ronald, "Executive Order 13282, National Security Telecommunications Advisory Committee," 1982

[2] Clinton, William, "Presidential Directive PDD-63, Protecting America's Critical Infrastructures," 1998

[3] Bush, George W., "Homeland Security Presidential Directive 3 (HSPD-3)," 2002

[4] Bush, George W., "Homeland Security Presidential Directive 7 (HSPD-7)," 2003

[5] "Infrastructure Resilience Requires All Hazards Plan, Panel Advises DHS," *Emergency Preparedness News*, March 21, 2006

[6] Vugrin, Eric D., Drake E. Warren, Mark A. Ehlen, and R. Chris Camphouse, "A Framework for Assessing the Resilience of Infrastructure and Economic Systems," in *Sustainable and Resilient Critical Infrastructure Systems: Simulation, Modeling, and Intelligent Engineering*, Kasthurirangan Gopalakrishnan and Srinivas Peeta, eds., Springer-Verlag, Inc., forthcoming, January 2010

[7] Fiksel, Joseph, "Designing Resilient, Sustainable Systems," *Environmental Science and Technology*, 37(23), pp. 5330-5339, 2010

[8] Eidson, Eric D., and Mark A. Ehlen, "NISAC Agent-Based Laboratory for Economics (N-ABLE^{TM}): Overview of Agent and Simulation Architectures." Sandia National Laboratories SAND Report SAND2005-0263, February 2005

6TH GLOBAL CONGRESS ON PROCESS SAFETY

Modeling Dispersion and Deposition of Smoke Generated from Chemical Fires

Shahryar Khajehnajafi, SAFER Systems, 5284 Adolfo Road, Suite 100, Camarillo, CA 93012, 805-383-9711, shah@safersystem.com

Reza Pourdarvish, SAFER Systems, 5284 Adolfo Road, Suite 100, Camarillo, CA 93012, 805-383-9711, reza@safersystem.com

Hardik Shah, SAFER Systems, 5284 Adolfo Road, Suite 100, Camarillo, CA 93012, 805-383-9711, hardik@safersystem.com

Prepared for Presentation at
American Institute of Chemical Engineers
2010 Spring Meeting
6th Global Congress on Process Safety
San Antonio, Texas
March 22-24, 2010

UNPUBLISHED

Modeling Dispersion and Deposition of Smoke Generated from Chemical Fires

Shahryar Khajehnajafi, SAFER Systems, 5284 Adolfo Road, Suite 100, Camarillo, CA 93012, 805-383-9711, shah@safersystem.com

Reza Pourdarvish, SAFER Systems, 5284 Adolfo Road, Suite 100, Camarillo, CA 93012, 805-383-9711, reza@safersystem.com

Hardik Shah, SAFER Systems, 5284 Adolfo Road, Suite 100, Camarillo, CA 93012, 805-383-9711, hardik@safersystem.com

Keywords: smoke, dispersion, deposition, chemical equilibrium composition, chemical release

Abstract

Smoke is a mixture of toxic gases and suspended particulate matter of solids and liquids that evolves from a fire of flammable materials. This paper presents real-time consequence modeling to track the concentration of individual species in smoke as well as its soot deposition. In the modeling process presented, the burning rate or vapor mass is fed into a combustion model in which the combustion of products has been identified and quantified along with the temperature of the fire. The output of the combustion model is the smoke that will be dispersed into the ambient. The fire geometry, which depends on the type of fire (e.g., pool or flare), is identified. A dispersion model with the capability of determining particulate deposition is then utilized for tracking the smoke plume. The combustion model provides the formation of the toxic species present in the smoke: carbon monoxide, halogen acids, organic and inorganic irritants, and soot, and their quantities as the fuel-to-air ratio is changed. The end product is an excellent what-if analysis tool for responders. Among its many benefits is the ability to determine whether to ignite or when to ignite a flammable, toxic release by studying its consequence in comparison.

1. Introduction

Smoke generated by fire can kill, injure or irritate humans by thermal damage and/or inhalation of poisoning byproducts of the fire. There is increasing concern about the impact of toxic gases and soot particles generated by combustion of chemicals that are dispersed in the ambient and environment. Refineries and chemical companies usually have a fire accident in the plant, which is a big concern for industries. Therefore there is a need to calculate the concentration of toxic gases and soot particles and predict the impact these gases have on the city and all the population near these industries.

This paper presents consequence modeling as a means to track the smoke composition and deposition caused by the release of flammable chemicals into ambient that subsequently catches fire. Depending on the mode of flammable material released, various types of fire scenarios are possible: flare, pool fire, jet fire, and tank top fire. This consequence modeling uses different modules, with each module providing information for the next module. At the time of ignition, the pool fire or jet fire module calculates the burning rate of a pool fire or mass of vapor emanating into ambient. The fire geometry, which depends on the type of fire (e.g., pool or flare), is identified. This data is fed into a combustion module in which the combustion of products is quantified along with the temperature of the fire. The output of the combustion model is the smoke that is dispersed into ambient. Finally, dispersion and deposition modules are then utilized to track the smoke plume and particulates deposition. This simulation software predicts dispersion and deposition of toxic gases and soot particles present in the smoke that are generated from hydrocarbons and chemical fires.

This paper addresses the consequence of hazardous material emanating from a fire, which will help firefighters, hazmat, or other emergency responders to study the impact of hazards associated with toxic smoke and particulate, devise a more informed decision-making process, and plan for evacuation, shelter in-place and other response actions.

1.1 Health Hazard from Particulate

The health effect of the particulate/soot depends on the size and mass concentration in the breathing air. The size determines where in the respiratory tract the particulate will penetrate and, depending on the location of the penetration—upper respiratory tract, lungs, or into alveoli—the health effect would be different. Fine particles, even if they are not toxic, can cause dryness of throat, coughing, and, in high concentration, tightness of the chest and extreme difficulty in breathing. The three exposure levels defined for particulate are inhalable, thoracic, and respirable. The mean particle diameter in each category is defined as 100, 10, and 2.5 micron. The soot particles with diameter of 2.5-5 microns are of respirable type that can penetrate into the alveolar region of the lungs. The deposition of soot in the bronchiole and alveolar region of the lung reduces its capacity of passing oxygen into the blood stream.

2. Modeling Modules

Simulation of dispersion and deposition of final products of fire is constructed on modules with each module consequently calculating certain information and input data for the next modules. The following sections explain each module.

2.1 Pool Fire Module

Pool fire is a common occurrence in chemical industries when chemical spills from a tank or pipe and creates a pool that gets ignited. This module computes real-time pool fire size, burning rate and geometry of the flame such as diameter, length and tilt.

2.1.1 Burning Rate

The burning rate ($\dot{m}$) at any time is obtained from the following equation:

$$\dot{m} = \dot{E} A_P \quad \text{[Eq. 1]}$$

where $\dot{E}$ is the flux of burning and A_P is the pool area, which can be a fixed area or changed dynamically. Mudan[1] proposed a correlated model to calculate the flux of burning that covers a wide range of chemicals as follows:

$$\dot{E} = 0.001 \frac{\Delta H_c}{\Delta H_v + \int_{T_i}^{T_b} C_P \, dT} \qquad \text{[Eq. 2]}$$

where $\dot{E}$ is the flux of burning (kg/m^{2}s), ΔH_c is the heat of combustion, ΔH_v is the heat of vaporization, C_P is the specific heat capacity, T_b is boiling temperature and T_i is initial temperature of liquid in the pool. The dynamic size of the pool is a function of ignition time, duration and rate of the spill into the pool and minimum height of the pool. The dynamic pool area is numerically calculated from the following equation by considering a mass balance on the pool:

$$\frac{d[\rho h A_P(t)]}{dt} = R_i - \dot{E} A_P(t) \qquad \text{[Eq. 3]}$$

where R_i is spill rate into the pool, A_P is the dynamic pool area, h is height of the pool, and ρ is density of the liquid. It is assumed that the size of the pool shrinks when it reaches the minimum height of the pool.

2.1.2 Flame Geometry

The diameter and height of the smoke release are required to employ as inputs in the dispersion module. It is assumed that flame diameter and height can be considered as geometry of the smoke release in the dispersion module. The diameter of the flame relies on the wind speed and can be different than the diameter of the pool. The flame drag diameter is given by the following equation [2]:

$$\frac{d_{drag}}{d_{pool}} = 1.5\, Fr^{0.069} \qquad \text{[Eq. 4]}$$

where the Fr number is defined as:

$$Fr = u_{10}{}^2 / gd \qquad \text{[Eq. 5]}$$

where u_{10} is the wind velocity at the height of 10 m.

The length of the flame is determined as follows[3]:

$$L_{flame} = 55\, d \left(\frac{\dot{E}}{\rho_a \sqrt{gd}}\right)^{0.67} (u^*)^{-0.21} \qquad \text{[Eq. 6]}$$

where u^* is the dimensionless wind velocity given by:

$$u^* = u_{10} \left(\frac{\rho_v}{\dot{E} g d}\right)^{1/3} \qquad \text{[Eq. 7]}$$

The following correlation to determine the flame tilt is proposed by AGA[4].

$$\cos(\theta) = \begin{cases} \frac{1}{\sqrt{u^*}} & u^* \gg 1 \\ 1 & u^* < 1 \end{cases} \qquad \text{[Eq. 8]}$$

The height of the smoke release is given by:

$$h_{flame} = L_{flame}\cos(\theta) \qquad \text{[Eq. 9]}$$

2.2 *Jet Fire Module*

This module obtains the geometry of the flame for flare or jet fire scenarios. The flame length (L_{B0}) in a wind-free environment is given by [5]:

$$\left(\frac{D_s\beta}{L_{B0}W}\right)^{2/3} = 0.2 + 0.024\left(\frac{g}{D_s^2u_j^2}\right)^{1/3} L_{B0} \qquad \text{[Eq. 10]}$$

where D_s is source diameter, u_j is gas exit velocity, and W is mass fraction of fuel in a stoichiometric mixture with air. β is approximately constant with a value of 2.85. The correlation for flame length L_B in a windy environment is given by:

$$L_B = L_{B0}(0.51e^{-0.4V} + 0.49)\left[1 - 6.07 \times 10^{-3}(\theta_{jV} - 90)\right] \qquad \text{[Eq. 11]}$$

where L_{B0} is flame length in still air, V is wind speed, and θ_{jV} is the angle between hole axis and wind vector. Flame length is the distance between the exit point to the tip of the flame OC line in Figure 11.

The flame angle or tilt α is given as:

$$\alpha\xi(L_{B0}) = 8000R + \xi(L_{B0})(\theta_{jV} - 90)(1 - e^{-25.6R}) \quad R \le 0.05 \qquad \text{[Eq. 12]}$$

$$\alpha\xi(L_{B0}) = 1726(R - 0.026)^{0.5} + 134 + \xi(L_{B0})(\theta_{jV} - 90)(1 - e^{-25.6R}) \quad R \ge 0.05 \qquad \text{[Eq. 13]}$$

where $\xi(L_{B0})$ is the Richardson number based on flame length in still air:

$$\xi(L_{B0}) = \left(\frac{g}{D_s^2u_j^2}\right)^{1/3} L_{B0} \qquad \text{[Eq. 14]}$$

where R is ratio of wind speed to jet velocity. The flame width at the frustum tip is calculated by:

$$W_2 = L_B(0.18e^{-1.5R} + 0.31)(1 - 0.47e^{-25R}) \qquad \text{[Eq. 15]}$$

2.3 Combustion Module

This module calculates the chemical equilibrium composition of a product of fire by assigning two thermodynamic state functions such as enthalpy and pressure. The chemical equilibrium is calculated by the minimization of Gibbs free energy based on the work of Zeleznik and Sanford [6, 7]. In this method, there is no need to specify a set of reactions and equilibrium constants.

The Gibbs energy using the Lagrange multiplier method is:

$$G = g + \sum_{i=1}^{k} \lambda_i \left(\sum_{j=1}^{N} a_{ij} n_j - b_i^o \right) \quad \text{[Eq. 16]}$$

with the following mass-balance constrain:

$$\sum_{j=1}^{N} a_{ij} n_j - b_i^o = 0 \quad \text{[Eq. 17]}$$

where g is Gibbs free energy per mass of mixture, λ_i are Lagrangian multipliers, a_{ij} are the number of atoms of element *i* per mole of species *j* , n_j are moles of species *j* per mass of mixture, b_i^o are the total number of atoms of element *i* per mass of total reactants, *k* is number of elements and *N* is number of species.

The chemical potential for species *j* is:

$$\mu_j = \left(\frac{\partial g}{\partial n_j} \right)_{P,T,n_{i \neq j}} \quad \text{[Eq. 18]}$$

At the equilibrium G is at a minimum:

$$\delta G = \sum_{j=1}^{N} \left(\mu_j + \sum_{i=1}^{k} a_{ij} \lambda_i \right) \delta n_j + \sum_{i=1}^{k} \left(\sum_{j=1}^{N} a_{ij} n_j - b_i^o \right) \delta \lambda_i = 0 \quad \text{[Eq. 19]}$$

Considering $\delta\lambda_j$ and δn_j as independent:

$$\mu_j + \sum_{i=1}^{k} a_{ij} \lambda_i = 0 \quad \text{[Eq. 20]}$$

$$\sum_{j=1}^{N} a_{ij} n_j - b_i^o = 0 \quad \text{[Eq. 21]}$$

The equilibrium compositions are calculated by solving Eq. 5 and 6 with the following thermodynamic state:

$$P = P_o \quad \text{[Eq. 22]}$$

$$h = h_o \quad \text{[Eq. 23]}$$

where P is pressure, which is considered 1 atm., h is the enthalpy of the mixture and h_o is the enthalpy of the reactants.

This module considers flame as a reactor, that flammable chemicals react with air (oxygen and nitrogen) and the final products of reaction can be gases and possible soot particles. The following assumptions and constraints are applied to the combustion module:

1. It is only applicable to chemicals with C, H, N, S, O and halogen (F, CL, Br, I).
2. It is applicable to a single gas or a gas mixture with ten species at most.
3. It is applicable to a gas mixture with existing inert gases such as CO2, N2, etc.
4. The final products of the fire are considered as:
   For hydrocarbon: CO_2, H_2O, H_2, N_2, CO, HCN, CH_4, C
   For halogen: HF, HCl, HBr, HI, Cl_2, I_2, F_2, Br_2
   For sulfur: SO_2, SO_3, COS, H_2S
   For nitrogen: NO, NO_2, NH_3
5. The temperature of the reaction is calculated from adiabatic flame temperature at atmospheric pressure.

By providing the air-fuel ratio and reactants as input data, this module calculates equilibrium composition, total smoke rate, smoke temperature and amount of soot particles due to uncompleted burning.

2.4 Dispersion/Deposition Module

In the dispersion/deposition module, it is assumed that flame behaves like a stack that is releasing smoke and soot particles. The dispersion model is an integrated Lagrangian puff model[8]. The dispersion model simulates the transport and dispersion of airborne particles with deposition onto the surface. Particles or soot are embedded into gaseous combustion product that plays the role of a carrier gas. The soot deposition takes place on a user defined grid domain with a user defined range and cell size. The deposition module considers smoke as a carrier gas for soot particles and defines the soot particle concentration (mg/cm^3) in the smoke and the amount of the deposition of soot particle onto the ground (mg/cm^2). The main assumption in this model is that the carrier gas and particles form a homogeneous cloud that moves at cloud speed. This assumption excludes particles with gravitational velocity high enough to fall out of the cloud. The particles with radii greater than about 80-100 μm have significant gravitational settling speeds. The Stokes law for the terminal settling velocity is valid for particles with small diameter. The Stokes formulation is used to calculate terminal or gravitational settling velocity as:

$$V_g = \frac{2gd^2\rho_p}{18\mu} \quad \text{[Eq. 24]}$$

where particle density is represented by ρ_p, air viscosity by μ, and particles diameter by d, and g is the acceleration due to gravity. The deposition of the airborne material onto the ground is caused by a combination of mechanical processes such as gravitational settling, turbulent diffusion, and surface impact. The deposition flux F_{dep} is assumed to be directly proportional to the local air concentration evaluated at a reference height z_{dep}, usually assumed to be one meter.

$$F_{dep} = V_{dep}C(x, y, z_{dep}) \quad \text{[Eq. 25]}$$

A typical range of the deposition velocity is between 0.001 to 10 cm/s. The deposition of the airborne material reduces the source soot content of the carrier gas. The deposition velocity is defined as Slinn and Slinn[9], Pleim et al[10]:

$$V_{dep} = \frac{1}{r_a + r_d + r_a r_d V_g} + V_g \quad \text{[Eq. 26]}$$

where V_g is gravitational velocity and r_a and r_d are aerodynamic resistance and deposition layer resistance, respectively. The aerodynamic resistance is defined as:

$$r_a = \frac{1}{ku_*} \ln\left(\frac{z_r}{z_0}\right) \quad \text{[Eq. 27]}$$

where u_* is the friction velocity and z_0 is the surface roughness. The deposition layer resistance is defined as:

$$r_d = \left[Sc^{-2/3} + 10^{-3/st}\right]^{-1} u_*^{-1} \quad \text{[Eq. 28]}$$

The Sc and St are the Schmidt and Stokes number.

$$Sc = \frac{v}{D} \quad \text{[Eq. 29]}$$

$$St = \left(\frac{V_g}{g}\right)\left(\frac{u_*^2}{v}\right) \quad \text{[Eq. 30]}$$

where v is air kinematic viscosity and D is diffusivity of pollutant in the air. The source depletion model of Van der Hoven[11] is used to account for this effect.

$$\frac{dQ(t)}{dt} = -\iint_A V_{dep} C(x, y, z_{dep}) dA \quad \text{[Eq. 31]}$$

Mass depletion due to dry deposition can be also be simplified to:

$$\left.\frac{dQ_s}{dt}\right|_{drydep} = -2\pi\sigma_y^2 C_s V_{dep} \quad \text{[Eq. 32]}$$

The trajectory of the puff is obtained as a solution to the vertical and horizontal momentum balance. Horizontal momentum balance:

$$\frac{d\left[(Q_s + Q_{cg})\, u \cos(\theta)\right]}{dt} = Q_a U_{air} \quad \text{[Eq. 33]}$$

The Q_s and Q_{cg} are the mass of soot and carrier gas in the cloud. Centerline velocity is represented by u, and Q_a is the air entrainment rate.

Vertical momentum balance:

$$\frac{d\left[(Q_s + Q_{cg})\, u \sin(\theta)\right]}{dt} = V_{tot}\, g\, (\rho_{air} - \rho_{puff}) \quad \text{[Eq. 34]}$$

V_{tot} is puff volume. The adiabatic heat balance on the puff is written as:

$$\frac{dH}{dt} = Q_a\, C_{pair}\, T_{air} - Q_s\, C_{ps}\, T_{cloud} \quad \text{[Eq. 35]}$$

where H, total enthalpy change of the cloud, is a function of air and water vapor entrained into the cloud along with particle and carrier gas heat transfer to the surrounding. Smoke composition, rate, temperature, geometry of the flame and metrological data are used as input information for the dispersion module. The dispersion model assumes constant molar ratio among the smoke gas constituents for dispersion. Gas concentration is displayed in ppm, and the output for soot plume concentration and deposition are in mg/m^3 and mg/m^2 respectively.

3. Case Studies

In the following sections we show the application of the model to three scenarios; tank top fire, pool fire, and sour gas ignition. The chemical(s) involve in the fire can be pure or a mixture of inert and flammable materials. The required inputs for case studies are shown. For each scenario the smoke composition, its temperature as well as plume impact for a few specific gaseous component of smoke is displayed. In case of sooty fire soot plume and its deposition on the ground is also provided.

3.1 Tank Top Fire

A fire occurs at a benzene (C_6H_6) floating roof tank due to an earthquake. The floating roof collapses, benzene floats above the roof and ignites due to an impact and friction from sloshing, creating a tank top fire. Tank top fire is one of the most common types of fire in refineries. Figure 1 shows a schematic of a tank top fire. Simulation is carried out to study the impact of smoke generated from the fire. Information about the storage tank and meteorological conditions is given in Table 1.

Table 1. Input Data for Tank Top Fire

Tank	
Tank geometry	Vertical cylinder
Tank height	50.0 (ft)
Tank diameter	50.0 (ft)
Tank temperature	70.0 (deg F)
Meteorology	
Ambient temperature	70.0 (deg F)
Wind Speed	2.2 (m/s)
Stability	2.0
Humidity	50 (%)
Surface roughness	1.0 (m)
Ceiling height	10000 (m)
Soot	
Soot density	1.8 (g/cm3)
Soot mean diameter	5 (micron)

It is assumed that smoke disperses from a stack with the height and diameter of the flame. During the fire, benzene reacts with air to produce new chemicals. The reaction in an unbalanced form can be represented as:

$$C_6H_6 + (O_2 + N_2) \rightarrow CO + CO_2 + H_2O + H_2 + HCN + CH_4 + N_2 + C$$

Table 2 shows the equilibrium composition and temperature of the smoke and geometry of the flame. Generally, the amount of the soot can range from fractions of percent to about 20% of the fuel mass in a regular pool fire [12]. We estimate Air to Fuel ratio of 4.0 due to the sooty fire of benzene. The amount of soot particles based on this Air to Fuel ratio is about 18% of the fuel mass.

Figure 2 demonstrates soot particle concentrations at three levels. Smoke carrying soot particles rises up quickly to higher elevation due to buoyancy before dispersing in the atmosphere. Figure 3 demonstrates the soot deposition on the ground at two levels.

Figure 1. Schematic of Tank Top Fire

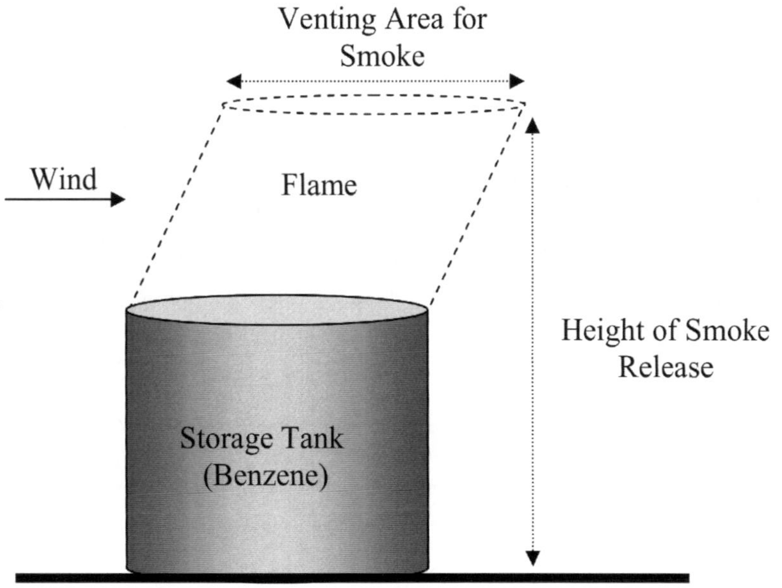

Table 2. Smoke Results for Tank Top Fire

Smoke Composition (weight %) for Air to Fuel ratio of 4.0 (lb/lb)	
N_2	64.20
CO	34.10
H_2	1.62
CO_2	0.03
H_2O	0.02
HCN	0.02
CH_4	0.01
Smoke Information	
Smoke rate	11669.2 (lb/min)
Soot rate	554.4 (lb/min)
Smoke temperature	2062.9 (deg F)
Smoke duration	120 (min)
Release height	137.5 (ft)
Source diameter	50 (ft)

Figure 2: Soot Plume Generated from Benzene Fire

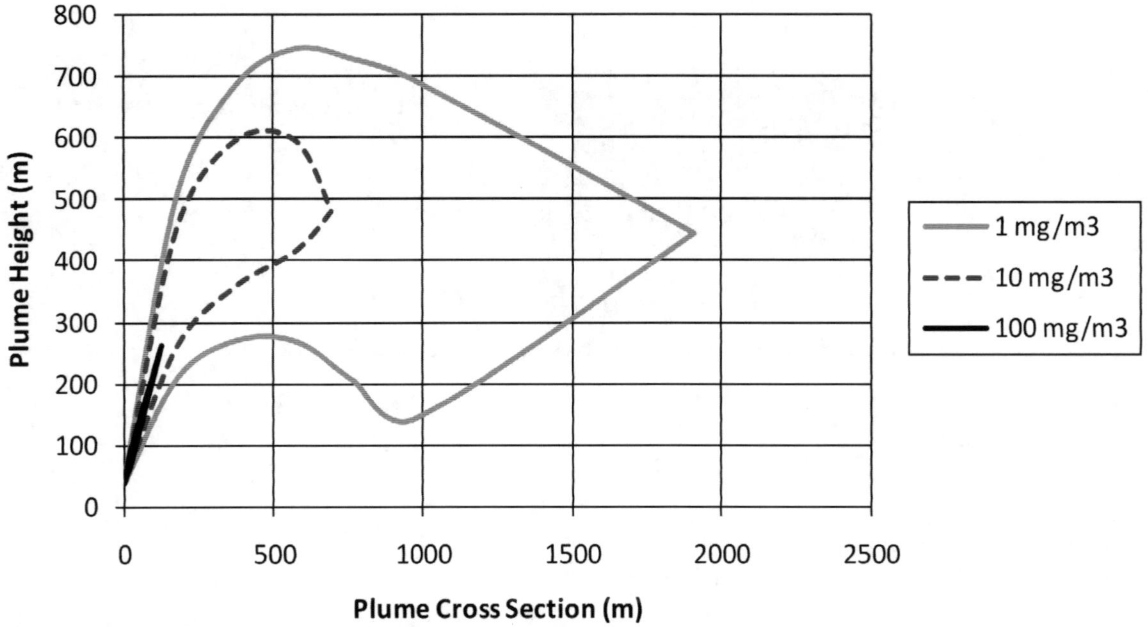

Figure 3: Soot Deposition from Benzene Fire

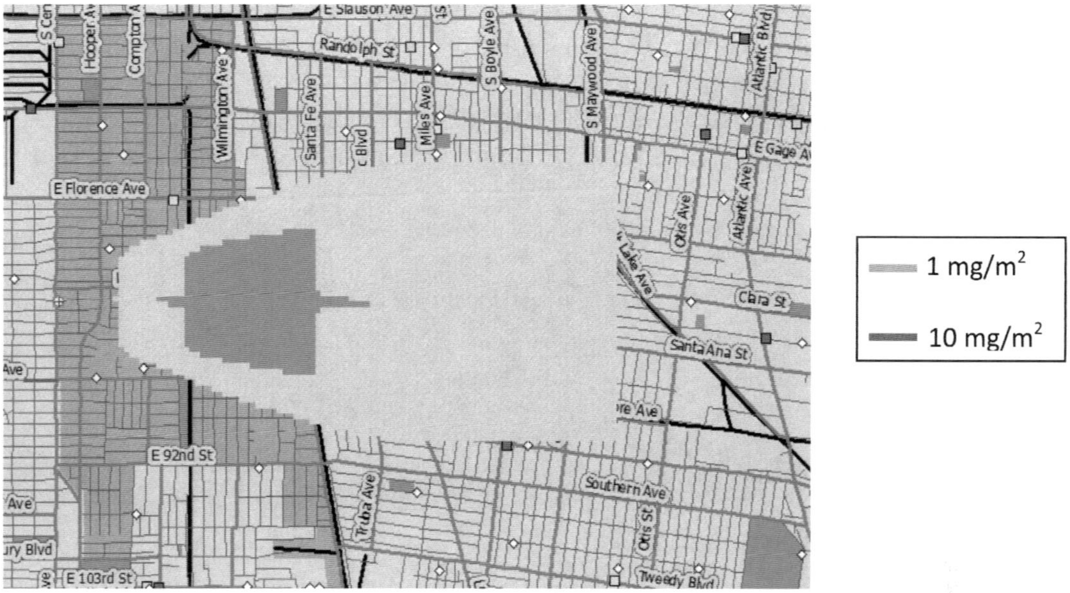

It is assumed that the molar composition of a smoke gas mixture is constant. This makes it possible to study concentration profiles of each individual species in the smoke. Figure 4 indicates the dispersion of carbon monoxide, which is one of the toxic products of benzene fire. Carbon monoxide concentration levels are based on the Emergency Response Planning Guidelines Handbook (ERPG).

Figure 4: CO Plume Generated from Benzene Fire

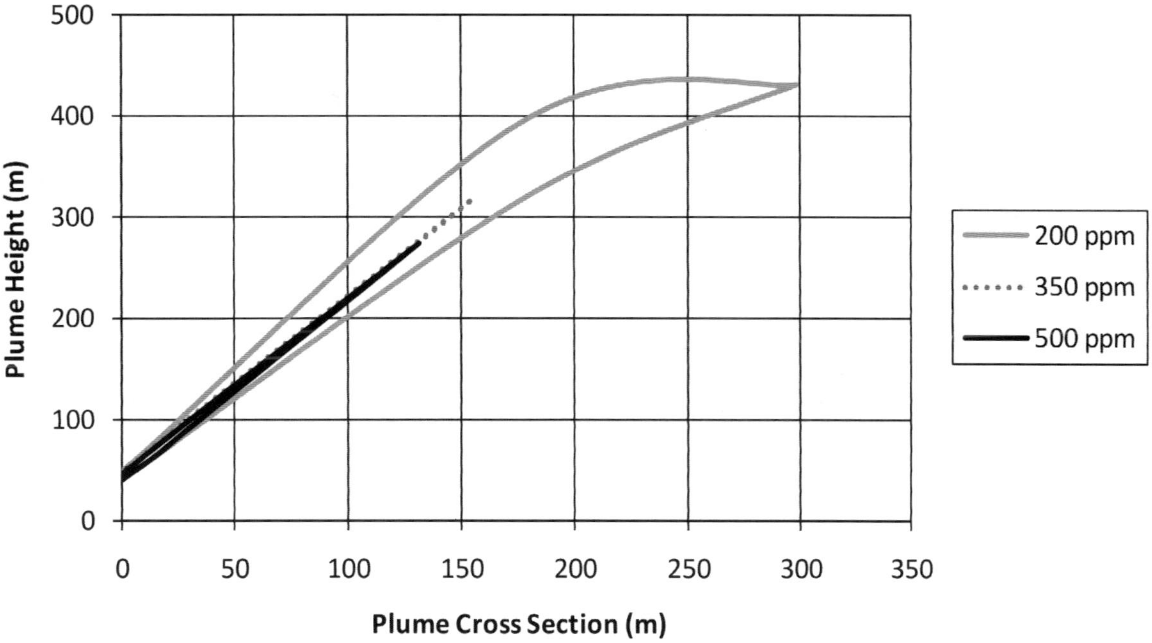

3.2 Pool Fire

A pipe carrying liquid octane from a tank is sheared. Liquid octane spills onto a concrete forming a pool. The flammable cloud emanating from the pool is transported downwind until it encounters an ignition source. The resulting flash fire burns back to the source, leading to an octane pool fire. Figure 5 shows a schematic of an octane pool fire scenario. Simulation is carried out to study the impact of smoke generated from the fire. Information of tank, pipe, pool, and meteorological conditions are given in Table 3.

Table 3: Input Data for Pool Fire

Tank	
Tank geometry	Vertical cylinder
Tank height	40.0 (ft)
Tank diameter	30.0 (ft)
Tank wall thickness	0.5 (in)
Tank temperature	70.0 (deg F)
Liquid level	32.0 (ft)
Leak	
Tank leak type	Pipe scenario
Pipe diameter	2.0 (in)
Pipe length	1.0 (ft)
Pipe elevation	4.0 (ft)
Pool	
Substrate type	Concrete
Substrate temperature	70.0 (deg F)
Pool area	Unconfined
Minimum pool depth	0.39 (in)
Meteorology	
Ambient temperature	70.0 (deg F)
Wind Speed	2.2 (m/s)
Stability	2.0
Humidity	50 (%)
Surface roughness	1.0 (m)
Ceiling height	10000 (m)
Soot	
Soot density	1.8 (g/cm3)
Soot mean diameter	5 (micron)
Ignition	
Ignition time at	5 (min)

Figure 5. Octane Pool Fire

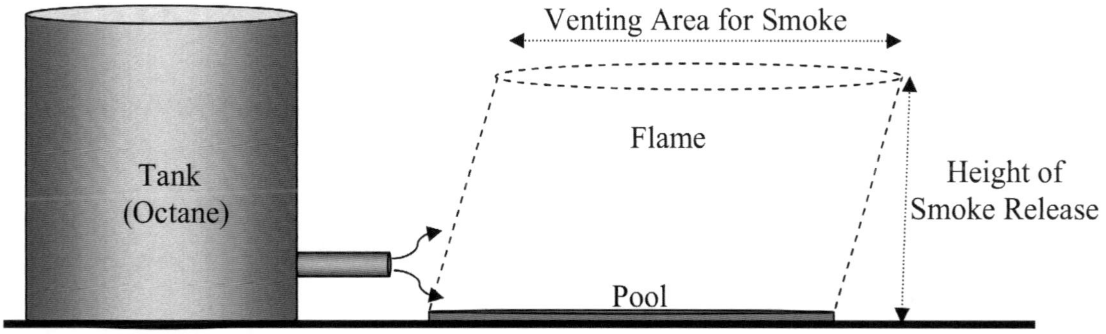

In this scenario, it is assumed that an ignition source occurs 5 minutes after the breach of the pipe. The total burning rate for the combustion module is obtained by the summation of octane vapor due to flashing and octane evaporation due to the fire. The area of the pool is assumed to expand and then shrink at the end of the fire. During the fire, octane reacts with air to produce the following gases (the reaction is represented in an unbalanced form):

$$C_8H_{18} + (O_2 + N_2) \rightarrow CO + CO_2 + H_2O + H_2 + CH_4 + N_2 + C$$

Table 4 demonstrates the equilibrium composition of the final product of octane fire. The Air to Fuel ratio is estimated at 4.0, which introduces soot particles in the final product. The amount of soot particles is about 15% of fuel mass.

Figures 6 and 7 show dispersion and deposition of soot, respectively. Figure 8 shows carbon monoxide plume impact generated from the octane pool fire.

Table 4: Smoke Results for Octane Pool Fire

Smoke Composition (weight %) for Air to Fuel ratio of 4.0 lb/lb	
N_2	63.8
CO	29.2
H_2	3.1
CO_2	2.4
H_2O	1.1
CH_4	0.4
Smoke Information	
Smoke rate	758.3 (lb/min)
Soot rate	31.3 (lb/min)
Smoke temperature	1419.7 (deg F)
Smoke duration	91.7 (min)
Release height	35.5 (ft)
Source diameter	18.0 (ft)

Figure 6: Soot Plume Generated from Octane Fire

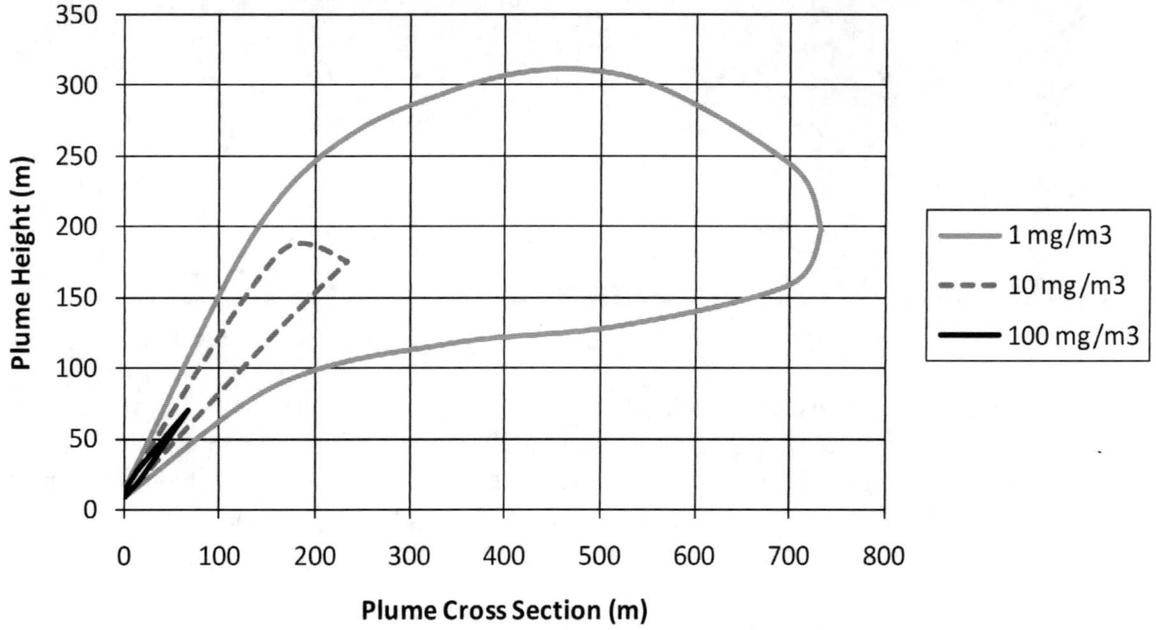

Figure 7: Soot Deposition for Octane Fire

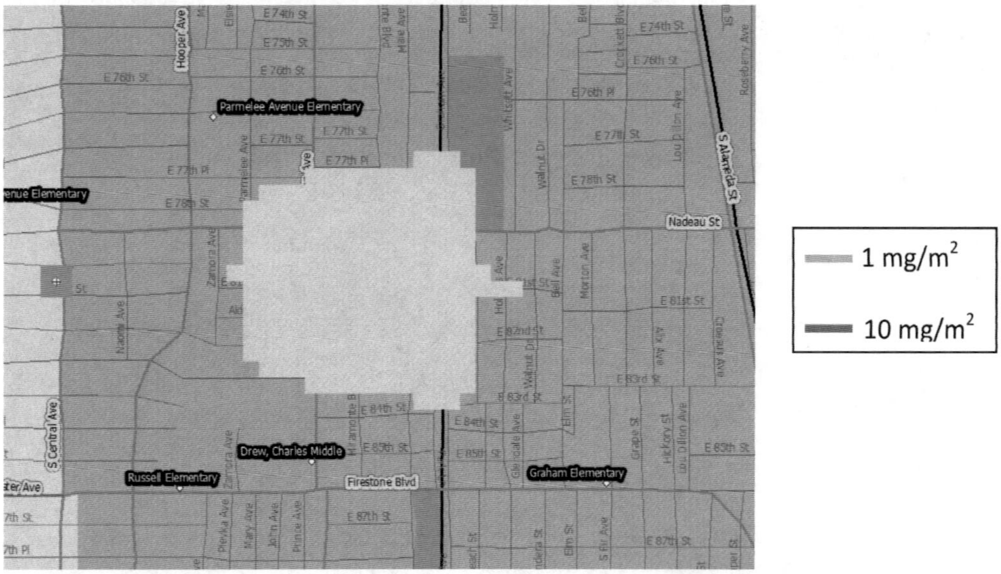

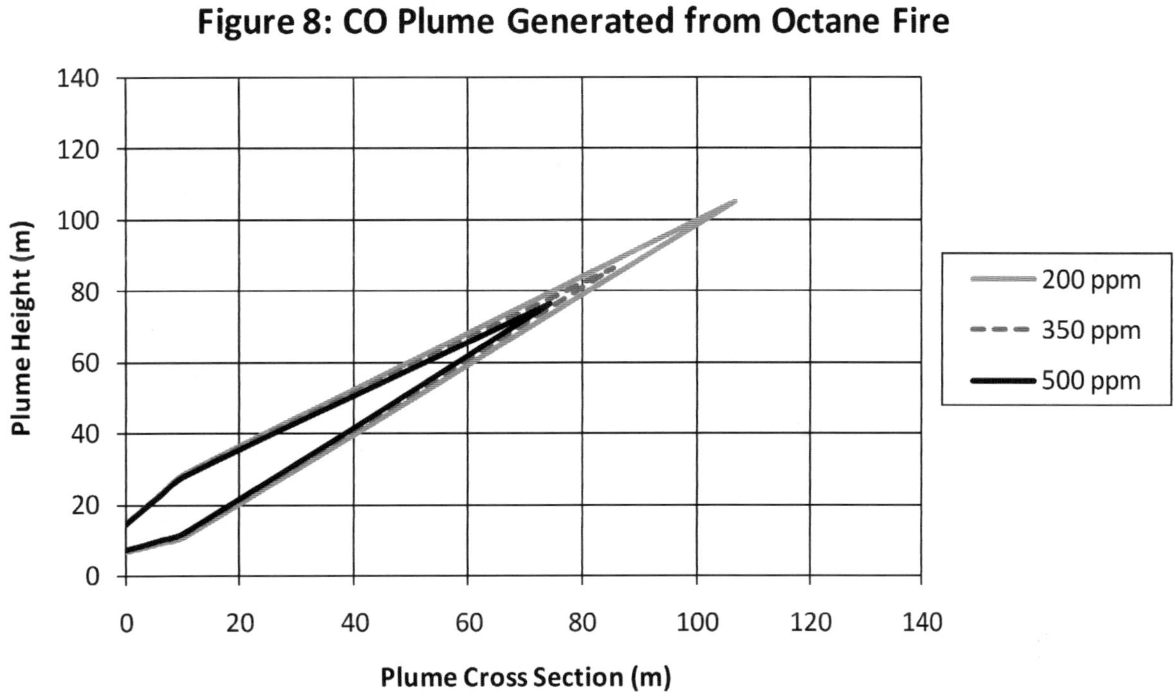

3.3 Stack /Flare

An operational upset at a refinery unit forces vent gases, that contain the extremely hazardous chemical of hydrogen sulfide (H_2S) to be dumped into the flare system to prevent fire or explosion. The flare is turned off. Management decides to do a quick study of the consequence of the un-ignited release to justify turning the flare on and burning the venting gases. The information regarding this sour gas composition and release condition are provided in Table 5. The meteorological conditions are the same as those used in the pool fire case study.

Table 5: Input Data for Stack/Flare

Stack	
Release height	100.0 (ft)
Release diameter	1.0 (ft)
Release rate	2500.0 (lb/min)
Release temperature	70.0 (deg F)
Release Gas Component (mole %)	
CH_4	90.0
C_2H_6	4.0
H_2S	3.0
CO_2	3.0

Figure 9 demonstrates the dispersion of H_2S from a stack before burning. This figure shows that the odor can be recognized at least 7 km from the source of the release based on an H_2S odor level of 0.1 ppm. Figure 10 also indicates the concentration of H_2S in the range between 30 ppm and 0.1 ppm that is touching the ground. For this range of concentration, it is believed that nearly all individuals could be exposed for up to one hour without experiencing serious health effects that could impair an individual's ability to take protective action. This means that the emergency responders must ignite the gas in a short time to prevent individuals from having any serious health effects. Based on the above results, management decides to ignite the cloud at the source to mitigate the hydrogen sulfide impact on the community. Therefore, the plume was ignited 15 minutes after the release.

Figure 9: H_2S Plume Impact before Ignition

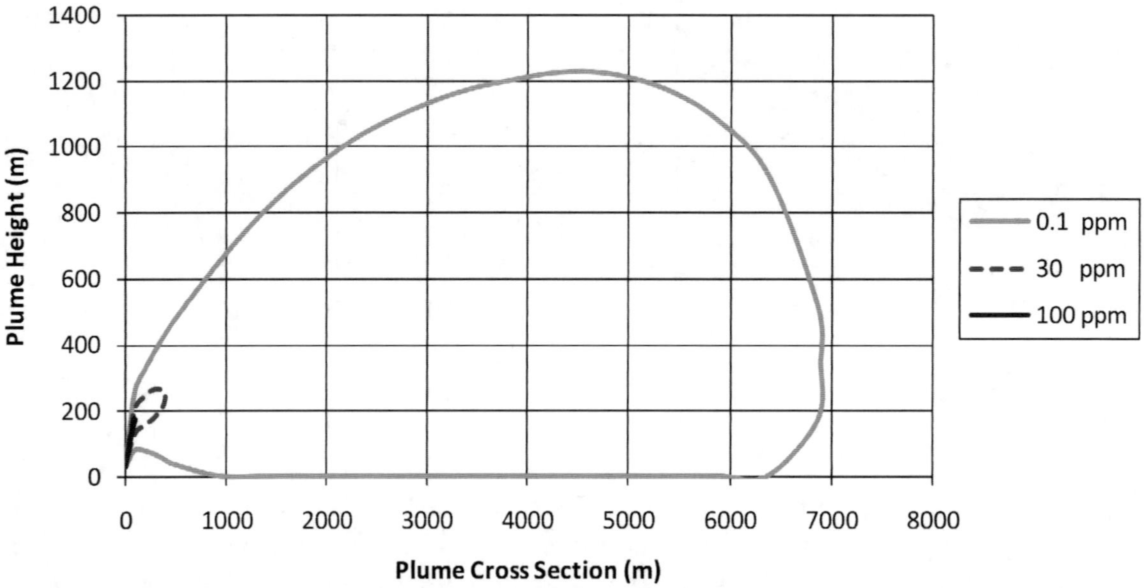

Figure 10: Contour Plot of Ground Level H_2S Concentration at 0.1 ppm

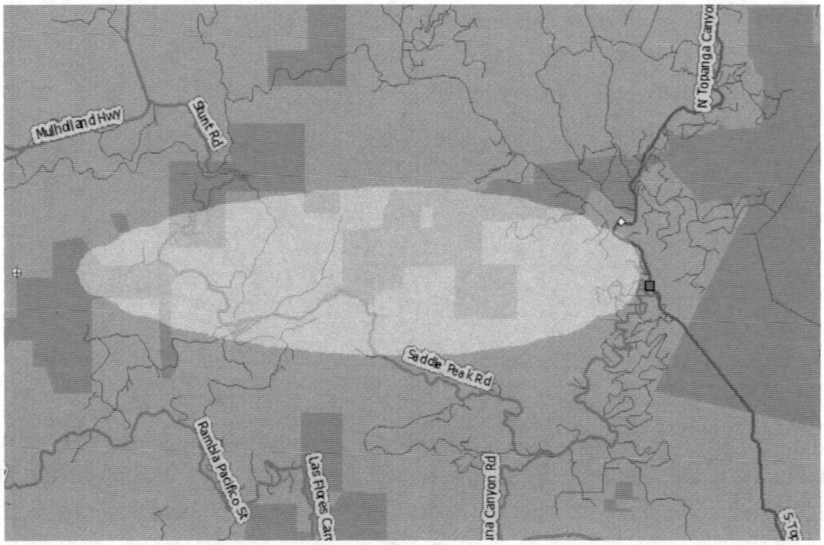

Figure 11 shows the flame due to release of a gas from a stack. We estimate an Air to Fuel ratio of 11.0 due to the non-smoky fire for the natural gas, which also can be tested for a range of 13.0 to 10.0. The equilibrium composition for burning of this sour gas is shown in Table 6. This table demonstrates characteristics of the smoke venting area, height, temperature and rate.

Table 6: Smoke Results for Sour Gas Fire

Smoke Composition (weight %) For Air to Fuel ratio of 11.0 (lb/lb)	
N_2	70.4
H_2O	12.3
CO_2	8.4
CO	7.8
SO_2	0.6
H_2	0.4
H_2S	0.04
Smoke Information	
Smoke rate	29867.0 (lb/min)
Soot rate	0.0 (lb/min)
Smoke temperature	3110.9 (deg F)
Smoke duration	120 (min)
Release height	220.9 (ft)
Source diameter	36.1 (ft)

Figure 12 indicates the concentration of H_2S after the ignition. This figure shows that the fire reduces the concentration of H_2S by dilution and reacting with air. Because of the high temperature of the smoke, unburned H_2S is dispersed into the atmosphere at a high elevation. Therefore, burning of the sour gas can make H_2S much less harmful for the population particularly if dilution is applied at the flare tip to increase the Air to Fuel ratio. However, SO_2 and CO are the other toxic gases that are produced from the burning of sour gas. Figures 13 and 14 show dispersion of SO_2 and CO in the atmosphere, respectively. These figures show that toxic gases are dispersing into the high elevation and are harmless for the population near the release site.

Figure 11: Schematic of Stack/Flare

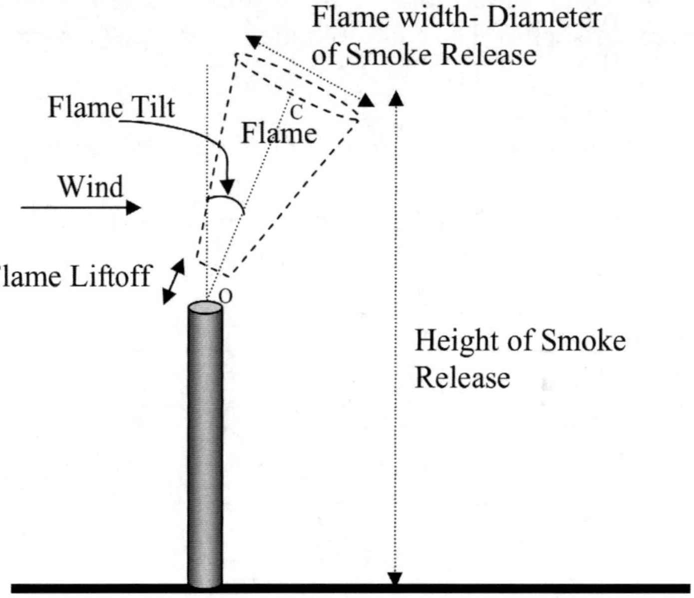

Figure 12: Unreacted H_2S Plume after Ignition

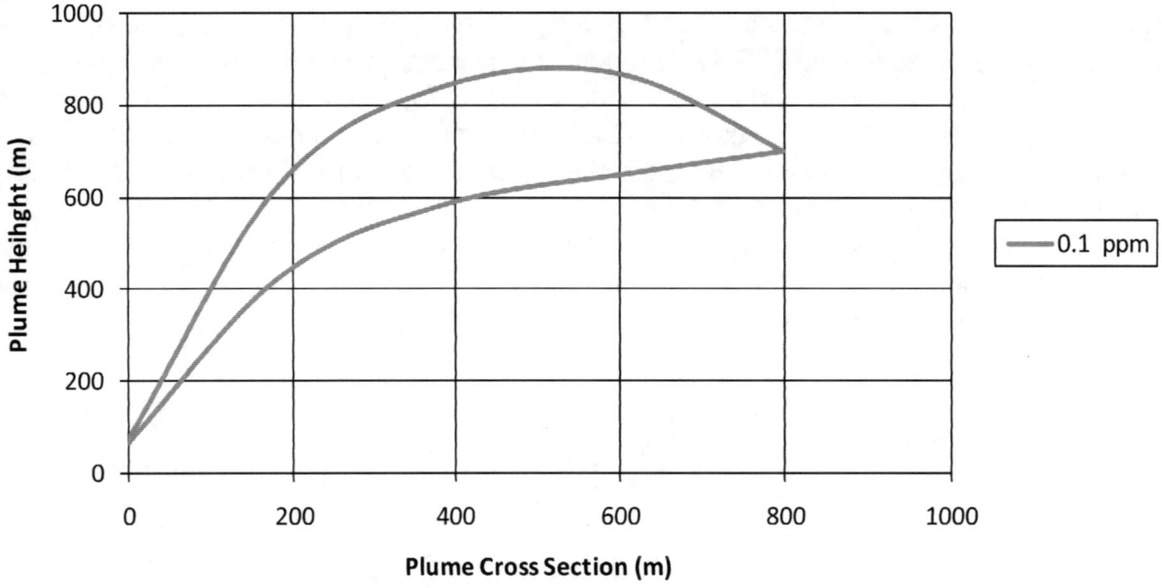

Figure 13: SO_2 Plume Generated from Sour Gas Fire

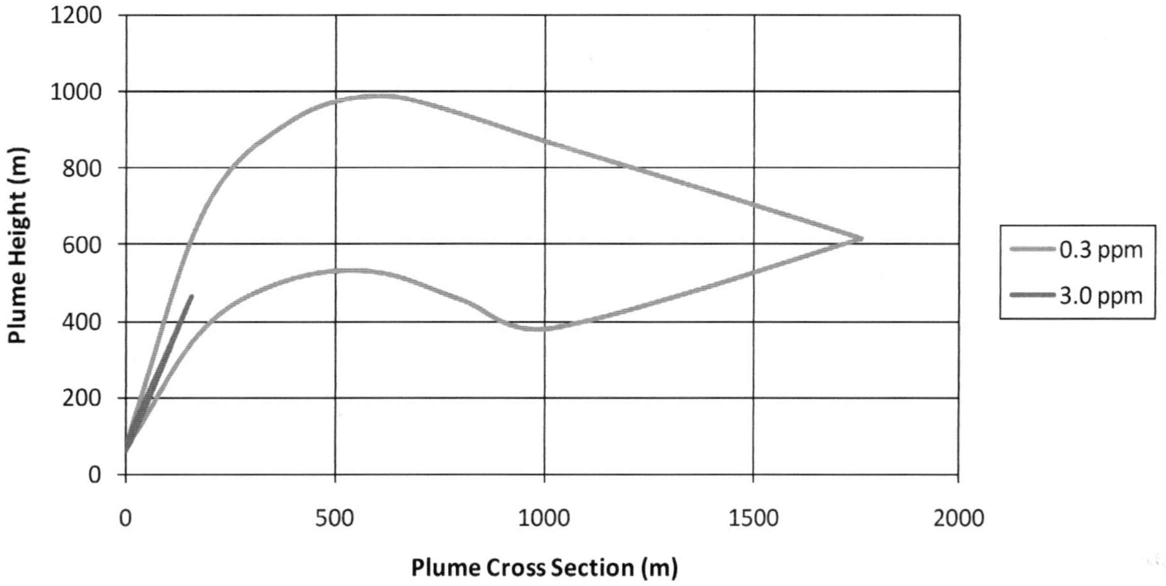

Figure 14: CO Plume Generated from Sour Gas Fire

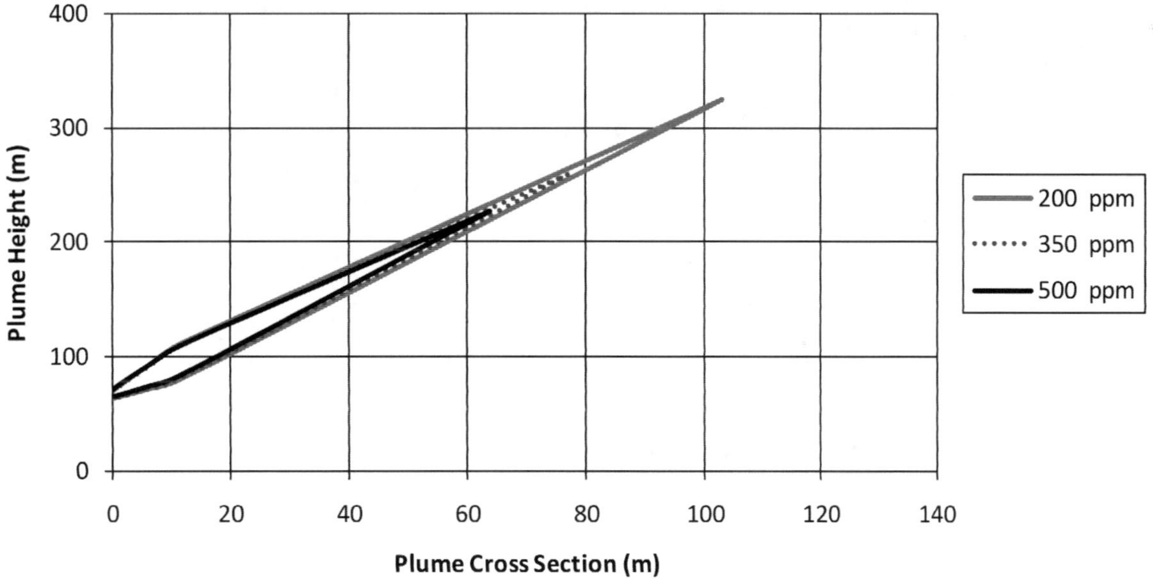

4. Conclusion

A comprehensive combustion model for emergency or risk management studies was presented. The model is capable of dealing with any type of fire resulting from the release of flammable materials: pool fire, jet fire, flares, and tank top fire. A combustion model based on the work of Zeleznik and Sanford made it possible to calculate the combustion material with a minimum amount of information. Through the three case studies, we showed the applicability of the model. Assumption of a constant molar ratio among smoke gas mixtures made it possible to look at concentration profiles of each individual species. The soot plume and soot deposition were also displayed for sooty smoke. In the flare example, we showed a screening hazard analysis of an un-ignited gas release versus an ignited one. The emergency response actions can reduce the exposure to the public through evacuation, which prevents exposure to toxic gas; sheltering; and ignition; to reduce exposure to H_2S by replacing it with exposure to a lower concentration of SO_2.

In conclusion, this model is capable of simulating the consequences of any flammable chemical release into ambient. The model provides the following type of information:
- Dispersion of toxic gases and liquids before ignition
- Equilibrium composition of final product of the fire
- Adiabatic flame temperature
- Dispersion of individual product of fire
- Dispersion and deposition of soot particles

5. References

[1] Mudan, Krishna S., "Thermal Radiation Hazards from Hydrocarbon Pool Fires," *Prog. Energy. Combust Sci.*, Volume 10, 59-80, 1984.

[2] Moorhouse, J., "Scaling Criteria for Pool Fires Derived from Large Scale Experiments," *I. Chem. E. Symposium Series*, Volume 71, 165-179, 1982.

[3] Thomas, P. H., "The Size of Flames from Natural Fires," *9th Int. Combustion Symposium Series*, 844-859, 1963.

[4] American Gas Association, "LNG Safety Research Program," Report 3-1, 1974.

[5] Chamberlain, G. A., "Developments in Design Methods for Predicting Thermal Radiation from Flare," *Chem. Eng. Res.*, Volume 65, 299-309, 1987.

[6] Zeleznik, F. J., Sanford, G., "Calculation of Complex Chemical Equilibria," *Applied Thermodynamics Symposium*, Volume 60, No. 6, 27-57, 1968.

[7] Zeleznik, F. J., Sanford, G., "An Analytic Investigation of Three General Methods Calculating Chemical-Equilibrium Composition," *NASA,* TN D-473, 1960.

[8] Sheih, C. M., "A Puff Pollutant Dispersion Model with Wind Shear and Dynamic Plume Rise," *Atmospheric Environment*, Volume 12, 1933-1938, 1978.

[9] Slinn, S. A., Slinn, W. G. N., "Predictions for Particle Deposition on Natural Waters," Atmospheric Environment, Volume 14, 1013-1016, 1980.

[10] Pleim, J., Venkatram, A., Yamartino, R., "ADOM/TADAP Model Development Program. Volume 4, The Dry Deposition Module," *Ontario Ministry of the Environment*, Rexdale, Ontario, 1984.

[11] Van der Hoven, I., "Deposition of Particles and Gases," *Meteorology and Atomic Energy*, 202-208, YID-24190, 1968.

[12] Butler, K. M., Mulholland, G. W., "Generation and Transport of Smoke Component," *Fire Technology*, Volume 40, 149-176, 2004.

Paper Title: Consequence Modeling of Chlorine Release

Presenter/Principal Author: Mr. Prakash Shah
Affiliation: Reliance Industries Limited (DMD)
Address: At & Po.: Dahej, Dist: Bharuch, Gujarat, India, Pin No.:392130
email address: prakash.a.shah@ril.com

Other Author(s):1. Mr. Chandrakant J Patel
Affiliation(s): Reliance Industries Limited (DMD)
2. Mr. Prabir Kumar Ghosh,
Affiliation(s): Advisor Safety, Centre for HSE Excellence, Reliance

Address(es): At & Po.: Dahej, Dist: Bharuch, Gujarat, India, Pin No.:392130
email address(es): 1. chandrakant.j.patel@zmail.ril.com, 2.Raja.Kalluri@ril.com

Prepared for Presentation at
American Institute of Chemical Engineers
2010 Spring Meeting
6th Global Congress on Process Safety
San Antonio, Texas
March 22-24, 2010

UNPUBLISHED

Consequence Modeling of Chlorine Release

Presenter/Principal Author: Mr. Prakash Shah
Affiliation: Reliance Industries Limited (DMD)
Address: At & Po.: Dahej, Dist: Bharuch, Gujarat, India, Pin No.:392130
email address: prakash.a.shah@ril.com

Other Author(s):1. Mr. Chandrakant J Patel
Affiliation(s): Reliance Industries Limited (DMD)
2. Mr. Prabir Kumar Ghosh,
Affiliation(s): Advisor Safety, Centre for HSE Excellence, Reliance

Keywords: Chlorine, Consequence Analysis, Hazard, Risk, Quantitative Risk Assessment, Safety Review, Chemical Exposure Indices.

Abstract

The paper describes the **consequence modeling of chlorine release** from a chlor alkali plant as a part of process hazard analysis. The objective was **to carry out Quantitative Risk Assessment (QRA) study of the chlorine system of the chlor alkali plant and build up in-house capability to carry out such studies**. QRA covers the following stages: Hazard Identification, Consequence Analysis, Frequency Analysis, Risk Analysis and Risk Management. QRA provides valuable insights into the features of the Industrial plant, highlighting those aspects where failure may result in harm to operators, members of the public, the environment and or the asset itself. **Hazard identification** was done in two phases by **Safety review of chlorine storage system** and using **Chemical Exposure Indices (CEI) method**. Based on the safety review and CEI rating, scenarios were selected for modeling and consequence analysis. **The results were summarized and scenarios were classified on basis of Onsite and Offsite Impact**.

A consequence modeling of losses of containment scenarios of chlorine bullet using PHAST software is explained in detail in the paper along with the risk management recommendations. Competency development initiatives such as this were helpful in understanding of process hazards by the plant operating staff and helped in implementation of mitigating measures. Thus a robust safety management culture can be instituted.

1.0 Introduction:

Industrialization is a vital part for the economic upliftment of vast population of India. Gujarat is one of the most industrialized states of the country and many of the industries including major chemical industries are concentrated in the Bharuch district of Gujarat. In Bharuch district, Ankleshwar and Dahej are the two important industrial areas, and in these two regions a total of six large size caustic chlorine industries are located, handling more than 7, 50,000 metric tonne per annum of chlorine out of which four units are located in the Dahej Industrial Region itself within a radius of 5 kilometers. Out of total chlorine production in India Dahej region produces 17% chlorine. M/s Reliance Industries Limited in Dahej handles about 400-430 tonne of chlorine daily.

To strengthen the overall safety approach M/s Reliance Industries Ltd. has decided to implement Process Safety Management (PSM) in a structured manner at Dahej facility. As a part of PSM implementation process hazard analysis (PHA) is a key study area. Hazard identification and consequence analysis are two important elements of a PHA study and the plant management decided to carry out these two studies in chlor-alkali plant. The Quantitative Risk Assessment consists of Hazard Identification, Consequence calculation, Frequency estimation, Risk Analysis and Risk Management and the main objective of study was to develop an in-house expertise in carrying out Quantitative Risk Assessment.

M/s Reliance engaged Mr. P.K. Ghosh as an Advisor who has carried out such studies before. Two teams one a QRA team and the other a Core team were formed to carry out the QRA study. QRA team was to carry out the study and the responsibility of the core team was to conduct a peer review particularly the assumptions and the final recommendations arrived. The QRA team consisted of members from various disciplines like Operation, Health, safety, environment and fire, Technical Services, and Group Manufacturing services and the members, all chemical engineers, were having experience from 2 to 20 years. The Core team to carry out the peer review consisted of the top level management including Plant head, Technical Service Head, Site Health, safety, environment and fire Head and Group Manufacturing Service PSM Leader having 15-25 years of industrial experience. All the team members were trained in a phased manner by the expert Mr. Ghosh. The training included the detail understanding of the basics of the QRA steps

On completion of training the study was carried out under the guidance of Mr. Ghosh.

2.0 Process Description

Chlorine is produced by electrolysis of Sodium chloride (brine) in electrolytic membrane based cell. The chlorine gas saturated with water leaves the cell is cooled down from 88 to 15 deg C in a chlorine cooler by means of polished brine, cooling water and chilled water. From the cooler gas enters the moist chlorine filter to separate brine mist carried along with gas. The moist chlorine gas coming from the filter dried by means of sulphuric acid to a final moisture content of 10% w/w ppm. The dry gas is passed from the dry chlorine filter to remove the traces of sulphuric acid. The dry chlorine gas is compressed in a chlorine compressor from 0.02 bar (g) to 7.0 bar (g). The compressor discharge is admitted into the chlorine liquefier under controlled pressure and liquefied by exchanging heat with the chilled Ethylene glycol. The liquid chlorine goes to chlorine evaporator and excess chlorine which is not evaporated goes to chlorine storage tank via liquid chlorine receiver by gravity. The overall process flow diagram for production of chlorine is shown in Fig.1. The evaporated chlorine is sent to downstream consuming unit via pipelines.

Fig-1: Overall Process Flow Diagram of Chlorine Production System

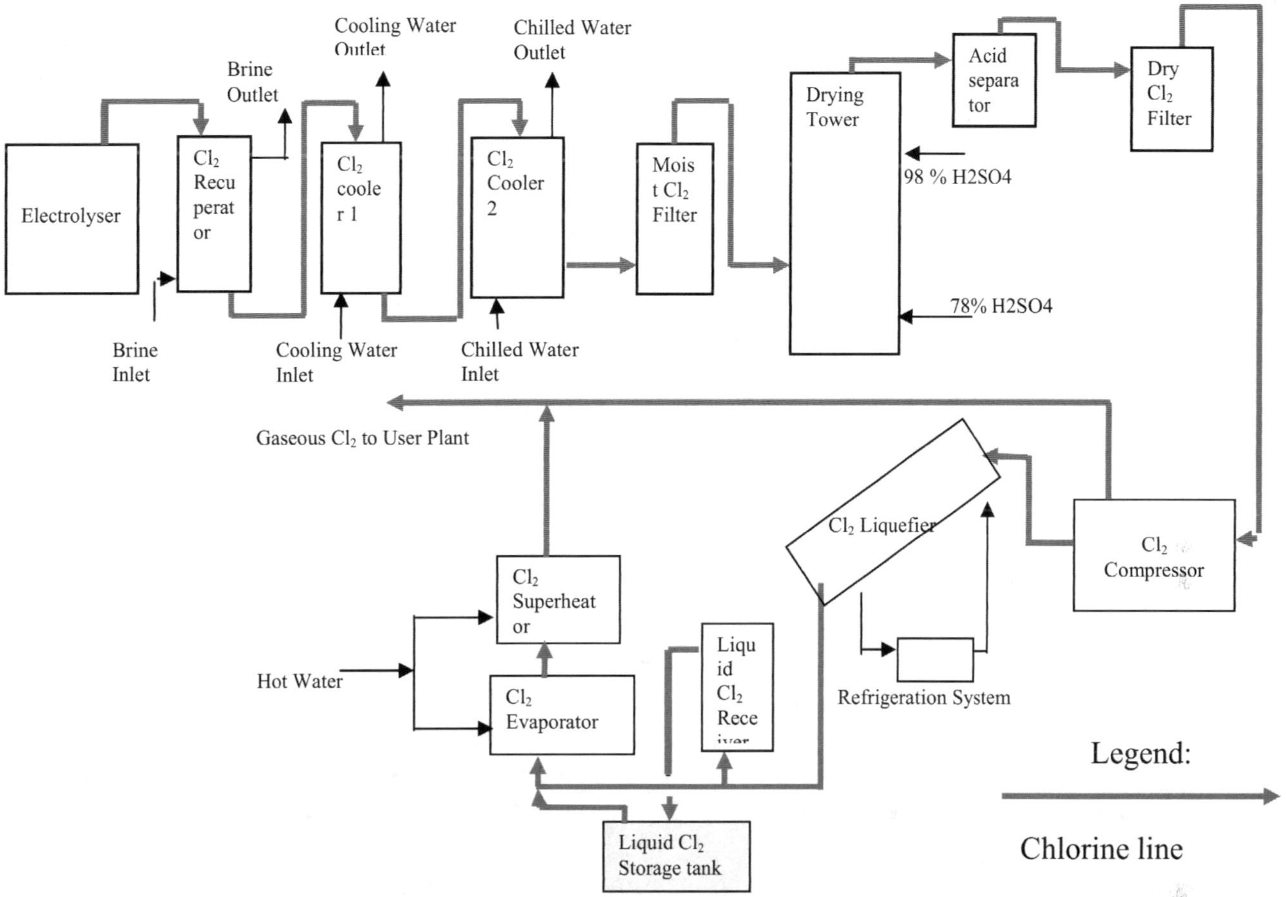

3.0 Hazard Identification

The hazard identification is the first step involved in the risk assessment. The main purpose is to identify hazards and failure cases which might initiate the hazards. The system boundary definition is important steps for the hazard identification. Out of all the chemicals handled at chlor alkali plant, the risk from chlorine release is considered worst as it is capable of spreading to large distances with significant gas concentrations and causing harmful effect to people. Hence, chlorine system of the plant was defined as boundary for the QRA study.. The techniques used for QRA study are as follows:

1. Process Safety Review of the Liquid Chlorine Storage System
2. Hazard Identification Using Chemical Exposure Indices (CEI)

3.1 Process Safety Review:

Process safety review is essential for hazard identification and focuses on major risk situations such as major loss of containment scenarios which can have offsite consequences. The Review is generally carried out based on some Standard and verifies whether the existing facilities conform to the Standard and making recommendations for specific actions that are needed, based on deviations from the Standard, with which it was compared, justifying the recommendations to eliminate the hazards.

3.1.1 Objective of the Process Safety Review:

The main objective of the Process Safety Review is to evaluate the nature and scale of hazards at the facility site by means of following:

1. To identify hazards based on deviations from Standard.
2. To postulate scenarios due to deviations from the Standard.
3. To select major loss of containment scenarios, this can have offsite consequences out of the various scenarios.

3.1.2 Review of existing Liquid Chlorine Storage System:

In this study, the process safety review study was carried out for Liquid Chlorine storage system. The standard used for the review was **"Safety Advice for Bulk Chlorine Installation HS (G) 28** which was prepared by the Working Party which met at the auspices of Health and Safety Executive of UK and is based on the Guidelines for Bulk Handling of Chlorine at Customers Installation published by the **Chemical Industries Association, UK**. Based on the standard following points of chlorine installations were reviewed during study.

1. Storage Vessel: Design criteria, connecting pipe work, chlorine inlet and outlet, vent, relief, instrumentation.
2. Protection of Storage Vessel against Overpressure: Relief valve, bursting disc, expansion vessel, pressure alarm
3. Inspection and Commissioning of Tank
4. Transfer of Chlorine to Consuming Unit, Transfer of chlorine (liquid & gas), precautions and transfer of liquid chlorine using vapor pressure, padding by N2.
5. Chlorine Vaporizer: Heating, types, throughput, general installation, safety hazards, isolation, PCV, corrosion
6. Chlorine Absorption System: Vent collection
7. Pipe work for Liquid Chlorine: Design
8. Protection of Liquid Pipelines against Over – Pressure: Expansion tank to relieve line
9. Valves: Types of valves, remotely controlled
10. Protective and Emergency Equipment: PPE, emergency tool kit, containment of leak
11. Operator's training & operating instructions
12. Maintenance
13. Emergency Procedures: Detector, alarm, on site/off site plan, control centre, team.

3.1.3 Results of Process safety Review and Postulation of Release Scenario:

From the process safety review study, team found the following deviation from the standard. Based on the deviation team has postulate the probable scenario for the deviation which are given in table-2.1.

Table-2.1: Safety Review findings and postulated scenario for deviation

Sr. No.	Item Description	Requirements:	Deviations	Postulated Scenario
1.1	Storage Vessel	The bund should be capable of taking the contents of the largest storage tank with adequate freeboard and a sump.	At present there is no sump.	Catastrophic rupture of storage tank
1.2		Provision should be included for removal of rain water over the bund wall, not via drain or through bund valves.	At present rain water in the bund is removed through drain valve.	
1.3		If the vessel operates at low temperature and lagging is required, the material should be fire resistant, chemically inert to liquid or gaseous chlorine and resistant to ingress of atmospheric moisture.	The material for lagging is not fire resistant.	
2	Liquid Chlorine Outlet	The pressure in the Cl2 storage tank is measured by gauges specially developed for use with liquid chlorine; these should have silver or tantalum diaphragms.	The pressure in the chlorine storage tank is measured by gauges having stainless steel diaphragms.	Leakage from pressure gauge due to diaphragm failure will result in gaseous chlorine release.
3	Protection of storage vessels against overpressure	The design requirements for the construction of the expansion vessel are similar to those for the storage tanks; the capacity should be about 10% of the largest storage vessel.	The expansion vessel is not of the specified capacity (10% of tank).	Lower capacity may lead to liquid chlorine carry over to Hypo system and can lead to chlorine release as the chlorine absorption system is not designed for liquid Cl2.
4	Inspection & Commissioning of Cl2 Tank	For lagged tanks, sufficient lagging should be removed to enable the condition of the external surface to be assessed.	Tank lagging is selectively removed for carrying out NDT. Thorough external inspection of the tank is not carried out.	Local corrosion at non-inspected areas may not be detected and can lead to liquid/gaseous Cl2 leakage.
5.1	Containment of chlorine leakage	Materials and equipment including plastic sheets to contain chlorine leakages should be readily available.	The materials and equipment to contain chlorine leakages are not available near storage area.	Liquid chlorine release from vessels and pipelines due to inadequate containment.
5.2			Materials to contain chlorine leak not available.	
6	PPEs: Emergency tool kit	Emergency tool kits (including spare valve keys) should be provided in the chlorine storage area.	There are no emergency tool kits available in the chlorine storage area.	May lead to liquid Cl2 release for an extended period due to delay in getting proper tools for leakage control.

3.2 Hazard Identification using Chemical Exposure Index (CEI) Method:

Chemical exposure index (CEI) is a simple method of rating the relative acute health hazard potential to people in the neighbouring plants or communities from possible chemical release incidents. Absolute measures of risk are very difficult to determine, but the CEI system will provide a method of ranking one hazard relative to another. The methodology utilizes a linearized equation for estimating air borne quantity released. The calculated index will be consequence based and independent of the frequency of events. Index will be used as a screening tool for selection of scenarios for consequence analysis. Scenarios, which have greater than 200 CEI, require consequence analysis study. For this study Dow's Chemical Exposure Index method was used.

3.2.1 Objective of Hazard Identification by CEI:

The main objective of the Hazard Identification by CEI is to evaluate and identify the possible scenario which could have potential of toxic hazards and ranking process equipment on the basis of their highest potential of toxic hazard and to identify scenario for further consequence analysis.

3.2.2 Scenarios Identification & CEI Calculation:

Based on the Dow's CEI guidelines the different scenario for the gaseous and liquid chlorine lines and equipments were considered and CEI calculated which are summarized in table 3.1. In each scenario single phase is considered for the calculation.

Table-3.1: Summary of CEI Results

Scenario No.	Scenario	Equipment Tag No.	Hole Dia. (mm)	Airborne quantity	CEI
1	Main Cell House header to Cl_2 Recuperator header leakage	NA	318	25.66	1106.19
2	Vacuum Pump Discharge line leakage	07P001	68.1	1.36	254.764
3	Vacuum Pump Discharge to CDT line leakage	NA	50.8	0.75	189.304
4	Anolyte tank top leakage	07T001A	68.1	1.22	241.502
5	Cl_2 Recuperator to Cl_2 Cooler-1 header leakage	NA	272.6	19.41	962.234
6	Cl_2 Cooler-1 to Cl_2 Cooler 2 header leakage	NA	272.6	20.01	976.802
7	Cl_2 Cooler -2 to Moist Cl_2 Filter header leakage	NA	272.6	20.68	993.144
8	Moist Cl_2 Filter to Drying Tower header leakage	NA	272.6	20.35	985.106
9	Drying Tower to Drying Cl_2 Filter header leakage	NA	227.1	13.33	797.214
10	Dry Cl_2 Filter to Cl_2 Compressor line leakage	NA	227.1	12.82	781.963
11	Cl_2 Compressor 4th Stage aftercooler outlet leak	22K001	90.87	10.41	704.639
12	Cl_2 evaporator Top Leak	26E001	113.6	19.23	957.635
13	Cl_2 receiver top leak	23V002	50.8	3.331	398.569
14	Cl_2 Bullet top line leak	24T001	50.8	3.150	387.621
15	Cl_2 Liquefier outlet line leakage	23E001	50.8	68.83	1811.64
16	Liquid Cl_2 receiver Bottom inlet leak	23V002	50.8	69.75	1823.7
17	Cl_2 Evaporator Bottom leak	26E001	50.8	76.23	1906.5
18	Cl_2 bullet bottom leak	24T001	50.8	69.93	1826.08
19	Cl_2 liquefier outlet line U seal bottom leak	NA	50.8	78.98	1940.59
20	Cl_2 Expansion Vessel bottom line leak	23V001	50.8	75.58	1898.38

4.0 List of Scenario Identified for consequence analysis:

Based on the hazard review and CEI rating following scenario were identified for the consequence analysis study.

Scenario No.	Scenario
1	Main Cell House header to Cl_2 Recuperator header leakage
2	Vacuum pump discharge line leakage
5	Cl_2 Recuperator to Cl_2 Cooler-1 header leakage
6	Cl_2 Cooler-1 to Cl_2 Cooler-2 header leakage
7	Cl_2 Cooler-2 to Moist Cl_2 Filter header leakage.
11	Cl_2 compressor 4th Stage after cooler outlet leak at the bend
12	Cl_2 Evaporator top leak

13	Cl_2 Receiver top Leakage
15	Cl_2 Liquefier Outlet line leakage
17	Cl_2 Evaporator Bottom leak
18	Cl_2 Bullet bottom leak
Based on Safety Review Study	Catastrophic rupture of Cl_2 storage bullet

5.0 Consequence analysis:

Consequence analysis consists of evaluating the direct, undesirable impact of potentially hazardous events, such as fires, explosions, and toxic releases, resulting from loss of containment due to failure of engineering and administrative controls for the process.

5.1 Objective of Consequence analysis:

The main objective of the consequence analysis study is to predict the physical behavior of the chlorine release under different identified failure scenario and identifying of population area affected under such scenario.

5.2 Consequence analysis for the Chlorine System

In this study various scenarios identified from the safety review and hazard identification using CEI method study are considered for the consequence analysis study. Consequence analysis includes the estimating release amounts of materials, conditions of release, and area affected by such releases. As the calculation of models is very complex, PHAST software developed by Det Norske Veritas (DNV) has been used to analyze the physical effects of release following tank or line rupture or leakage and the subsequent damaged posed by the release to the surrounding people. The detailed modeling of the Worst case scenario catastrophic rupture of chlorine bullet and credible scenario 0.25" leak in chlorine bullet are given as illustration.

5.2.1 Assumption for Scenario Modeling:

There were various assumptions made for the scenario modeling to decide the inputs, calculation of release quantity, release duration etc which are as follows.

1. Isolation Philosophy: For the calculation of release quantity and duration the leakage is identified and isolated based on the available process control option and strategy including the equipment isolation, plant shut down procedure and gas scrubbing system. The release duration is calculated based on the time required for leakage detection and immediate action which are considered based on the practical experience, manpower availability, gas detection system, Operating practices, PPEs availability.
2. The release quantity is calculated based on the equipment dimension, isolation philosophy, corrective action time, control strategy and options.
3. The ambient condition of the facility was considered for the scenario modeling which are as follows: ambient temperature: 35 ^{o}C, Facility is surrounded with large obstacles like columns and buildings.
4. The various scenarios were modeled for different hole sizes for the 5/D and 1.5/F weather condition.

5. Based on the organization plot plan the boundaries of organization from chlorine facility are as follows: East: 900 m, West: 2400 m, North: 1100 m, South: 800 m. This data would help in identifying the offsite consequences by plotting the results on plant lay out drawings
6. Since wind direction is arbitrary, the minimum distance of 800 m is considered as boundary and scenario identified having dispersion results of distances greater than 800 m.

5.2.2 Scenario Modeling:

Modeling of scenario is complex process; hence PHAST software was used for Consequence Analysis. The software is using the Unified Dispersion Model (UDM) models for the dispersion calculation following a ground level or elevated two phase pressurized release. It effectively consists of Jet Dispersion, Droplet evaporation and rainout, touchdown, Pool spread and vaporization, Heavy gas dispersion, Passive dispersion, Pool Vaporization. To find out the toxic effects PHAST software designed based on probit models and calculates the toxic doze, probit number, probability of death and integrated probability of death for a user defined height and user defined offset in the crosswind direction.

5.2.2.1 Scenario: Catastrophic rupture of chlorine bullet having 25 MT liquid chlorine inventory under 1.5 / F weather condition – Worst Case

(A)Initial Conditions for Discharge Modeling:

Type of event : catastrophic rupture
Phase : Liquid
Type of Vessel : Padded
Discharge Pressure : 7.6 bar
Discharge Temperature : 8 °C
Elevation : 0.5 m
Status of Bund : Bund present
Area of dike : 365 m^2
Type of Bund surface : Concrete
Bund Height : 1m
Bund Failure Modeling: Bund can not fail
Mass inventory of material to discharge : 25000 kg

(B) Initial Conditions for Dispersion Modeling:
Ignition location : No ignition location
Mass inventory of material to disperse : 25000 kg
(Quantity of material release is found from the discharge modeling results. In this scenario there is catastrophic rupture and entire quantity released considered for the dispersion)
Atmospheric temperature: 35 °C
Relative Humidity : 85%
Wind Speed : 1.5 m/s
Pasquill Stability : F
Surface Roughness length: 183.156
Surface Roughness parameter : 0.0999

(C)Toxic Effect Modeling:
The toxic effect modeling uses the probit method for the defining the toxic effects. Dispersion modeling results are considered for the toxic effects calculation.

These inputs are provided into PHAST software and scenario run for the discharge, dispersion and effect models.

(D) Results of Modeling:

The evaluation and analysis of results obtained from the software is as follows:

(D-1)Summary of results of Discharge Modeling:

Liquid Fraction	: 0.88	Final Temperature	: -33.99 °C
Final Velocity	: 79.19 m/s	Droplet Diameter	: 233.45 um

(D-2) Summary of Results of Dispersion Modeling:

Distance (m) to Concentration results:

Concentration (ppm)	Distance (m)
ERPG 3 (20 ppm)	6267.6
IDLH (35 ppm)	3407.79

Concentration at various distances Results:

Distance (m)	Concentration (ppm) at ERPG average time of 3600 s
100	30687.5
300	8934.7
800	596.231

Dispersion for ERPG- 3 and IDLH Concentration limits

Weather Condition	Distance and Time for Concentration			
	35 ppm		20 ppm	
	Time (Min)	Distance meter	Time (Min)	Distance meter
1.5 F	56.	7300	83	12500
5 D	9	3520	11	5200

Discussion: The following graphs show the maximum concentration footprint for the ERPG-3 and 35 ppm (IDLH value) concentration limits. It shows that the 35 ppm concentration can reach up to the distance of 7300 m, 20 ppm concentration can reach up to 12500 m.

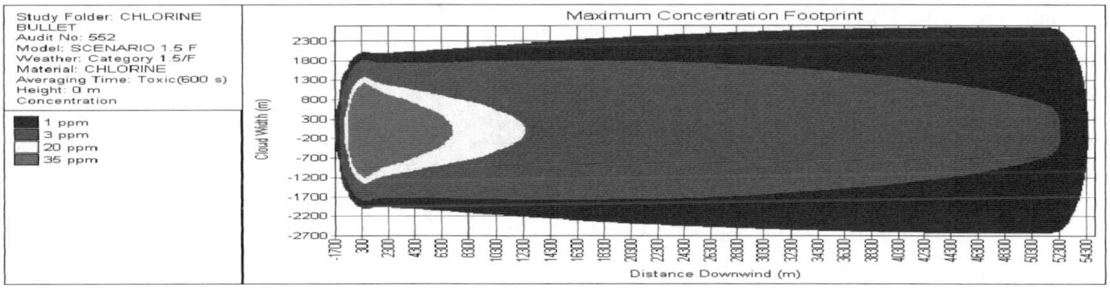

(D-3) Summary of Results of Toxic Effect Modeling:

Distance to Equivalent toxic doze:

Concentration(ppm)	Reference Time (s)	Distance (m)
IDLH (35 ppm)	3600	3115.66
ERPG 3 (20 ppm)	3600	7414.27

Time to Equivalent toxic doze at various distances

Concentration(ppm)	Reference Time	Time (s)		
		100 m	300 m	800 m
IDLH (35 ppm)	3600	12.0	34.9	229.4
ERPG 3 (20 ppm)	3600	11.7	32.6	199.3

Probit model calculation results

Distance (m)	-250	-125	-50	100	300	500	800
Integrated Probability of fatality (1.5 F)	0.82	1.0	1.0	1.0	1.0	0.91	0.27
Integrated Probability of fatality (5 D)	0	0	0.19	1.0	0.99	0.84	0.19

Discussion: The results show that the probability of fatality is 1.0 for the upstream distance of -123.07 m to downwind distance of 328.18 m. It means the all the personal present in this distance and plume width area for this distance will die. The study location is located in a petrochemical complex and surrounded by the other plants, this area covers almost more than 100 people during the day time and about 50 people during silent hrs of operation. In addition to the above large numbers of people may get injured and requires immediate medical treatment in case of scenario occur. The probability of fatality at different distance is given above. The graphic presentation of probit, probability of fatality, toxic doze, equivalent concentration vs. downwind distance are shown in table-5.2.2.1

Table-5.2.2.1: Graphic presentation of probit, probability of fatality, toxic doze, equivalent concentration vs. downwind distance

Probit No. vs. Distance downwind (m)	Probability of fatality vs. Distance downwind (m)
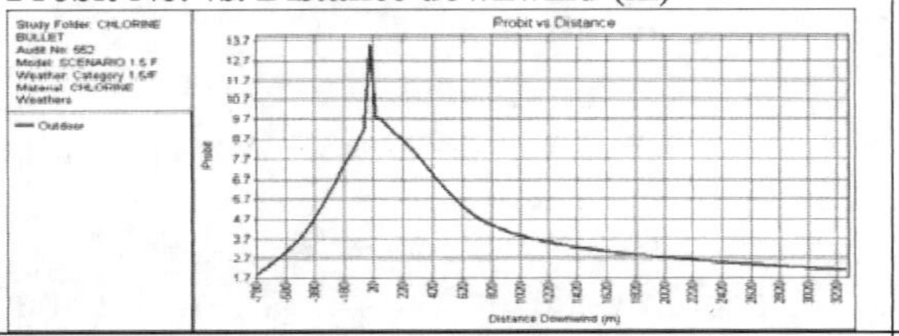	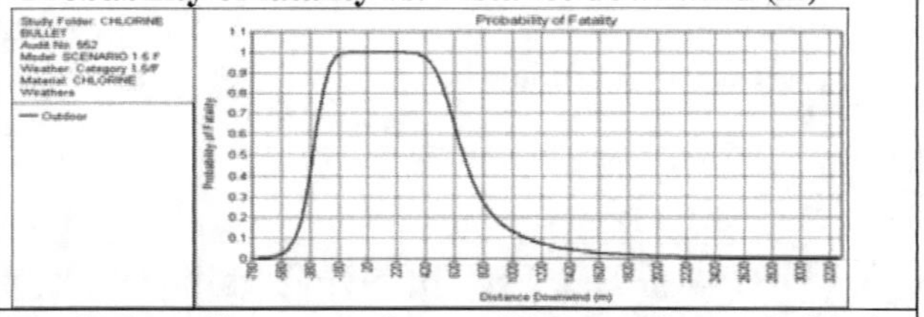
Toxic Doze vs. Distance downwind (m)	**Equivalent Concentration vs. Distance downwind (m)**
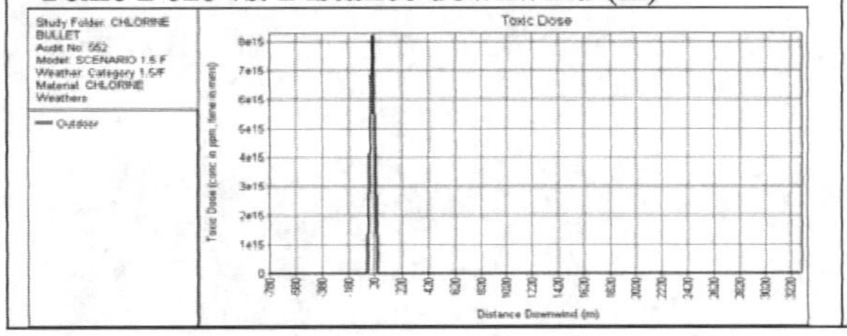	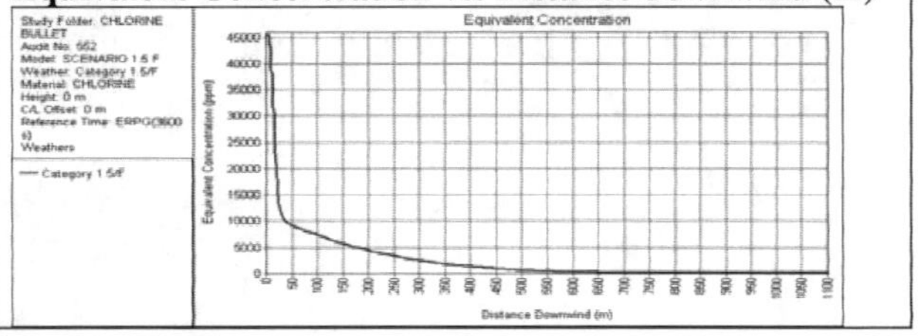

5.2.2.2 Scenario: Chlorine bullet leakage (0.25” diameter hole) –Credible Scenario

(A) Initial Conditions for Discharge Modeling:

Type of event	: Vessel leakage	**Phase**	: Liquid
Type of Vessel	: Padded	**Hole Size**	: 0.25”
Discharge Pressure	: 7.6 bar	**Discharge Temperature**	: 8 °C
Elevation	: 0.01 m	**Status of Bund**	: Bund present
Area of dike	: 365 m^2	**Type of Bund surface**	: Concrete
Bund Height	: 1m	**Bund Failure Modeling:**	Bund can not fail

Mass inventory of material to discharge : 25000 kg
Outdoor Release Direction: Down Impinging on Ground.

(B) Initial Conditions for Dispersion Modeling:

Ignition location : No ignition location
Atmospheric temperature: 35 °C
Relative Humidity: 85%
Surface Roughness length: 183.156
Surface Roughness parameter : 0.0999
Mass inventory of material to disperse : 25000 kg
Wind Speed & Pasquill Stability: Modeled for two weather condition (5/D and 1.5/F)

(C) **Toxic Effect Modeling:**
The toxic effect modeling uses the probit method for the defining the toxic effects. Dispersion modeling results are considered for the toxic effects calculation.

(D) Results of Modeling:
The evaluation and analysis of results obtained from the software is as follows:

(D-1) Summary of results of Discharge Modeling:
Liquid Fraction : 0.88
Final Temperature: -33.99 °C
Final Velocity : 84.14 m/s
Droplet Diameter : 41.81 um

(D-2) Summary of Results of Dispersion Modeling:
Distance (m) to Concentration results:

Concentration (ppm)	Distance (m)	
	1.5 F	5D
ERPG 3 (20 ppm)	1424.26	317.176
IDLH (35 ppm)	933.647	225.767

Concentration (ppm) at ERPG average time of 3600 at various distances Results:

Distance (m)	Concentration (ppm)	
	1.5 F	5D
100	354.48	130.525
500	75.482	8.81278
800	42.625	3.68

Dispersion for ERPG-3 and IDLH Concentration limits

Weather Condition	Distance and Time for Concentration			
	35 ppm		20 ppm	
	Time (Min)	Distance meter	Time (Min)	Distance meter
5 D	1.2	250	1.5	330
1.5 F	28	1100	34	1900

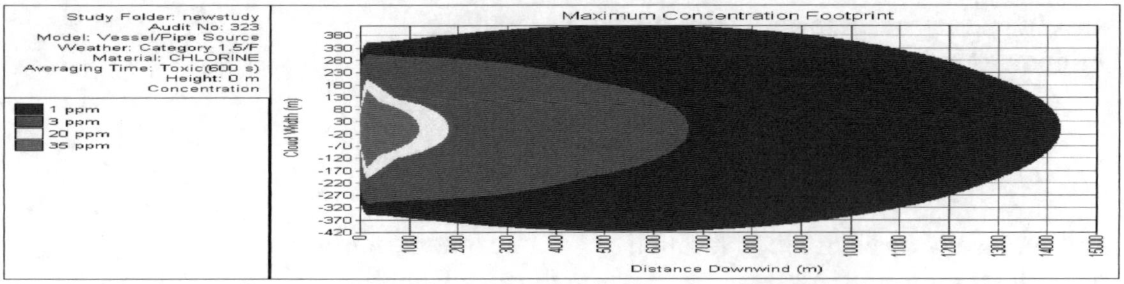

Graph: Maximum Concentration Footprint (1.5/F Weather Condition)

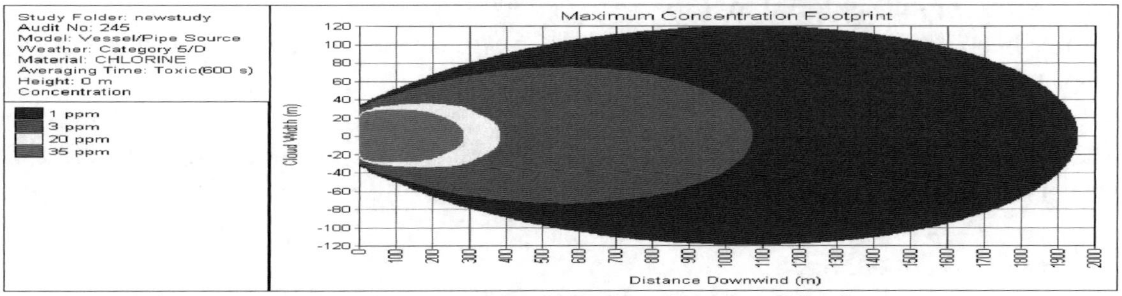

Graph: Maximum Concentration Footprint (5/D Weather Condition)

Discussion: For the 5 D weather condition the maximum concentration footprint shows that the 35 ppm concentration can reach up to the distance of 250 m, 20 ppm concentration can reach up to 330 m.

For the 1.5 F weather condition the maximum concentration footprint shows that the 35 ppm concentration can reach up to the distance of 1100 m, 20 ppm concentration can reach up to 1900 m.

(D-3) Summary of Results of Toxic Effect Modeling:

Distance to Equivalent toxic doze:

Concentration(ppm)	Reference Time (s)	Distance (m)	
		1.5 F	5 D
IDLH (35 ppm)	3600	818.925	288.679
ERPG 3 (20 ppm)	3600	1342.99	389.549

Time to Equivalent toxic doze at various distances for 1.5 F weather condition

Concentration(ppm)	Reference Time	Time (min)		
		100 m	500 m	800 m
IDLH (35 ppm)	3600	3.63	14.0	65.2
ERPG 3 (20 ppm)	3600	3.60	11.8	18.2

Time to Equivalent toxic doze at various distances for 5 D weather condition

Concentration(ppm)	Reference Time	Time (min)		
		100 m	500 m	800 m
IDLH (35 ppm)	3600	1.8	No Hazard	No Hazard
ERPG 3 (20 ppm)	3600	0.86	No Hazard	No Hazard

Probit model calculation results

Distance (m)	25	75	150	300	500	1000
Integrated Probability of fatality (1.5 F)	1.0	0.73	0.28	0.08	0.02	0
Integrated Probability of fatality (5D)	0.99	0.56	0.07	0.0	0.0	0.0

Discussion: The results show that the probability of fatality is 1.0 for the downwind distance 25 m. It means the all the personal present in this distance and plume width area for this distance will die. The study location is located in a petrochemical complex and surrounded by the other plants, this area covers almost more than 20-30 people during the day and 10-15 people during silent hrs of operation. In addition to the above large numbers of people may get injured and requires immediate medical treatment in case of scenario occur. The graphic presentation of probit, probability of fatality vs. downwind distance for 1.5/F and 5/D weather condition is shown in table-5.2.2.2A & 5.2.2.2B respectively.

Table-5.2.2.2.A: Graphic presentation of probit & probability of fatality vs. downwind distance for 1.5/F Weather condition

Probit No. vs. Distance downwind (m)	Probability of fatality vs. Distance downwind (m)
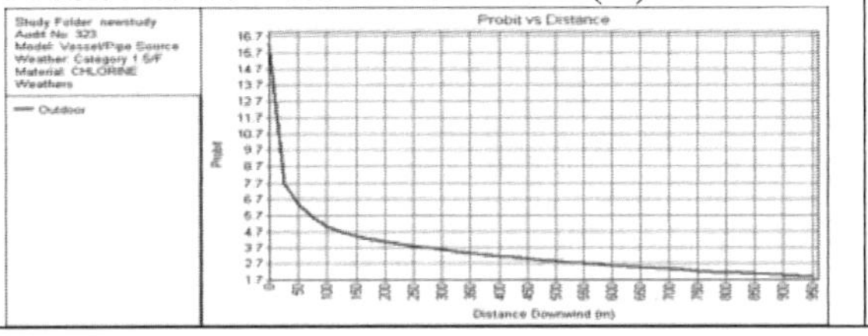	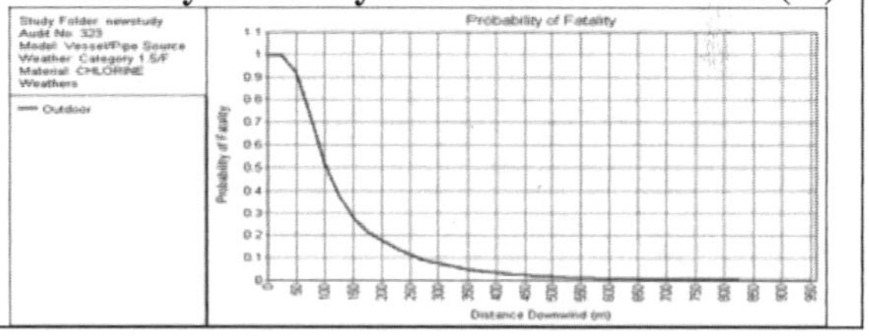

Table-5.2.2.2.B: Graphic presentation of probit & probability of fatality vs. downwind distance for 5/D Weather condition

Probit No. vs. Distance downwind (m)	Probability of fatality vs. Distance downwind (m)
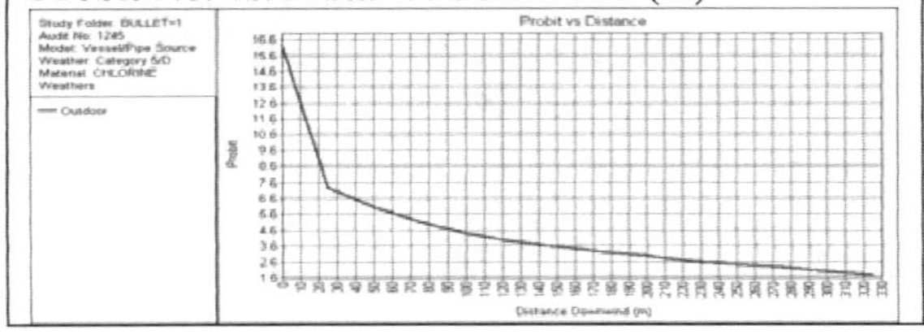	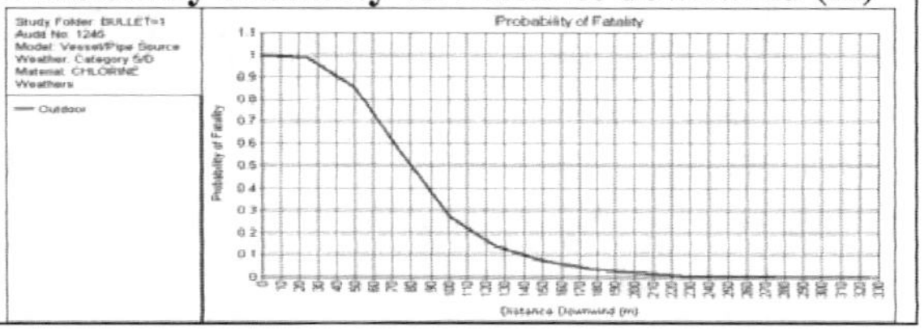

6.0 Recommendation:

Based on the detailed safety review study, CEI and consequence analysis study, following modifications were recommended for the system.

1. Chlorine inventory should not be more than 25 MT/Tank.
2. Provision of a sump in the dyke with a capacity of 10 M3 with suction hood connected to Hypo system
3. Provision should be made for removal of rain water over the dyke wall using a pump and not through drain valve in the dyke.
4. At present there is provision of tackling a large release of liquid chlorine. HAZMAT foam for liquid chlorine leak containment can be explored.

5. The isolating valves for the bursting disc system may be mechanically interlocked so that one pair of discs is always operative or isolating valves may be individually locked.
6. The pressure in the chlorine storage tanks should be measured by gauges specially developed for use with liquid chlorine; and these should have silver or tantalum diaphragms instead of stainless steel used.
7. In the storage area where there is a potential of release of liquid chlorine self contained breathing apparatus (SCBA) should be employed and since in case of emergency the timely action is of utmost importance provision of two SCBA's in the storage area is needed.
8. Emergency tool kits (including spare valve keys) should be provided in the chlorine storage area. The equipment supplied should be regularly inspected and maintained.
9. The main risk of chlorine escape to environment is from incorrect operation of the plant and hence licensing system for operators may be introduced.
10. Regular practices and mock drill of the Emergency Plan must be carried out.
11. Chlorine Emergency handling team must be identified and trained to combat any emergency situation.

7.0 Conclusion:

In the context of the risk arising out of handling of chlorine at organization, the paper discussed the hazard identification using safety review and CEI and consequence analysis results of the scenario developed from the process safety review of the existing facility and CEI calculation, effect zone at onsite and off site are identified for the worst case and credible scenarios. Hence, the outcome of the study gives valuable inputs to the organization management to augment its On-site and Off-site emergency Plan for the chlorine handling. The facility can be made safer than the earlier by implementing the recommendation. This in-depth study has developed the capability of the team for carrying out such studies. The team learning shared with the other sites of the organization and studies are replicated which helps in building the safe and robust safety system and culture at the organization.

References:

1. PHAST software, DNV
2. Dow's Chemical Exposure Index Guide, AICHE 1994
3. Guidelines for Bulk Handling of Chlorine at Customers Installation published by the Chemical Industries Association, UK
4. Publication Series on Dangerous Substances (PGS3): Guidelines for Quantitative Risk Assessment, Yellow Book
5. Publication Series on Dangerous Substances (PGS3): Guidelines for Quantitative Risk Assessment, Purple Book
6. Perry's Chemical Engineers Handbook
7. Loss Prevention in the process Industries Volume1,2 & 3, Second Edition by Frank P. Lees

8. Quantitative Risk Assessment by DNV Technia
9. Guidelines for consequence analysis of chemical releases –AICHE ,1999
10. Guidelines for chemical process quantitative risk analysis – CCPS,2000
11. Evaluating Process Safety in the chemical industries – a user guide to quantitative risk analysis-CCPS,2000
12. Plant Operating Manual

Process Safety Management in the Energy Sector

Marc McBride
Group Head of Process Safety
Centrica plc
Millstream
Maidenhead Road
Windsor
Berkshire, UK
SL4 5GD
marc.mcbride@centrica.com

Harry Chen
Director, Environment & Compliance - North America
Direct Energy
Suite 600
12 Greenway Plaza
Houston, TX 77046
harry.chen@directenergy.com

Prepared for Presentation at
American Institute of Chemical Engineers
2010 Spring Meeting
6th Global Congress on Process Safety
San Antonio, Texas
March 22-24, 2010

UNPUBLISHED

Process Safety Management in the Energy Sector

Marc McBride
Group Head of Process Safety
Centrica plc
Millstream
Maidenhead Road
Windsor
Berkshire, UK
SL4 5GD
marc.mcbride@centrica.com

Harry Chen
Director, Environment & Compliance - North America
Direct Energy
Suite 600
12 Greenway Plaza
Houston, TX 77046
harry.chen@directenergy.com

Keywords: process hazard analysis, leadership, culture, accountability, stakeholder engagement

Abstract

This paper describes experience of application of the principles of process safety management within Centrica plc, an integrated energy company with operations in North America and Europe. The paper describes the key challenges for implementation of process safety requirements in this sector, the approaches which Centrica has developed and the successes and lessons learned to date.

The elements Centrica has identified as critical to success in process safety are presented under the categories of 'people', 'plant' and 'processes'. A case study is then described of application of one of these elements - process hazard analysis - to Centrica's US gas-fired power plants and Canadian upstream gas business. The focus of the paper then broadens to the 'people' elements of process safety management - including leadership, culture, stakeholder engagement and accountability - as key to maintaining the effectiveness of risk controls. Specific issues examined include leadership behaviors, clarity of expectations around process safety, the role of internal and external stakeholder engagement and organizational design. The paper brings an international dimension to these topics, describing approaches which Centrica has adopted based on industry best practices and accident learnings.

The conclusion of the paper is that the principles of process safety management have much wider application than the traditional process industries, indeed that there are common issues and important learnings to be gained across a variety of high hazard industries. The paper underlines human, corporate culture and organizational issues as key to delivering sustained, high levels of performance in process safety.

1. Introduction

Centrica plc is an integrated energy company, formed from the demerger of British Gas in 1997, with around 33,000 employees worldwide. Centrica is involved at most stages of the energy supply chain, from sourcing new supplies of gas, through gas processing and storage, to power generation (gas-fired and renewable) and supply of energy and energy-related services to customers. Process safety is not a term that is used widely throughout the energy sector, yet there is considerable scope for its application and learning from the chemical process industries.

Indeed it is vital that such cross-sector learning takes place if we are to overcome the challenges that face us today in process safety. Three incidents in the US and UK in 2005 highlighted this challenge. In March, an explosion and fire occurred at BP's Texas City refinery in the US in which 15 people died and 180 were injured. This led to record fines and far-reaching recommendations from both the Chemical Safety and Hazard Investigation Board (CSB, 2007) and an independent safety review panel which looked at BP's five US refineries (Baker, 2007). In April there was a notification of a leak of radioactive liquor at the Thermal Oxide Reprocessing Plant (THORP) at Sellafield in the UK, leading to an extensive investigation and major lessons for the operator and the nuclear industry generally (HSE, 2007a). Finally, in December, a massive explosion and fire occurred at an oil depot near London operated by Hertfordshire Oil Storage Limited, a joint venture between Total and Chevron, in which 43 people were injured. This led to a major investigation, again with far-reaching recommendations for the petroleum industry, the regulator and government in the UK (MIIB, 2008).

A number of common themes emerged from these accidents, including failures of leadership, poor safety culture, ineffective arrangements for monitoring and audit, inadequate management oversight, failure to learn from past incidents, ineffective management of change and failure to ensure integrity of safety-critical systems. Unfortunately these are not the only instances in which such themes have been identified. Accidents before and since, eg the Longford gas plant fire in Australia in 1998 (Longford Royal Commission, 1999) and the incidents in the Statfjord offshore field in Norway in 2007 (PSAN, 2008a) and 2008 (PSAN, 2008b) have raised similar serious concerns. Recent regulatory interventions in the UK and US have also shown that there is no room for complacency. For example, the UK Health and Safety Executive's 'KP3' offshore asset integrity program (HSE, 2007b) found that the state of plant on over 50% of the installations inspected was poor, with underlying failings in leadership, the engineering function and learning. The preliminary findings of the US Occupational Safety and Health Administration's refinery National Emphasis Program (Lay et al, 2009) show that citations were issued for an average of 17.4 violations per refinery.

Considerable effort is now being put into addressing these issues. The question therefore is whether this will be sufficient to bring about the required improvements in a way that is sustainable going forward.

2. Centrica's approach to process safety management

Three factors are often cited as key to process safety - people, processes and plant. (Broadribb, 2008). Get these right and the fourth 'P' - performance - will look after itself. In this approach the 'people' element refers to the behaviors or culture within an organization. The 'processes' element refers to how things are done, in other words systems and procedures. The 'plant' element is, in a sense, what distinguishes process safety from other aspects of safety, ie the major hazards that are often inherent to the businesses we are in. Such hazards can manifest themselves in a variety of forms including inventories of hazardous substances, high energy sources (eg high pressure, high voltage equipment and rotating machinery) and operation in extreme and/or sensitive environments.

In Centrica we have taken this model and populated it with the specific elements that we regard as key to delivering sustained, high levels of performance in process safety, as shown in Figure 1.

Figure 1: Centrica's approach to achieving excellence in process safety

This paper focuses on a subset of these elements, ie hazard identification and risk assessment, leadership (including leadership behaviors and clarity of priorities and expectations), external stakeholder engagement, and accountability (in particular organizational design). These are some of the areas where Centrica, in common with other companies, is prioritizing its efforts. It is not surprising that three of these elements fall within the 'people' category. The past few decades have seen rapid advances in technology and increasing sophistication in safety management systems. However it is the 'people' element that has proved most challenging, yet is vital if the other two elements are to function effectively. This illustrated by Reason and Kletz respectively below.

"[There is a] *widely spread misconception . . . that somehow systems sit apart from culture. It is this belief that drives managers' over-reliance on systems on the one hand, and insufficient understanding of, and emphasis on, workplace culture, on the other. They believe, mistakenly, that compliance with such rules and procedures can be achieved simply by the imposition of systems, while ignoring the crucial cultural dimension. Yet it is the latter that ultimately determines the success or failure of such systems.*" [Reason in Hopkins, 2005]

"*It is easy to buy safety equipment. All we need is money and if we make enough fuss we will probably get it in the end. It is much more difficult to make sure that the equipment is kept in full working order when the initial enthusiasm has faded. Procedures, including testing and maintenance procedures, are subject to a form of corrosion more rapid than that which affects steelwork and can vanish without trace in a few months once a manager loses interest.*" [Kletz, 2001]

3. Hazard identification and risk assessment

It is fundamental for businesses to understand the hazards and risks that they create. This is why hazard identification and risk assessment appear as one of the core elements of Centrica's approach to process safety management, in common with most companies and industry guidelines (CCPS, 2007). Centrica requires its upstream businesses to formally identify and assess process safety risks. This section of the paper presents a case study of the application of process hazard analysis (PHA) to Centrica's gas-fired power plants in Texas and upstream gas business in Alberta, Canada. In neither case are PHA studies required by law.

The technique chosen for the PHA studies was a hazard-based, semi-quantitative approach, following relevant industry guidelines, eg CCPS (2008) and EPSC (2003). In this approach the facilities are firstly split into their component systems and activities. The hazard scenarios associated with each system and activity are then identified by the study team, together with the potential causes, with the aid of checklists. Next, the consequences and safeguards are documented and a semi-quantitative assessment of risk made using a risk matrix. On the basis of the assessed risk, the adequacy of the existing safeguards is then judged by the team and actions raised if appropriate.

Table 1 shows how the gas-fired power plants and upstream gas operations, including well drilling and completion activities, were split into their component systems and activities. Table 2 shows the PHA template and an example entry, whilst Table 3 shows examples from the checklist of hazard scenarios and potential causes.

Table 1: Breakdown of facilities into systems and activities for PHA studies

Facility		**Component systems and activities** (Notes 1, 2)
Gas-fired power plants		Fuel gas system Gas turbines and generators Hydrogen system High voltage electrical systems Heat recovery steam generator, steam cycle systems and cooling towers Steam turbine and generator Water treatment plant Fuel oil system
Upstream gas	Operations	Well sites Pipelines Compressor stations
	Well drilling and completion	Drilling Completion Workover Suspension/abandonment

Note 1: For the purpose of the PHA, each system was further split into sub-systems, eg the fuel gas system at one of the gas-fired power plants was split into the incoming gas pipeline and manifold station, the pig receiver, the buried section of pipeline on-site and the filter/separator vessels and associated aboveground pipework.
Note 2: In the case of the upstream gas business, a two tier approach to the PHA was adopted, addressing the individual assets and activities (as listed above) but also the generic, overarching management controls.

Table 2: PHA template and example entry

System:	1. Fuel gas				
Description:	*< Description of fuel gas system >*				
Sub-system	**Scenario**	**Consequence** (Note 1)	**Causes**	**Existing safeguards** (Prevention, control, mitigation) (Note 1)	**Action** (Note 1)
1.1 Incoming gas pipeline and manifold station	Gas release, fire/explosion	Potential for serious injuries or fatality Severity Rating = □	- Accidental excavation - Corrosion - etc	- Pipeline operator's planned preventive maintenance program - Pipeline owner's audit program - etc Probability Rating = □	Risk Level = □

Note 1: Severity and probability ratings assigned according to 5x5 risk matrix giving risk level of 1 (low) to 5 (high)

Table 3: Checklist of hazard scenarios and potential causes (examples)

Hazard scenarios	**Potential causes** (Note 1)
Uncontrolled release of hazardous substance Uncontrolled release of steam/hot water Fire Explosion Asphyxiation Release of contained pressure	'Internal' causes: • Incorrect material specification • Corrosion • Internal pressure outside design limits • Overfilling

Hazard scenarios	Potential causes (Note 1)
Arc flash/electrocution Catastrophic failure of rotating equipment Contaminated surface water run-off Well blowout Pipeline leak or rupture	• Human error • Dropped load or impact • Loss of well control 'External' causes: • Vehicle impact • Sabotage/vandalism/terrorism • Extreme weather: • Flooding • Subsidence • Hazardous neighbor activities

Note 1: Internal causes are those within the direct control of the plant. External causes are those outwith the direct control of the plant

Centrica's experience of undertaking the PHAs was overwhelmingly positive. The PHA technique itself proved a cost-effective way of identifying and assessing risk for a variety of types of plant and activity. In this case it took one week to carry out the PHA for each power plant and two weeks for the upstream gas business, approximately one third of which was formal study time. The team-based approach, involving frontline personnel, engineers, superintendents, managers and directors, was valuable in drawing out a variety of experience and providing the opportunity of knowledge-sharing with the team.

Another key success factor was ensuring that the PHAs were operationally-focussed and site-specific, drawing on actual experience of operation including any plant modifications and incidents which had occurred. This helps translate what might otherwise be a theoretical study into a useful reference for ongoing operations. Of particular value was the opportunity to compare outputs from PHAs for similar types of facility. This enabled common issues to be identified and sharing of best practice from one facility to another.

A critical factor for the success of the whole study was gaining the buy-in of senior management at the outset to the study objectives and approach. This ensured the availability of the necessary resources to complete the study and commitment to the actions which arose.

The PHA study record provides a valuable record of the key controls in place - both hardware and procedural - to prevent major accidents. However the effectiveness of these controls ultimately relies on people and it is this topic which we turn to the following sections.

4. Leadership

4.1 Leadership behaviors

Following the Texas City incident, leadership is now commonly regarded as key to delivering success in process safety, and health, safety and the environment (HS&E) generally. The behavior of leaders is a significant determinant of the culture within an organization, which, as we've seen above, is a prerequisite to effective functioning of processes and plant. This has been put by Schein as follows:

"*Leaders create and change cultures, while managers and administrators live within them ... Leaders create cultures by what they systematically pay attention to. This can mean anything from what they notice and comment on to what they measure, control, reward and in other ways systematically deal with.*" [Schein in Hopkins, 2005]

Centrica also recognizes leadership as a key ingredient of the success of the business generally. To this end we have developed a model for leadership behavior, the key elements of which are summarized in Table 4 below, with examples of specific behaviors.

Table 4: Centrica's leadership behaviors

Leadership behavior	What does it mean ?	Examples	
		More of ...	Less of ...
Being one team, where everyone counts	We achieve more by listening to, and being honest with, each other	*Connects with people through active listening and clear communication*	*Creates a silo mentality*
Positive environment, where people want to work	We give people the confidence and freedom to succeed	*Builds trust and acts with humility*	*Creates complexity and fails to keep things simple*
Putting our customers first, where they define excellence	We always wear our customers' shoes	*Drives innovative solutions to emerging customer needs*	*Ignores frontline feedback*
Seizing responsibility, to deliver on our commitments	We keep our promises and act with integrity	*Takes action and gives other the confidence to take action*	*Avoids ownership when things go wrong and blames others*
We see beyond our goals	We keep people safe and care for the society and the environment	*Seeks out, learns from and transfers best practices internally and externally*	*Compromises long term value for short term gain*

The attributes in Table 4 were developed by the senior leadership team within Centrica and the detail behind each behavior was put together by a project team of over 70 leaders from around the Group worldwide. These behaviors are supported by an online 360° feedback tool and 'story telling' initiatives, which are key to reinforcing positive behaviors. The attributes described above are similar to those which other companies have identified, eg BP's leadership model (Flynn, 2008).

The reason for mentioning these generic models is that the leadership behaviors which are relevant to process safety are likely to be those which are relevant to the business generally. For example the behavior '*builds trust and acts with humility*' is as relevant to process safety as it is in the general business context. Furthermore if, as process safety advisers, we can tap into these generic models (rather than invent or own) then we will achieve much greater leverage for process safety objectives.

The approach we have taken within Centrica therefore is to use our generic leadership behaviors as a starting point and develop from them specific behaviors for HS&E leadership (including leadership in process safety). The specific behaviors we have arrived at are as shown in Table 5.

Table 5: Centrica's HS&E leadership behaviors

No.	Behavior description
1	*Lead by example - what you do is more important than what you say*
2	*Reinforce HS&E as a top priority for Centrica, integral to the success of our business*
3	*Engage with key stakeholders, particularly those at the frontline of HS&E delivery*
4	*Be mindful of our HS&E impacts*
5	*Take the initiative on emerging HS&E issues and drive delivery of effective solutions*
6	*Share our HS&E successes and learn from our mistakes*
7	*Actively seek and give honest feedback on HS&E performance*

Leadership of course is not just an attribute we expect of those at the top of the organization but which promulgates throughout all layers of management, including frontline supervisors. Figure 2 shows an example of how Centrica has cascaded the leadership behaviors within its businesses and established straightforward expectations for people working for us.

Figure 2: Continual mindfulness - Centrica Storage

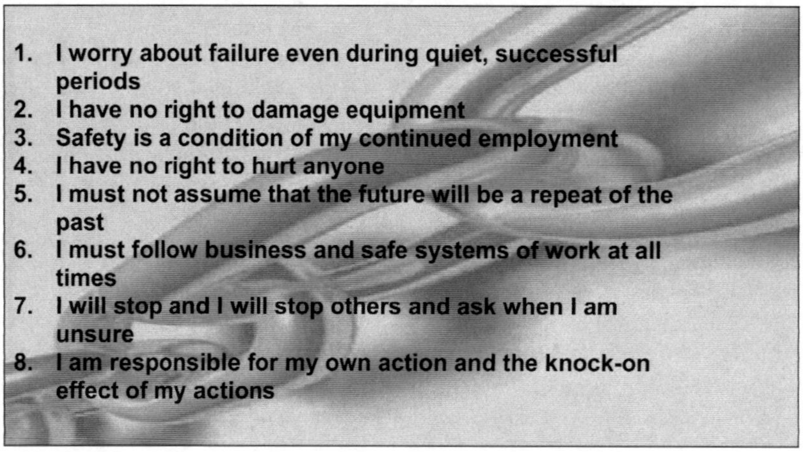

The essential premise behind behavioral safety is that accidents stem largely from the unintended consequences of people's action (or inaction) rather than any malicious act, hence the need to understand the drivers behind this action (or inaction) and correct them. In our safety leadership training we've therefore emphasized the goal of influencing this 'well intentioned' majority.

4.2 Clarity of priorities and expectations

4.2.1 Background

Another important attribute of leadership is clarity of priorities and expectations, and the special emphasis that's required in relation to safety, and process safety in particular. Companies have phrased this in different ways, for example BP emphasize their number one priority as '*safe, compliant and reliable operations*' (BP, 2008) while Du Pont expects '*adherence to the highest standards to ensure the safety and health of its employees, its customers and the people of the communities in which it operates*' (Du Pont, 2008). British Energy states safety as '*its overriding priority*' (British Energy, 2009)

In Centrica we stress that safety is our top priority, fundamental to the success of our business. We state our specific expectations on process safety within a Group standard. This links to our expectations on HS&E management generally which are expressed in a separate Group standard aligned to BS EN ISO 14001 (BSI, 2007) and BS OHSAS 18001. By structuring our standards in this way it has enabled us to be specific about what's required in terms of process safety management over and above HS&E management generally. Thus, for example, we have stressed the need for suitably rigorous and systematic means of identifying and assessing process safety risks and the need for proper management of change, neither of which feature strongly in OHSAS 18001.

4.2.2 Risk acceptability criteria

This section focuses on one specific aspect of process safety management, namely the criteria for acceptability of risk. In ISO 14001 and OHSAS 18001 the stated objective is 'continual improvement' and this is the criteria we have adopted within Centrica at the broad HS&E level. In terms of process safety, however, we consider that this does not go far enough and have therefore adopted the standard of 'ALARP' ('as low as reasonably practicable'). This is the UK approach but is mirrored in other countries, eg the Netherlands and Australia.

We think ALARP is a more appropriate standard to apply in the field of process safety because, whilst it captures the important principle of continual improvement, it also implies an absolute standard which must be achieved. Moreover, as a process, it drives continual review and challenge of process safety performance.

It is important however to be clear what ALARP means in practical day-to-day terms. Much has been within the last ten years about ALARP, eg HSE (2001, 2003a, 2003b, 2004, 2006). The simplest way in which the UK Health and Safety Executive (HSE) have phrased the ALARP requirement is '*tell me what you're not doing and why you're not doing it*', emphasizing that the burden of proof is on the operator.

Within Centrica, the way we approach ALARP is threefold:

1. To apply the principles of inherently safer design;
2. To apply relevant good practice; and
3. To identify and assess options for minimizing risk throughout the lifecycle of facilities, and implement those that are reasonably practicable (balancing cost and risk benefit).

The above indicates the fundamental role the discipline of process safety has to play in decision making within a business, whether in terms of budgeting, planning, conceptual design, engineering, operations or decommissioning. We discuss below how organizational design can assist in ensuring that this influence is effectively brought to bear.

Ironically one of the advantages of the ALARP model (its flexibility) is also its undoing in the sense that it can lead to inconsistent decision-making between (or within) different industries, operating companies or the regulator. This is also discussed further below in the context of the Buncefield incident.

5. Stakeholder engagement

5.1 Internal engagement

In the preceding section we identified engagement with those at the frontline of HS&E delivery as one of the positive HS&E leadership behaviors which Centrica has highlighted. This means engagement with all those who have a role to play in process safety, especially the workforce, supervisors and line management. Amongst the most importance characteristics for leaders in this context are listening, follow-up and feedback.

This approach was exemplified by the "Boots off for Safety" campaign in the UK offshore industry in 2006-7 in which industry leaders visited offshore installations to create an industry-wide dialogue on safety, an outcome of which has been the development and implementation of a minimum industry safety training standard. Weick and Sutcliffe (2007) have also identified sensitivity to the frontline as a key characteristic of high reliability organizations.

"*High reliability organizations (HROs) are sensitive to operations. They are attentive to the front line, where the real work gets done. The "big picture" in HROs is less strategic and more situational than is true of most other organizations. When people have well-developed situational awareness, they can make the continual adjustments that prevent errors from accumulating and enlarging. Anomalies are noticed while they are still tractable and can still be isolated.*"

Whilst engagement with the frontline may not be expected to provide all (or even most of) the answers to achieving a step change in process safety performance, it nevertheless provides a measure by which the likely effectiveness of initiatives to improve safety can be judged. The importance this frontline engagement is that it also provides leaders with a means of sense-checking information provided by their direct reports. We mention follow-up and feedback as key characteristics of leaders because unless follow-up and feedback takes place then frontline personnel will eventually stop raising issues.

Like most companies, Centrica employs a variety of means to gather frontline intelligence, including daily operations meetings and plant walk-arounds, fortnightly safety and environmental awareness meetings and quarterly worker safety representative meetings and management meetings. Increasingly we have made process safety part of this engagement, for example by conducting periodic 'ALARP workshops', at which ideas for safety improvement are solicited from representative groups of the workforce, evaluated and included as part of formal improvement programs.

5.2 External engagement

External engagement is likely to become increasingly important if we are to achieve a collective step change in process safety performance. History suggests that significant steps forward in safety have tended to follow major accidents, such as Flixborough (1974), Three Mile Island (1979), Alexander Kielland (1980), Bhopal (1984), Chernobyl (1986), Piper Alpha (1988) and Longford (1998).

Flixborough was one of the accidents which prompted the UK (as part of an EU-wide initiative) to introduce specific legislation to control major accident hazards. Bhopal had a global impact and, together with the Pasadena chemical plant explosion in 1989, triggered the introduction of the Emergency Planning and Community Right-to-know Act (EPCRA), the Process Safety Management (PSM) standard and Risk Management Program (RMP) in the US. The accidents at Three Mile Island and Chernobyl were watershed events in the nuclear industry, demonstrating the importance of defense in depth, including human factors and emergency preparedness considerations, and the need for globalized coordination on nuclear safety issues and improved understanding and management of the safety risks of nuclear activities (IAEA, 2003). Piper Alpha prompted a change from prescriptive to goal-setting legislation in the UK offshore industry and, along with the capsizing of the Alexander Kielland in Norwegian waters and other accidents, eg the sinking of the Ocean Ranger offshore Newfoundland in 1982, triggered a transformation in offshore safety practices around the world. The Longford incident had a similar impact on the onshore major hazard industries in Australia leading to the introduction of the Occupational Health and Safety (Major Hazards Facilities) Regulations in the State of Victoria.

The question therefore is how we can break this cycle and drive the necessary improvements in process safety to prevent, or at least reduce the chance of, further major accidents occurring. A clue to this could be in the increased collaboration which has occurred, to varying degrees, in different parts of the high hazards sector in the aftermath of accidents such as those mentioned above.

This includes for example chemical industry's worldwide Responsible Care program, strengthened following Bhopal, and initiatives within the offshore industry such as 'Step Change in Safety' and hydrocarbon release reporting in the UK, the tripartite approach of the Norwegian Petroleum Safety Authority, contractors initiatives, such as those of the International Association of Drilling Contractors (IADC) and collaboration amongst regulators, eg the International Regulators Forum (IRF). Of all the high hazard industries, the nuclear industry is perhaps characterized by the most collaboration, which includes legally-binding international conventions and knowledge management services, through the International Atomic Energy Agency (IAEA), and peer review programs, through the World Association of Nuclear Operators (WANO).

An unprecedented degree of collaboration is also now evident in the UK downstream petroleum industry following the Buncefield incident, with the industry and regulator collaborating through the Buncefield Standards Task Group (BSTG), now the Process Safety Leadership Group (PSLG), to set new standards for petroleum storage and timescales for implementation. This approach addresses one of the weaknesses of goal-setting regulatory regimes, such as the UK's ALARP approach introduced above, by establishing a more consistent, authoritative approach to safety for similar types of facility. Furthermore the Buncefield, Texas City and THORP incidents together have prompted the formation of a cross-sector process safety forum by the UK Petroleum Industry Association (UKPIA) and Oil & Gas UK in association with the UK Chemical Industries Association (CIA) and the Nuclear Industry Association (NIA).

It appears therefore that if increased collaboration is a desirable characteristic post accident then it is even more desirable as a preventative measure. Table 6 indicates some of the forms which this collaboration could take and the possible topics for the collaboration. Table 7 then gives examples of schemes which are already in place, based on the authors' experience. Undoubtedly there are many more collaboration schemes in place than those identified in Table 7, however this approach could be used to build up a more complete picture to identify where there are gaps and potential synergies. The appropriate collaboration structures could then be put in place as a pro-active means of minimizing the chance of further major accidents occurring.

Table 6: Possible models of collaboration in process safety

Form of collaboration		Possible topics for collaboration	
1	Cross-industry	A	Development and sharing of good practice
2	Cross-sector	B	Common incident reporting and sharing protocols
3	Supply chain	C	Common performance metrics
4	Cross-regulator	D	Peer review
5	Industry-regulator	E	Personnel secondments
6	Inter-governmental	F	Incident investigation
7	Industry-workers	G	Research and technology development
8	Industry-insurers	H	Knowledge management
9	Contractors	I	Competence assurance and training
		J	Provision of advice or consultancy services
		K	Consultation
		L	Information sharing

Table 7: Examples of existing models of collaboration

Collaboration model	Examples
1-A	Chemical industry's Responsible Care program, UK offshore oil and gas industry's 'Step Change in Safety' program, professional bodies (eg American Petroleum Institute, UK Energy Institute and UK Engineering Equipment and Materials Users Association)
1-B	UK offshore oil and gas industry's Incident Alerts Database
1-C	CCPS Guidelines for Process Safety Metrics (CCPS, 2009), API RP 754 project, UK offshore industry's common asset integrity metrics (HSE, 2009)
1-D	WANO peer review program
1-H	UK Oil and Gas Academy; ENFORM (Canada)
2-D	UK Institution of Occupational Safety and Health (IOSH) Hazardous Industries Group (HIG) pilot peer review scheme
3-L	GE Technical Information Letters
4-L	Offshore International Regulators Forum
5-G	UK Joint Industry Projects, eg on Blast and Fire Engineering for Topsides Structures (following Piper Alpha); Buncefield Explosion Mechanism Advisory Group
6-B	IAEA International Nuclear and Radiological Event Scale
6-H	IAEA knowledge management program
7-A	Norwegian oil and gas industry 'Working together for safety' project
7-I	UK offshore industry "Boots off for Safety" campaign resulting in new standards of industry safety training
8-L	Marsh Risk Consulting 100 largest Losses (Marsh, 2003)
9-I	International Association of Drilling Contractors training accreditation schemes

As a footnote to the above, it almost goes without saying that collaboration works best when industry (and other interested parties) take a collective view of their future, ie that an incident affecting one party affects all, a notable characteristic of the nuclear industry.

6. Accountability

Highlighted above is the fundamental role process safety has to play in decision-making within businesses in the high hazards sector. This section examines the influence organizational design has in determining whether this role is effectively exercised. The context for this discussion is a large company comprising a number of business units each of which may have a number of significant assets. One can consider two extremes of organizational design for such a company - one in which the process safety function is centralized (Figure 3) and the other in which it is distributed within the business units or assets of the organization (Figure 4).

Figure 3: Example of centralized process safety function

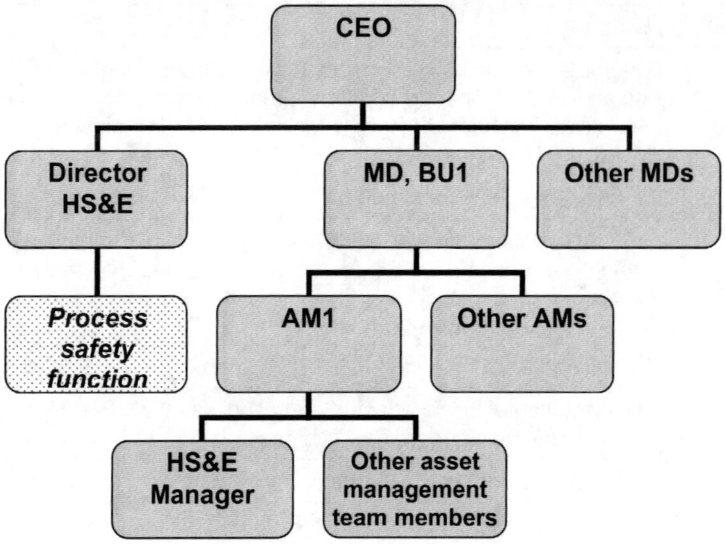

Key: CEO = Chief Executive Officer, MD = Managing Director, BU = Business Unit, AM = Asset Manager

Figure 4: Example of distributed process safety function

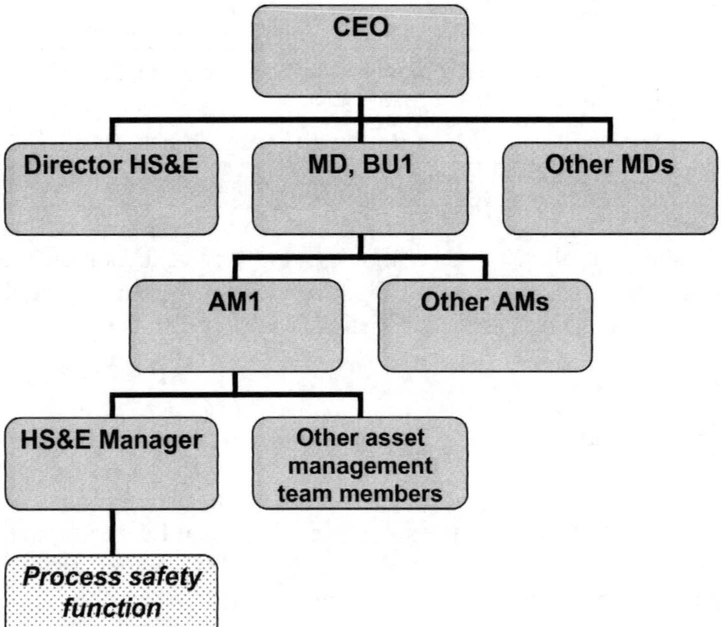

The advantage of the structure shown in Figure 3 is that it provides a strategic focus on process safety at the top of the organization, however this may not translate to practical, day-to-day impact at the operational level. The advantage of the structure shown in Figure 4 is that there is close connection between operational decision-making and process safety requirements but there is no overall strategic direction or "pull from the top", so implementation within different businesses may be inconsistent and/or reactive in nature.

Recognizing the above dichotomy, most large companies opt for a hybrid structure in which process safety specialists are present both at a Group or corporate level and within individual businesses or assets. However this does not necessarily overcome the problems identified above. This has been illustrated by Hopkins' analysis of the BP Texas City incident.

"*The existence of group vice-presidents for HSE and technology meant that, theoretically, process safety had a champion or champions at a very high level within the corporate structure. In practice, however, these people wielded little influence at site level.*"

"*... there was no way for the process safety manager [at Texas City] to raise his concerns at a higher level. In his view, the site manager did not understand the distinction between personal safety and process safety and was not giving sufficient attention to the latter. But there were no lines of reporting between him and the technical people at higher levels in the organization that would have enabled him to raise this issue. In particular, there was no direct line of reporting to the functional units mentioned earlier. Had there been, he might well have been able to wield more influence.*" [Hopkins, 2008]

In hybrid organizational structures conflict can arise where business units and the Group have different objectives and priorities on process safety. Within Centrica we have added in features to our organizational structure (see Figure 5) to ensure commonality of objectives and priorities. These comprise our HS&E Committee, reporting to our Executive Committee, which sets the strategy for HS&E (including process safety) across the Group, and the HS&E Leadership Team, which implements the strategy. We also maintain a strong functional link between the Group process safety function and specialists within the businesses.

In our organizational design, we have also found value in the process safety function reporting in at the same level as (rather than a sub-set of) the general safety function. This ensures that our process safety advisers are part of the senior management decision-making team within assets and that the process safety 'message' is delivered in as simple and direct a way as possible. An important feature of our organizational structure is also that the process safety adviser reports to a manager who is independent of the asset management team. In devising the structure in Figure 5 we have also recognized the importance of a strong functional link between our process safety advisers (who are part of our team of technical authorities) and our engineering contractor (via their design authorities).

Figure 5: Centrica organizational design for process safety (simplified example)

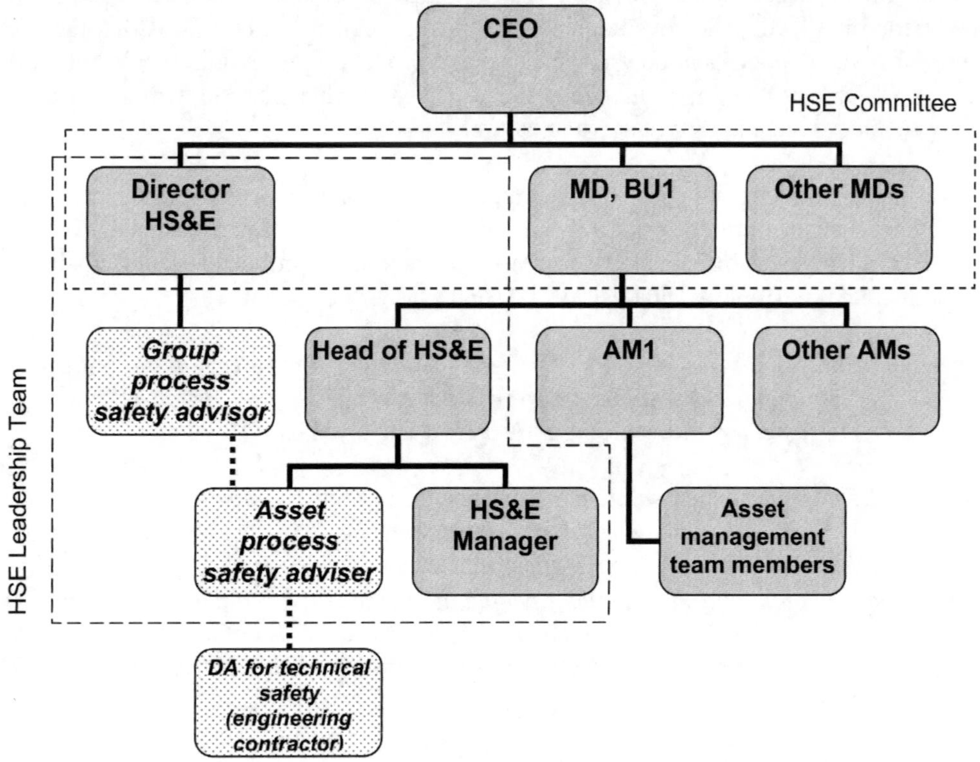

Key: .... indicates functional link, DA = design authority

7. Conclusion

The energy sector operates simple processes compared to the chemical process industries yet, as this paper has shown, the principles of process safety management apply equally. Effective management of process safety requires a balanced approach in the areas of people (behaviors), processes (how we do things) and plant (operational and asset integrity). Of these elements, the people element has proved the most challenging to address, yet is vital if we are to achieve the step change in safety performance we desire, and regulators and the public demand.

Accidents such as those mentioned in this paper will continue to occur unless we embed learning and drive continual review and challenge of process safety performance. The opportunities for learning exist across exist not just within particular industries but across the high hazard sector as a whole.

8. References

Baker, J. (2007), The Report of the BP US Refineries Independent Safety Review Panel, James A Baker (Chair).

BP (2008), Annual Review (www.bp.com).

British Energy (2009), Safety and Security Policy (www.british-energy.co.uk).

Broadribb, M. (2008), 3 Years on from Texas City, AIChE Global Congress on Process Safety.

BSI (2007), Occupational health and safety management systems - requirements, BS OHSAS 18001, British Standards Institution.

BSI (2004), Environmental management systems - requirements, BS EN ISO 14001, British Standards Institution.

CCPS (2007), Guidelines for Risk Based Process Safety, Center for Chemical Process Safety, American Institute of Chemical Engineers, Wiley.

CCPS (2008), Guidelines for Hazard Evaluation Procedures, 3rd Ed, Center for Chemical Process Safety, American Institute of Chemical Engineers, Wiley.

CCPS (2009), Guidelines for Process Safety Metrics, Center for Chemical Process Safety, American Institute of Chemical Engineers, Wiley.

CSB (2007), Investigation Report - Refinery Explosion and Fire, US Chemical Safety and Hazard Investigation Board.

Du Pont (2008), Du Pont Code of Conduct (www.dupont.com).

EPSC (2003), Hazard Identification Methods, European Process Safety Centre, UK Institution of Chemical Engineers.

Flynn, S (2008), Leading from the top in BP, UK HSE 'Leading from the top - avoiding major incidents' conference.

Hopkins, A. (2005), Safety, Culture and Risk - The Organisational Causes of Disasters, CCH Australia Ltd.

Hopkins, A. (2008), Failure to Learn - the BP Texas City refinery disaster, CCH Australia Ltd.

HSE (2001), Principles and guidelines to assist HSE in its judgements that duty-holders have reduced risk as low as reasonably practicable, UK Health and Safety Executive.

HSE (2003a), Assessing compliance with the law in individual cases and the use of good practice, UK Health and Safety Executive.

HSE (2003b), Policy and guidance on reducing risks as low as reasonably practicable in Design, UK Health and Safety Executive.

HSE (2004), Hazardous Installation Directorate's operational policy on the role of good practice in demonstrating compliance - applying the ALARP suite guidance in the context of an ALARP demonstration, SPC/Permissioning/22, UK Health and Safety Executive.

HSE (2006), Offshore Information Sheet No 2 - Offshore Installations (Safety Case) Regulations 2005 - Regulation 12 Demonstrating compliance with the relevant statutory provisions, UK Health and Safety Executive.

HSE (2007a), Report of the investigation into the leak of dissolver product liquor at the Thermal Oxide Reprocessing Plant (THORP), Sellafield, notified to HSE on 20 April 2005, UK Health and Safety Executive.

HSE (2007b), Key Programme 3 - Asset Integrity Programme, Offshore Division - Hazardous Installations Directorate, UK Health and Safety Executive.

IAEA (2003), Nuclear safety: a maturing discipline, International Atomic Energy Agency statement.

Kletz, T. (2001), Learning from Accidents, 3rd Ed, Gulf Professional Publishing.

Lay, J., Long L. and Marshall, M. (2009), Preliminary findings on OSHA's refinery NEP, Process Safety Progress (online article), American Institute of Chemical Engineers.

Longford Royal Commission (1999), The Esso Longford Gas Plant Accident, Report of the Longford Royal Commission.

Marsh (2003), The 100 Largest Losses 1972-2001 - Large Property Damage Losses in the Hydrocarbon-Chemical Industries, 20th Ed.

MIIB (2008), The Buncefield Incident 11 December 2005 - Final Report, Major Incident Investigation Board.

PSAN (2008a), Oil discharge from Statfjord OLS-A on 12 December 2007 - Investigation Report, Petroleum Safety Authority Norway.

PSAN (2008b), Hydrocarbon leak in utility shaft on Statfjord A 24 May 2008 - Investigation Report, Petroleum Safety Authority Norway.

Weick, K. and Sutcliffe, K. (2007), Managing the Unexpected - Resilient Performance in the Age of Uncertainty, 2nd Ed, John Wiley & Sons.

6TH GLOBAL CONGRESS ON PROCESS SAFETY

Analyzing Historic Process Data to Identify Near Misses and Warning Signs: Examples from the Buncefield Incident

Timothy J. Myers
Exponent, Inc.
9 Strathmore Road, Natick, MA 01760
(508) 652-8572
tmyers@exponent.com

Harri K. Kytömaa, Alfonso F. Ibarreta, Nicolas F. Ponchaut
Exponent, Inc.
9 Strathmore Road, Natick, MA 01760

Prepared for Presentation at
American Institute of Chemical Engineers
2010 Spring Meeting
6th Global Congress on Process Safety
San Antonio, Texas
March 22-24, 2010

UNPUBLISHED

Analyzing Historic Process Data to Identify Near Misses and Warning Signs: Examples from the Buncefield Incident

Timothy J. Myers
Exponent, Inc.
9 Strathmore Road, Natick, MA 01760
(508) 652-8572
tmyers@exponent.com

Harri K. Kytömaa, Alfonso F. Ibarreta, Nicolas F. Ponchaut
Exponent, Inc.
9 Strathmore Road, Natick, MA 01760

Keywords: tank overfill, process data analysis, near misses, alarms

Abstract

Process data is often analyzed during the investigation of industrial fires and explosions. Such analyses sometimes reveal prior near misses or warning signs that foreshadow the eventual causes of a catastrophic event. In some cases the facility may have been unaware of these near misses or not realized their frequency. Proactive analysis of process data prior can be a tool to allow facilities to identify additional near misses and warning signs, and address problems before a catastrophic event occurs. Such techniques are currently being used in a voluntary safety program in the commercial aviation industry.

In this paper we examine historical process data from the Buncefield Fuel Terminal. A December 2005 explosion and fire at the facility was caused by the overflow of approximately 300 m^3 of premium gasoline from a storage tank. The historical process data from the facility identified multiple reoccurring near misses or warning signs that preceded the December 2005 event. These include sticking of tank level gauges and reliance of operators on alarms. With the benefit of hindsight, it can be argued that, had the facility proactively evaluated their near misses prior to the incident, these warnings signs may have been more fully understood and addressed.

In this paper we describe the ways in which this facility could have recognized the significance of the frequent occurrences of equipment failure and operator reliance on alarms in the months prior to the incident and acted upon it.

system failures and operating practices that ultimately led to a catastrophic failure had occurred multiple times in the months preceding the incident.

2. Description of Buncefield Incident

At approximately 6:00 am on Sunday 11 December 2005, an explosion and fire occurred at the Buncefield fuel terminal. The incident caused significant damage to the facility and the surrounding area, but no fatalities. The facility received fuel from multiple pipelines, stored fuel in tanks, and distributed fuel to outgoing pipelines and tanker trucks. The initial explosion was caused by the overfilling of Tank 912, which caused approximately 300 m^3 of premium gasoline to spill from roof top vents on the tank; creating a vapor cloud over a large portion of the facility [1]. Figure 1 shows the recorded liquid level for Tank 912 in the hours preceding the incident.

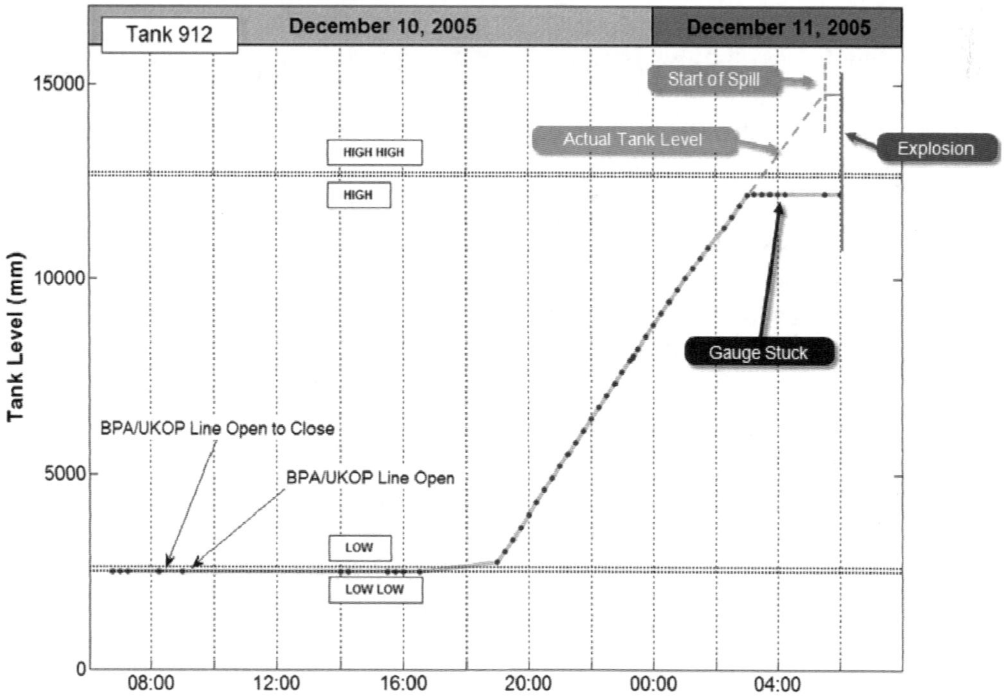

Figure 1 Liquid level in Tank 912 in the hours prior to the explosion.

At approximately 3:05 am, while Tank 912 was being filled from a pipeline, the level gauge became stuck and failed to record the increasing tank level even though the tank continued to fill. The data from the tank level gauges located in multiple tanks were displayed on a computer monitor in the control room. HIGH and HIGH HIGH level alarms were designed to activate and alert operators before a tank overflowed, but did not activate in the case of Tank 912 because the gauge that activated the alarms had become stuck. Additionally, an independent ultimate high level alarm and overfill prevention system, designed to shutdown the pipeline filling the tank before the tank overflowed, failed to shut down the pipeline.

The fuel tank was expected to reach its HIGH alarm level at 3:29, or about 24 minutes after the gauge became stuck. During normal operations, the pipeline flow to Tank 912 would have been diverted to another tank soon after the time the gauge became stuck. Instead, fuel began overflowing from the tank at around 5:35 – 5:40 am and approximately 300 m^3 overflowed, creating a vapor cloud of approximately 80,000 – 120,000 m^2. The vapor cloud was ignited at 6:01 am, leading to a vapor cloud explosion with estimated peak overpressures above 200 kPa, causing damage to buildings and structures in the surrounding area [2].

3. Monitoring System

The Buncefield fuel terminal contained a variety of tanks that were monitored by electronic tank gauges and a centralized tank gauging system. A single computer terminal simultaneously monitored and controlled 23 main fuel storage tanks.

The computer monitoring system was used to monitor tank levels and temperatures, control tank level gauges, and control and monitor the opening and closing of tank valves at Buncefield. The level gauge measured the tank level automatically and displayed the tank level in the control room.

The computer system was also configured to log data including tank level and temperature, alarm data, valve actuation, and other operator actions. The computer system also activated audible and visual alarms when certain conditions were exceeded. Historical logs of process data were kept in the computer system for several months.

3.1 Level gauge monitoring

The tank 912 level gauge was a servo gauge, which measured the level of liquid in the tank with an accuracy of ± 1 mm. The level sensor was mounted on the top of the tank and consisted of a drum that raised or lowered a displacer on a wire. The displacer moved up and down with the liquid level in a stillwell. The gauge was designed to maintain the displacer at the fuel / air interface.

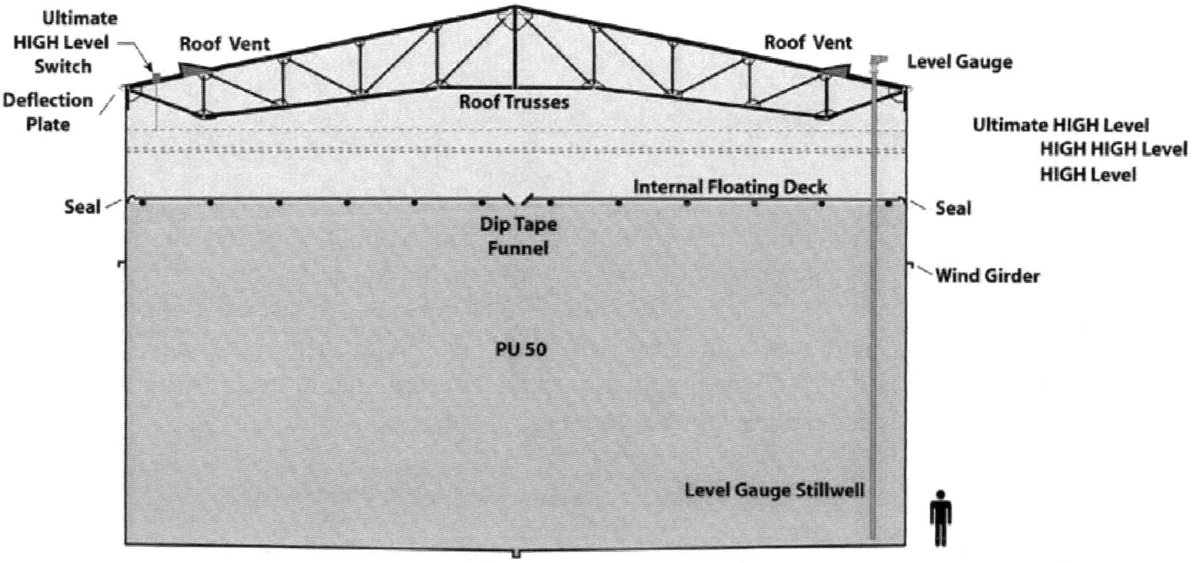

Figure 2 Schematic of tank 912 showing location of level gauge and ultimate HIGH level switch, as well as the level activation of the HIGH and HIGH-HIGH alarms.

The software was equipped with alarms that activate when the tank level and/or temperature reach specific values. Some alarms, such as the HIGH and HIGH HIGH level alarms, were predetermined, based on the tank safe operating levels. Other level alarms, such as the USER HIGH and USER LOW, could be manually set by the operator or by the control system based upon the calculated end points of fuel movements.

Figure 2 shows the location of the level gauge and ultimate high level switch in Tank 912. The schematic also shows the activation heights for the HIGH and HIGH HIGH level alarms, as well as the liquid level of activation of the independent shutdown switch, also known as the Ultimate HIGH level.

3.2 Fuel Movement Planning and Monitoring

The software kept track of both the expected fuel movement direction as entered by the operator, and the actual movement direction based on level changes measured by the tank gauge. This allowed the operator to visually observe whether the actual (measured) fuel movement direction was consistent with the expected fuel movement direction.

During a filling movement, the software assisted the operator in monitoring the progress of a fuel movement by displaying the length of time a tank had been filling with fuel and the total amount of fuel that had been transferred into the tank. The system also calculated the amount of fuel filled or drained into a tank from changes in the tank level. However, the tank gauging system did not have an independent measurement or display of fuel flow rates in pipelines feeding the tank. Such flowrates were not available real-time in the control room.

4. Analysis of Historical Data

The software settings, tank levels, alarms, and operator inputs were logged in separate files that were archived over varying periods of time, with alarm data available for 5 years prior to the incident and tank level data available for the previous 6 months. In the incident analysis, Exponent used several software tools to analyze the data contained within the computer files. Matlab was used to combine several historical files and reconstruct the tank level and temperature data along with alarm, valve activation and other command data. Algorithms were also developed to identify the fuel movements and faulty tank level measurements.

4.1 Failure of level gauge system

Analysis of the historical data indicated that some of the Buncefield level gauges often "stuck". During these sticking events the level gauge displacer became submerged in fuel when the tank was filling or remained hanging in air when the tank was draining.

Stuck points were identified by analyzing the tank level record. During a sticking event while filling, the level gauge displacer stops rising as the tank continues to fill. At this point the gauge displacer becomes submerged in the fuel and the gauge is stuck. When the gauge eventually unsticks, the displacer rapidly returns to the actual liquid surface, and once again accurately measures the tank level. The rapid change in reported tank level that occurs when the gauge unsticks typically corresponds to flow rates that are not possible in practice, when compared to typical filling or emptying flow rates. The apparent filling or draining flow rate can therefore be used as a simple means to identify the occurrence of sticking events.

A sample of multiple sticking events for Tank 912 in a single day is shown in Figure 3. Analysis of the data clearly shows that the recorded gauge level does not follow the true liquid level for much of the time between 00:00h and 16:00h.

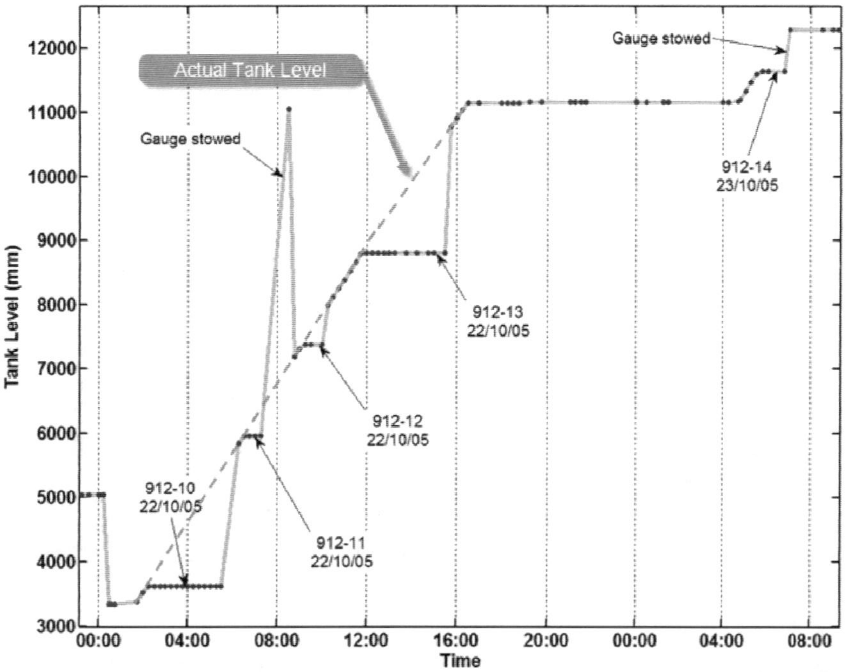

Figure 3 Plot of Tank 912 liquid level in October 20, 2005, showing multiple stuck points.

The level histories of all the main tanks during the 180 days before the explosion were analyzed to identify stuck points. A total of 21 sticking events were detected in the level data for the 23 main fuel tanks. Out of the total 21 sticking events, 18 were associated with Tank 912. Prior to the December incident, two Tank 912 sticking events occurred in September, thirteen in October, and two in November.

The level data and operator action log data also showed that gauges were often remotely controlled via the software to be stowed or reset during sticking events (see "gauge stowed" in Figure 3). These commands involved manually raising the gauge displacer to the gauge housing and then sending it back to find the liquid level. These actions typically caused the gauges to unstick.

Through analysis of the level data, the gauge on Tank 912 was found to stick at repeatable locations that are spaced by a constant interval. The height differences between all of the stuck points correspond to multiples of 473 mm (18.6 inch). This distance corresponds to the approximate circumference of the 150 mm (6 inch) diameter gauge drum. This indicates that each time the gauge stuck, the gauge drum was at the same angular position. This behavior suggests that friction at a specific angular position of the wire drum or the gear shaft on the servo side of the gauge probably caused the sticking.

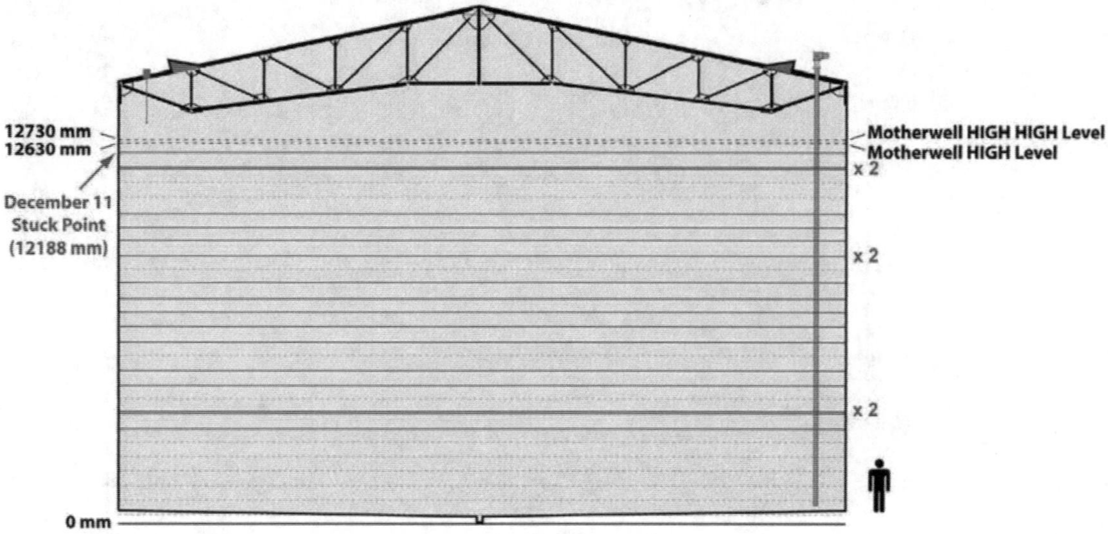

Figure 4 Levels at which the gauge in tank 912 got stuck are indicated by solid blue lines. The dashed lines are separated by one gauge drum revolution (~473 mm). The "x 2" symbol indicates the gauge stuck at that height on two separate occasions.

The post-incident analysis of data indicated that the failure rate of the tank level gauges was high, particularly on Tank 912. The failure rate was much higher than the assumed failure rate in risk analyses performed at the facility. Additional layers of protection were therefore necessary to meet the level of risk determined in pre-accident risk analyses.

4.2 Reliance on HIGH and HIGH HIGH level alarms

According to the API Recommended Practice 2350, "High-level detectors and/or automatic shutdown/diversion systems on tanks containing Class I and Class II liquids shall not be used for control of routine tank filling operations. These devices are intended to signal a potential emergency and initiate certain manual responses or activate automatic response mechanisms." Analysis of the historical data logs, however, shows that the HIGH and HIGH HIGH level alarms were often activated during tank filling, particularly when multiple tanks were being filled from a single fuel movement.

The data was analyzed using automated algorithms to compare the alarm activation times to the number of fuel movements. Fuel movements were identified from tank level increases. Analysis of the data identified a total of 1,136 tank fillings during the 180 days prior to the incident, which were compared with the rate of alarm activation and operator command usage. During fuel parcels where more than one tank was filled, it is expected that the operator would fill a tank to near maximum capacity before switching over to another tank. It is not expected that the last tank filled during a fuel movement would be filled to capacity. Analysis of the data indicated that HIGH alarms activated relatively frequently (21%) during these multiple tank fillings when tanks were expected to be filled to capacity.

In some cases, multiple tanks in a fuel group were filled to or above the HIGH level alarm one after the other. Such a case is shown in Figure 5.

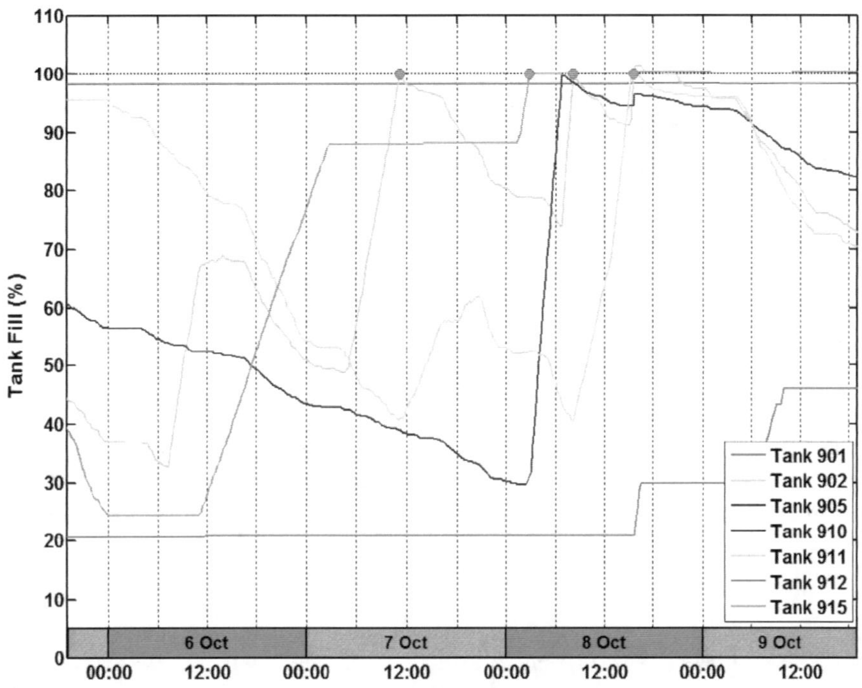

Figure 5 Plot showing the filling of several tanks in sequence. Tanks were filled to over 100% capacity causing an alarm (orange dots) before flow was switched to another tank.

The frequent activation of alarms during tank filling is not consistent with guidance provided in API Recommended Practice 2350. Operator's reliance on alarms resulted in a lower level of protection than if the operator was independently monitoring and anticipating the time when tanks would approach their full level. This was especially problematic given the frequency of tank level gauges sticking.

Had this increased alarm activation frequency been addressed prior to the incident, operators would have been more likely to detect that the gauge had become stuck. This is particularly true since the Tank 912 level gauge of tank 912 stuck just 24 minutes before the tank was expected to reach its full level and continued to fill for over 2 hours before fuel began overflowing from the tank. Had the operators independently anticipated the time the tank became full, they would have checked the tank level shortly after the gauge became stuck, well before the tank began overflowing, and would have been able to communicate with the pipeline or use other means to determine whether the tank was actually still filling.

4.3 Lack of fuel movement planning and monitoring

Scheduling commands including changing the movement direction, calculating the stop dip level, volume, or weight of a fuel movement and setting a variable user level alarm were logged in an operator action database. This data was only available for the 30 days prior to the incident but still provided valuable information. The operator action log entries were analyzed using algorithms to identify how often the movement calculator and movement planning function commands were used when tanks were filled. Analysis of the operator commands shows that the calculation tools provided in the software system were rarely used to plan fuel movements into a tank.

Analysis of the data in the 30 days prior to the incident indicated that operators had in many cases not updated the expected movement direction during tank fillings. At the time of the explosion most of the main fuel storage tanks did not have the expected movement direction set to the actual expected direction. The expected movement direction for the incident tank (912) had been incorrectly set to "outflow" for about two and a half days before the explosion.

Analysis of the operator actions indicated that the software tools used to calculate fill stop times and end levels were rarely used. The software calculator function was only used 5% of the time on the premium unleaded tanks like Tank 912. Had some of the tools been used by the operators, they would have provided a quick comparison to allow the operator to determine that the Tank 912 level plateaued even though it was expected to be rising.

Calculators would have provided the operator with estimates of the time at which the tank was expected to become full. Had the operator independently anticipated the time the tank became full, they could have checked the tank level after the gauge became stuck and well before the tank began overflowing. There was ample time for Buncefield operators to communicate with the pipeline operators or use other means to determine whether the tank was still filling had they become aware the gauge was stuck.

5. Conclusions

Post-incident analysis of the available data provided evidence of prior gauge failures and of only limited planning and monitoring by operators. These factors played an important role in the December 2005 incident at Buncefield. Had the facility proactively and critically evaluated these events prior to the incident, these warning signs may have been more fully understood and addressed. In this case the process data identified three factors that increased the risk of an overflow event at the facility:

- Tank level gauges on tanks frequently became stuck and remained stuck for long periods of time, particularly on Tank 912. The failure frequency of the level gauges was much higher than values used in pre-incident risk analyses. As a result, the actual risk of overflow at the facility was much higher than anticipated.

- High level alarms often activated during multiple tank fillings. This is not consistent with API Recommended Practice 2350, and provides a lower level of protection than if operators independently monitored and anticipated the time tanks would become full.
- Available planning tools included in the tank level and control software were rarely used. These tools would have allowed operators to more easily identify times when tanks were not filling or draining as expected, and would have also allowed them to estimate times when tanks would become full. Such planning tools allow operators to independently compare the progress of a fuel movement to the expected progress.

Although the analysis presented in this paper was carried out with the benefit of hindsight, it demonstrates that the proactive review of historic process data can reveal additional near misses and warning signs at facilities. In most operating facilities, such near misses can and often are identified by other means, such as maintenance logs or daily alarm logs. The methodology presented in this paper provides an additional method to identify near misses that may have originally occurred undetected. Irrespective of the method of identification of near misses, the Buncefield experience clearly demonstrates how important it is to address these rigorously.

Most of the analysis performed by Exponent utilized automated algorithms to detect discrepancies in liquid level recordings, alarm activations and usage of planning functions. Similar algorithms could be created for other types of processes to identify near misses and warning signs. Similar reviews of historical data are regularly performed as part of a voluntary safety program in the commercial aviation industry aimed at "determining the root causes of systemic problems that need correction" [3]. Historical process data is widely available in modern industrial facilities. The analysis of historical data using automated systems could provide an additional method to screen and identify past near misses.

6. References

[1] Buncefield Major Incident Investigation Board Health "The Buncefield Incident - 11 December 2005: The final report of the Major Incident Investigation" Health & Safety Executive (HSE) Volumes 1, 2 & 3 (2008) http://www.buncefieldinvestigation.gov.uk/reports/index.htm#final

[2] Steel Construction Institute "Buncefield Explosion Mechanism: Phase I" Health & Safety Executive (HSE) Volumes 1 & 2 (2009)

[3] DOT – Federal Aviation Administration (FAA) "Flight Operational Quality Assurance" Advisory Circular, AC No. 120-82, 4/10/2004.

6TH GLOBAL CONGRESS ON PROCESS SAFETY

Paper Title
Journey towards PSM Excellence

Presenter/Principal Author
Ravi Ramaswamy
Reliance Industries Ltd,
RCP, 7A, First floor,
Thane Belapur road,
Ghansoli, Navi Mumbai
Email: ravi.ramaswamy@ril.com

Other Authors
Ruskin Damani,
Debendra Das,
Prakash A Shah
Vishal Patel,
Prasad Chaitanya,
Ramesh Solanki,
Nemsingh Verma,
Vikas Shori

Reliance Industries Ltd,
RCP, 7A, First floor,
Thane Belapur road,
Ghansoli, Navi Mumbai

Email: Ruskin.damani@ril.com
Debendra.das@ril.com
Prakash.a.shah@ril.com
Vishal.patel@ril.com
Prasad.chaitanya@ril.com
Ramesh.solanki@ril.com
Nemsingh.verma@ril.com
Vikas.shori@ril.com

Prepared for Presentation at
American Institute of Chemical Engineers
2010 Spring Meeting
6th Global Congress on Process Safety
San Antonio, Texas
March 22-24, 2010

UNPUBLISHED

Paper Title

Journey towards PSM Excellence

Presenter/Principal Author
Ravi Ramaswamy
Reliance Industries Ltd,
RCP, 7A, First floor,
Thane Belapur road,
Ghansoli, Navi Mumbai
Email: ravi.ramaswamy@ril.com

Other Authors
Ruskin Damani,
Debendra Das,
Prakash A Shah
Vishal Patel,
Prasad Chaitanya,
Ramesh Solanki,
Nemsingh Verma,
Vikas Shori

Reliance Industries Ltd,
RCP, 7A, First floor,
Thane Belapur road,
Ghansoli, Navi Mumbai

Keywords: PSM, Process Safety Management, Reliance Industries Limited, RIL, HSE Management framework, safety culture, operational discipline

Abstract

This paper describes the experience, learning and major challenges in implementing Process Safety Management at Reliance Industries Limited (RIL) India.

The Reliance Group, founded by Dhirubhai H. Ambani (1932-2002), is India's largest private sector enterprise, with businesses in the energy and materials value chain.

Over the last two years, RIL has taken several strategic steps to make a quantum jump in Health, Safety & Environment (HSE) management systems, practices, performance and culture. RIL had tied-up with the DuPont Safety Resources who are the world leaders in HSE management to help in the transformation process.

The major risk and focus area for RIL is Process Safety. Process Safety, however, needs a strong integration with the general safety principles which include behavior influencing. RIL adopted a comprehensive strategy to address all aspects of Process Safety Management and radically transform the way process safety was managed. Top down approach of coaching all leaders, building strong systems of governance, review and controls is adopted along with the bottom up approach of involving all employees, building employee's capabilities on safety through Training and Coaching.

Visible Management commitment is at the core of Process Safety management and RIL's top leadership has committed itself to implement the processes safety management in its true spirit by motivating, reviewing, supporting, guiding and providing resources on priority for PSM implementation. Detailed unit wise plans and PSM metrics are developed and reviewed by each level of leadership in the organization.

The first step in implementing effective PSM is setting clear accountability and ownership of line management for the Process Safety performance. Line managers performed their role in the management and implementation of process safety by participating in subcommittees, standards / procedures development, trainings and actual implementation activities. Subcommittees were formed to manage, govern and facilitate the whole process of implementing the elements of Process Safety. Networks were formed to benefit from each other's learning.

A strong behavior safety program is implemented to establish a culture that is sensitive to safety, specifically, process safety.

Building PSM competencies in the organization through training, coaching and handholding is an important aspect to implement PSM. It is critical for the line management that the employees are aware of, adequately trained and made to understand Process Safety. A critical mass of employees got trained on PSM competencies and became the torchbearers.

Management expectations on PSM were spelled out clearly through developing and rolling out PSM standards at corporate level demanding world class level of performance. The manufacturing sites clearly defined the ways to achieve those expectations through their site procedures. The process of training on the standards and procedures and communicating to the employees is the core of effective implementation of PSM.

A robust Auditing process has been established to identify and improve the gaps in PSM and the audit also checks the efforts being taken to improve the Operational discipline at the organization. A progressive motivation process has been established to drive the organization towards excellence.

RIL has established a strong Process Safety Management system and further strengthens it by learning from its experiences during this process. Implementation of Process safety management involves an integrated approach of addressing all the elements with strong management commitment and involvement of all employees.

RIL's learning and experience are of relevance to all process industries who want to implement Process Safety Management and this paper provides insight into:

- The Approach to implement Process Safety Management
- The Management role in implementation of PSM
- The Governance and Review process to implement PSM
- The Challenges in implementing a successful Process Safety Management.

1. Introduction

Companies world over have realized the criticality of safety and its inter linkage with operational excellence. Resources invested to improve Safety are no longer viewed as a cost but as an opportunity to achieve superior business performance. Thus safety has been adopted as a core value by the many companies across the world.

Hazardous chemical release poses significant threat to employees and general public and provides impetus internationally and nationally to authorities to develop or consider developing regulations to eliminate or minimize potential occurrence of such event. Process Safety management is critical and has become essential to the industry handling hazardous chemicals.

The above are the two key factors driving the companies to implement the process safety management programme.

Across the World, companies have adopted various approaches to improve process safety management and governments have come up with number of regulations.

The Reliance Group, founded by Dhirubhai H. Ambani (1932-2002), is India's largest private sector enterprise, with businesses in the energy and materials value chain. Group's annual revenues are in excess of US$ 28 billion. The flagship company, Reliance Industries Limited, is a Fortune Global 500 company and is the largest private sector company in India.

Backward vertical integration has been the cornerstone of the evolution and growth of Reliance. Starting with textiles in the late seventies, RIL has grown to become a world scale multisite petrochemicals and refining player in just over 30 years. RIL's management always scans the best practices in the world and implements them to attain world class performance. Though not demanded by Government regulations in India, RIL has embarked on the journey of PSM excellence to be a world class leader, true to its vision of "Growth is Life". Understanding the importance of Process Safety in the businesses it operates, RIL management has taken several strategic steps to become world class in HSE management. In order to make a rapid quantum jump in HSE management systems, practices, performance and culture, RIL has learnt from experiences and knowledge available on process safety through CCPS and has also partnered with one of the global leaders in safety management

This paper outlines RIL's experience in its journey towards achieving Process Safety excellence. It also covers the challenges of Technological diversity, People diversity and Resources. RIL manufacturing assets consists of petrochemicals plants established in 1970s to the latest refinery complex commissioned in 2009.Some of these assets were acquired inorganically and the world scale refinery complex was built from grassroots. People diversity included employees who worked in older manufacturing units to the employees who worked in manufacturing units employing cutting edge technologies of tomorrow. RIL's journey of process safety excellence coincides with commissioning of two world scale projects of RIL i.e. a Refinery complex at Jamnagar and the Oil and Gas facilities at Kakinada.

It is also important to understand, though the principles to implement process safety remain the same, every organization is unique and has to develop its own methodologies to implement process safety principles based on the strengths and challenges of the organization. RIL HSE management framework focusing on achieving excellence in Process Safety Management is built on unique principles to guide the organization and created a process safety culture that is shared by everyone from the CEO to the frontline employees.

2. Approach to implement Process Safety Management

RIL has taken up the Excellence Journey on a four pronged approach of Drivers, Enablers, Operational Elements and Control and Governance. The major thrust area for safety of RIL manufacturing sites handling hazardous chemicals and flammable materials is Process Safety. Process Safety, however needs a strong integration with general safety principles which include influencing behavior, systematic & effective training and systematic incident investigation and auditing. This integration of principles of Process Safety & General Safety at the line management level is essential for an organization to achieve sustained high levels of safety performance.

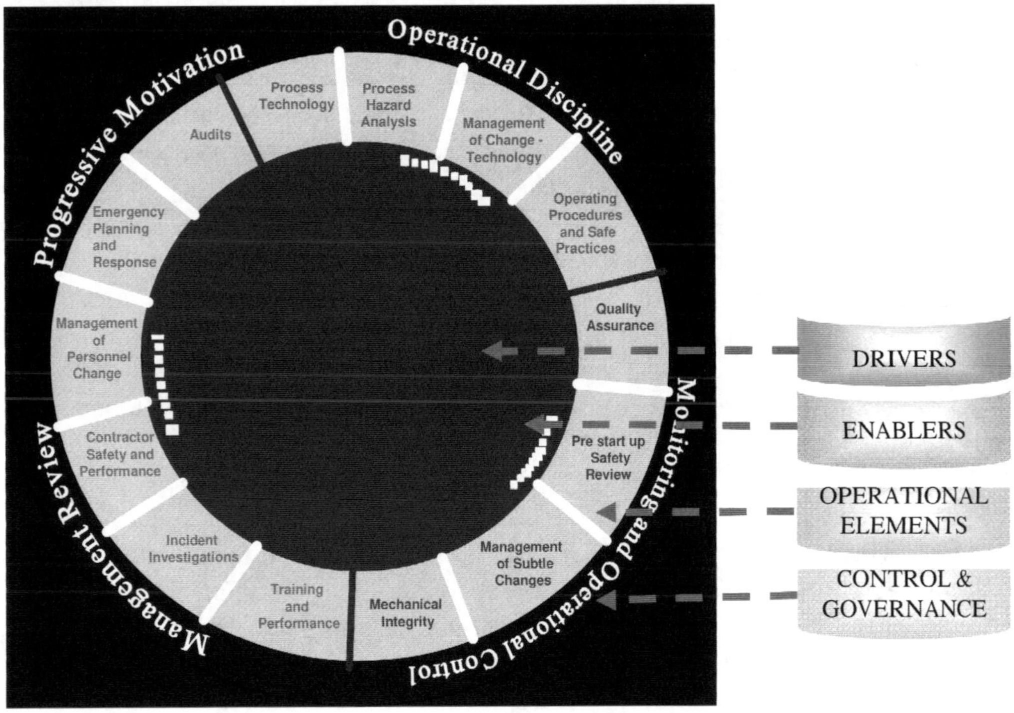

2.1. DRIVERS:

These are the key elements which provide the strategic direction for the management of Health, Safety and Environment in the Group:-

- Demonstrated Management Commitment
- Policies and Principles
- Challenging safety goals and objectives
- Health, Safety and Environment as a Line Organisation Responsibility

2.1.1. Management commitment

Commitment is the basic component of any successful safety programme and it is the key success factor of the RIL's journey towards Process Safety Excellence. RIL's board led by Chairman of the company Mr. Mukesh D Ambani committed itself to establish safety as a core value of the company and formed the HSE Committee of Directors to oversee, guide and drive the safety excellence journey. Since the commencement of the journey, top leadership was involved in regularly reviewing and enabling the manufacturing leadership in this transformation process. The Chairman made his expectations clear to all the leaders, the need to rapidly transform in order to reach the level of excellence in process safety. The commitment of the RIL Management was undeterred and top management ensured that the speed and quality of the programme was never compromised. A sense of urgency was demonstrated throughout the programme implementation.

The Management ensured that the best possible resources were made available for this transformation. One of the global leaders in safety was brought in to asses the current strength and weaknesses of the safety management system in the organization. A detailed evaluation of safety management systems was done across the organization's manufacturing facilities. Based on its findings, a high level strategic plan was established which involved development and implementation of standards, building capabilities in various facets of safety management and establishing managing system for safety in the organization.

A Centre for HSE Excellence was formed with the objective of supporting the HSE committee of directors in governance and supporting sites and businesses on various safety competencies through competency leads and teams. It was one of the critical steps to establish a seat of knowledge in HSE related matters for sustaining and subsequently leading the process safety management journey. A clear mandate was given to establish the Centre for HSE Excellence by sourcing the best available talents..

It was important to bring everybody on board for the transformation and to send a clear message across the organization. The Chairman addressed all the leaders across the manufacturing sites through a video conference and took commitment of each one of them himself. The communication was clear and consistent across the organization – **"A complete radical transformation in Safety Management"**. The same message, recorded and sent across the manufacturing sites was imbibed by the top to the lower most level of management personnel in the organization. The message from the Chairman of the company and the top management of

RIL and the commitment given by their leaders made a huge impact and motivated everybody in the organization and brought in the commitment at each level.

A "HSE Council" comprising of Top leadership of Manufacturing and key functions supporting manufacturing, led by Executive Director Mr. Hital R Meswani was formed and committed itself to the goal of complete radical transformation in safety. This council is the apex implementing organization and meets once in every forty-five days. The council led the transformation by reviewing progress and addressing the key challenges at every stage. Decisions were taken speedily in this forum and resources allocated immediately to catalyze the program. .

Building on its strength of project execution, Top leadership decided to take up this transformation as a mission with a defined timeline of 18 months. Having articulated the mission of safety transformation a detailed plan was chalked out with defined milestones and clearly defined roles and responsibilities. The Management committed the best resources available for this transformation by appointing operation heads themselves or the next in line person at each of its site as the project leaders. RIL has a programme for identifying high potential employees and developing them as future leaders by putting them through management programme called Management Programme for Reliance Engineers (MPRE) at one of the best management institutes in India. Such high potential MPREs were handpicked from across the organization and put in fulltime to assist the Project leads at the six manufacturing sites and at Centre for HSE Excellence to execute this transformation project and build a foundation for a robust sustainable process safety management system. Management sent a clear message to all employees in the organization that safety is the core value of the company and as important as production itself by establishing a strong execution team when other projects critical to businesses were also underway.

Such a demonstration through actions of the management was the one which built the foundation of the RIL process Safety excellence journey and everyone was enthused and were on board. The Safety Management framework had to be tailor-made to suit the organization's need and match its culture. RIL's four pronged approach of Drivers, Enablers, Operational Elements, Control and Governance was built on the Process safety management principles advocated by CCPS and DuPont. RIL customized, developed and implemented the RIL Safety Management Framework for managing safety, in particular the Process Safety. The first driver i.e. the demonstrated management commitment piloted the company in the Journey towards Process Safety Excellence.

This commitment is then demonstrated by all levels of management throughout the organization in their behavior. All levels of Management actively participated in incident investigations, in subcommittees functioning, conducting safety Observations, auditing, etc. All meetings started with safety contacts and safety became the top agenda in management review meetings. Resources were committed to achieve the long term goal of Zero injuries and incidents. Whole team of HSE council traveled to the manufacturing sites and reviewed the progress in field by carrying out Safety Observations Rounds and talking to employees and motivating them.

The period of this transformation was marked with deep economic recession throughout the world and companies had to put many projects on hold. However the RIL Management demonstrated the resilience and commitment to this transformation in safety by not only continuing with the resources allocated but also provided additional resources whenever a need arose. This again gave the message down the line that the safety is indeed a value for the company and made a huge impact in inculcating the safety culture

2.1.2. Policies and Principles

The next driver guiding the journey is the Policy and Principles. Inputs from internal and external stakeholders are gathered and used as and when appropriate in the planning process for HSE policies, goals, targets and resource needs. Several groups within RIL ensure that stakeholder perspectives are collected, understood, communicated and considered by the appropriate individuals and teams. The centre for HSE Excellence evaluated the stakeholder perspectives and made recommendation to the board. Core human values were the cornerstone for the development of HSE principles. The RIL HSE principles thus developed guided the RIL actions from development of HSE policy to the performance at work. The key challenge was how to make these HSE principles understood and remembered by all employees in the organization. A special communication package was developed and the acronym of **PROACTIVE- M** was publicized through pocket cards, posters, special campaigns, etc. and made popular. It became the catch word across the organization and the fundamental principles of HSE got imbibed in the organization's DNA and became the guiding force for employee behavior.

2.1.3. Challenging Safety Goals and Objectives:

The third driver for implementing process safety Excellence is the Challenging Safety Goals and Objectives. Managing Safety has to be like managing other aspects of a business by setting performance goals and objectives and making people accountable for achieving these goals. It is important to set aggressive and clear performance goals and objectives which motivate performance and drive improvements. Though, based on experiences of global safety consultants, safety transformation is a slow process and takes 3-5 years, RIL management set the challenging target for building the foundation for process safety management excellence within 18 months.

HSE goals supported by workable plans were prepared with clear responsibilities. Group leaders were made accountable for achieving group goals. The transformation journey was divided into 150 milestones and the HSE council reviewed the progress on milestone achievements every month and addressed the constraints.

2.1.4. HSE as a Line Organization responsibility

The Fourth driver is the **HSE as a Line Organization responsibility.** Effective implementation of Safety can only be achieved when the line organization is actively engaged. Line managers must manage safety as any other aspect of business like production, quality, etc. They are made accountable for their personnel safety and the safety performance of the organization through performance appraisals and introducing a concept of Personal Safety Action Plans.

Reliance Safety Observation programme (ReSOP) was introduced to build a safety culture and subsequently focused primarily on PSM to drive a PSM culture. The programme objective is to uncover the unsafe acts through proactive safety dialogue, reinforce safe behaviors, create awareness and build a strong safety culture. Safety Observations is an important tool to demonstrate management commitment and motivate people to be committed to PSM.

2.2. ENABLERS

These expectations underpin the line organisation's efforts in the delivery of Health, Safety and Environment performance. They provide the necessary support to the line organisation in implementation of the Health, Safety and Environment management system and thereby ensuring continual enhanced Health, Safety and Environment performance.

- Supportive Safety Personnel
- World Class Standards for Performance
- Integrated organisation for Health, Safety and Environment
- Effective Planning and Communication

2.2.1. Integrated Organization for Safety:

The first key enabler for process safety excellence journey is to establish a structured organization to administer safety effectively- one that cascades from the top to the lowest level at the sites. The organization of committees and subcommittees is the structure which involves employees at every organization level. The HSE committee of Directors and HSE council governed the transformation at the corporate level. HSE council is chaired by Executive Director responsible for the RIL manufacturing in the Board of Directors and has all site Presidents as its members. At each site, the Apex safety committee comprising of all subcommittee chairmen became the governing body. The subcommittees have representations from the plant level working committee. One of the key advantages in this approach is the overlapping representation which ensured a smooth two way communication across the organization and helps a rapid transformation process.

The Committees put forth policy and principles, formulate goals and objectives, discuss safety related matters and support line organization to achieve set goals and objectives. It is important to set aggressive and clear performance goals and objectives which motivates performance and drive improvements.

2.2.2. World Class Standards for performance

The next key enabler for implementing a process safety management programme is the " Safety Standards". They are the means to communicate the management expectations in process safety. Without written standards people are left to determine their own methods of performing activities: they may not always be safe and productive. The biggest challenge was to develop and implement the world class standards for performance at a rapid pace across all the six major manufacturing sites. Multisite teams of subject matter experts from all sites and Centre for HSE Excellence formed and developed these standards with support of external experts through standard development exercises. This helped in addressing the variations in practices and brought everyone on same platform with same understanding. These standards are then implemented by the line management at each site with the help of these subject matter experts and under guidance of the subcommittees.

A three level documentation structure has been created for implementing Policies. Level I is the Policy and principles and position statements which is a combination of group key statement and aspiration of the group. Level II are the group standards providing details on mandatory requirements in specific functional areas. Level III consists of documents which enable the implementation of the Group standards at the site level consisting of Site operating procedures and work instructions written to provide directions on how to accomplish a task or sequence of operations.

2.2.3. Supportive Health, Safety and Environment Personnel

The third key enabler in Process safety excellence journey is the Supportive Health, Safety and Environment personnel. While it is important for line function to take responsibility for HSE management, HSE personnel shall act as a supportive resource for smooth discharge of that responsibility. HSE personnel should be competent to perform the role of a resource for the line function. A thorough assessment of each HSE personnel across the six manufacturing sites and the corporate was done and key gaps identified. A detailed plan of development of HSE department and the Centre for HSE Excellence was chalked out and presented to HSE Council. HSE Council assigned dedicated senior level competent person at each site as PSM coordinators and a Corporate PSM leadership team was formed to drive the process safety management at sites. These are separate positions created in the organization to improve process safety. This effort was further strengthened by forming similar teams at plant levels. Capability in process safety management was built throughout the organization through training sessions.

Competencies team for each of the PSM element was, formed led by the competency lead at the CHSEE. Each competency team had a charter that describes the objectives, deliverable goals and critical success factors of the team. These charters were reviewed at least once a year to determine its suitability, adequacy and effectiveness. Competencies teams monitor and drive improvements in all aspects of respective competencies.

2.2.4. Planning and Communication

The fourth key enabler was the planning and communication process. Planning and communication is the next key success factor after demonstrated management commitment. The strategic direction is provided by the HSE Committee of Directors by providing goals and those goals converted into objective and then to detailed action plans at each level. The Site HSE plans are prepared inline with corporate objectives and further broken into subcommittee action plans based on which plant wise implementation plan is developed and the accountabilities rests with the plant heads. The plans gets reviewed by the subcommittees and Apex Safety committee at the sites and constraints are addressed appropriately at site level or raised to HSE Council level.

Realizing communication as one of the keys to the successful and uniform implementation of process safety in the process safety excellence journey, the management ensured that the message and methodologies should be consistent across the organization. The standards issued by the CHSEE were accompanied by the training and communication package to ensure every employee is clear of the management expectations.

It is important to maintain a two way communication so the employees are kept motivated and management be kept abreast with the issues in implementation and address them upfront. A separate communication subcommittee is established at the sites to look into the communication issues and ensure a consistent understanding across the organization. The Executive Director sent out communications regularly across the organization to provide the direction and drive for process safety management. The FY 2009-10 was declared the year of PSM by the Executive Director- manufacturing in one of his communication to employees.

External resources were inducted on board as consultants to learn and implement the best practices available in the country, notable being the Ex-Safety director of the Atomic Energy

2.3. OPERATIONAL ELEMENTS – PROCESS SAFETY MANAGEMENT MODEL

Our approach to Process Safety Management Model treats the 14 Operational elements as part of one holistic gamut. These operational elements are interrelated to one another. .

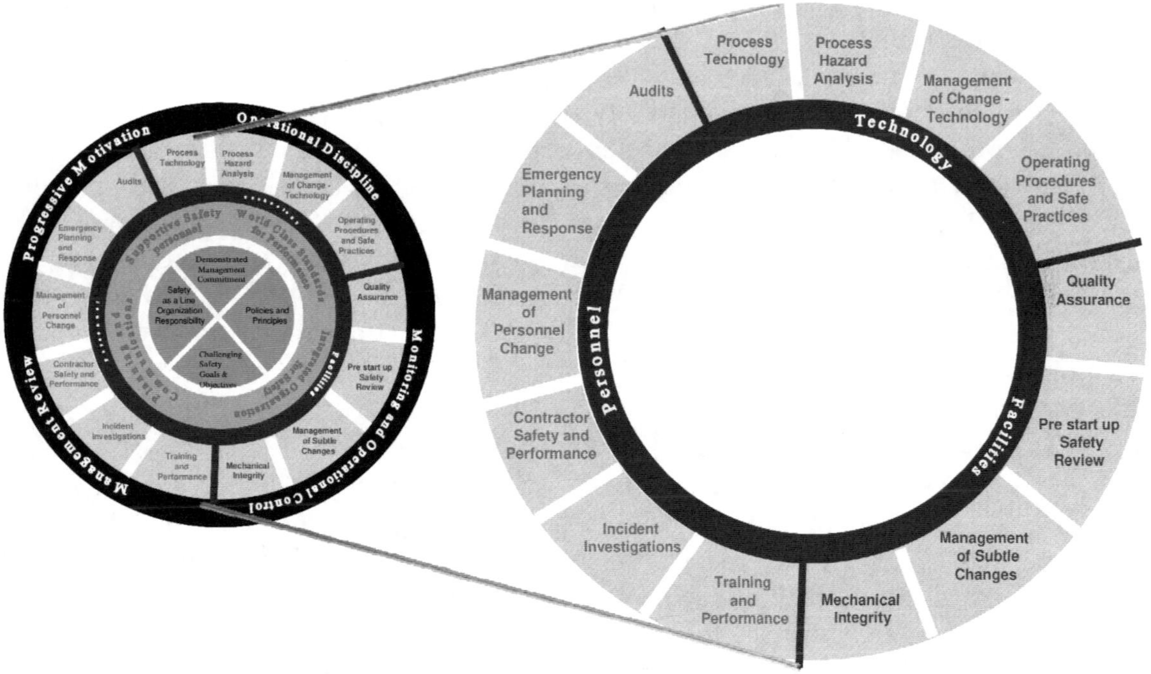

Management expectations on PSM are spelled out clearly through developing and rolling out PSM standards at corporate level, demanding world class level of performance. Manufacturing sites clearly defined the ways to achieve those expectations in their procedures. The process of training on the standards and procedures and communicating to the employees is the core of effective implementation of PSM

The implementation of Process Safety management faced key challenges due to the differences in the age of the plants, people competencies, and cultural differences and competed for human resources when other capital projects for new facilities were under construction and commissioning. Management commitment remained undeterred and it was clear that there should be no dilution and standards should match the world class level and every plant shall work towards complying with the same standards.

2.3.1. Process Technology

Process Technology is the first step in Process safety Management effort. It describes the process hazards so that operators can confidently work in the envelope that describes safe operation. PT element comprises of

1. Hazards of Material
2. Process Design Basis
3. Equipment Design Basis

For the new refinery site complying with requirements of PT element was not much of an issue as while constructing grass root plants, RIL always went for the best available technology. However for the old sites, process technology became a major issue due to non availability of the design documents and lack of competent people in the organization for developing such design documents. Massive efforts across the sites were started to update the process technology documents. The gaps are identified at each plant level and in many cases technology licensors or external support is explored to fill up the gaps. In some cases, Technology Licensors were not available to support. Process technology networks across the group work together to resolve the site issues with inter-site supports and unresolved constraints are referred to Apex committees at sites and HSE Council at Corporate level. Process Technology is one of the biggest resource gurgler and major pain area. The Management decided to build in-house technical capabilities for addressing the Process Technology issues and hence a specialized programme called Reliance Certified Process Engineers has been initiated with support of leading Technical institutes like IIT and external consultants.. The Gaps identified in Process Technology package provided an opportunity for the company to define requirement of Process Safety information upfront for the upcoming new projects

2.3.2. Training - Capability building in PSM

Implementing process safety management needs PSM competency in the organization. There were two key challenges; one was building a team of specialists in Process safety as PSM demands persons of high skills and competency. Second the training of more than 10000 employees on PSM. This called for huge no. of trainers. The specialized PSM capabilities were built through training and guidance from an external consultant resource and training packages were developed for all levels of employees. The cascade system of training is now established through these specialized persons who train another set of trainers in each of the plants and then the employees down the line get trained. Thus the training effort covers the whole organization. Currently, to increase competency on PSM, a comprehensive scheme for training, validation and certification is being launched to ensure a sustained and deep-rooted culture of PSM with a stream of specialists on PSM.

Imparting awareness and training on PSM elements to a large workforce including contractors conversant only in regional languages was a challenge and to overcome this, element specific campaigns were taken up with more focus on shop floor employees. These campaign which includes poster, exhibitions, experience sharing, quiz, lectures, competitions, demonstrations, street plays not only made it interesting but demonstrated the management commitment and enhanced employee's participation which is very critical for success of PSM.

Enhancement of process safety knowledge at all levels of organization was taken as a challenge and supplemented by subscribing to CCPS guidelines, Process safety magazines and imparting training and coaching through competent individuals and organizations.

2.3.3. Process Hazard Analysis:

Process Hazard analysis generated lot of debate on prioritization and time commitment. There were not many resources available as again this is a high skill area. To run a PHA programme simultaneously with so many activities in parallel was the tough task. PHA demanded very high caliber specialists and subject matter experts for specific areas where the competent persons were not available easily. It was decided to initiate PHA studies at twelve plants selected on priority and then subsequently initiating it at other plants when sufficient PHA resources are available. These selected PHA studies were supported by external experts who coached the teams and trained them on PHA. More than 100 personnel were trained on PHA to support further PHA studies. Experts on specific areas like Qualitative Risk Assessment were brought in at the Centre for HSE Excellence to support sites. All the plants are now covered under PHA studies. PHA experts are being developed in the organization through these studies and PHA training process is well established for competency development.

2.3.4. Incident Reporting and Investigation

Incident reporting and investigation was the main focus area of RIL management. The main challenges were to develop a culture of open reporting system. This called for a strong process safety culture and clarity on process safety incident. The management's expectation from incident reporting and incident investigations was to go beyond the global standards so that we have every opportunity for learning. Incident investigation standards for reporting and classification and investigation were rolled out which called for stringent timelines and quality of incident reporting and investigation process. A massive training effort was put in to train employees on reporting of incidents. The Management made it mandatory to investigate every incident by a six member team to ensure quality of incident investigation. A critical mass of people is trained on Incident Investigations and sample of some investigations conducted were reviewed with support of external experts. Subsequently the time limit to investigate an incident was also brought down to seven days to ensure that the critical information should not get lost due to the delays and learning from incidents can be shared at the earliest. A process of sharing learning from incidents is established to ensure that the recommendations, if relevant for other plants should get implemented quickly. The incident reporting and investigation system was IT enabled to bring in operational controls. All significant incidents during the month are presented to HSE council by the respective site head to appraise the actions taken up to prevent reoccurrence.

.

The analysis of recommendations helped in identifying focus areas to allocate the resources for strengthening PSM. Incident Investigation process not only helped in identifying weak areas in PSM for improvement but also built confidence, created openness and opened huge potential for learning process safety from our own incidents which were easy to understand by shop floor level employees.

One of the key challenges was ensuring100 % reporting of incidents. This was addressed by constituting teams of engineers from technical service, engineering service and plant operation who look at daily logs of plant, DCS systems, work orders and ensure that no incident goes unreported, howsoever minor it is. The intention was to identify and address issues at the initial stage itself before those issues result into major incidents.

2.3.5. MIQA: Mechanical Integrity and Quality Assurance

Mechanical Integrity and quality assurance was the major challenge area and in fact the first priority element of PSM on which RIL management focused upon. Much before the adoption of process safety as an integrated approach, RIL had realized that Mechanical Integrity is the most challenging area especially with older plants handling highly corrosive chemicals. MIQA is the backbone of any PSM programme and required an integrated approach with stronger operational controls. A separate external consultant of global repute on Plant Maintenance systems was brought in and all the maintenance business processes that were identified and developed by RIL teams were reviewed by external experts. These maintenance business processes (MBP) which included the operational and administrative controls became the foundation for MIQA programme. Considering the criticality of these maintenance business processes, they were again reviewed by external Safety experts on MIQA and modified to include additional safety requirements. A top level MIQA guidance manual was developed to bring in the clarity on the MIQA requirement for the company as a whole. These processes were prioritized and IT enabled and controls are put in to ensure compliance to world class standards.

There are bigger challenges involved in the area of management skills for MIQA due to overlapping responsibility areas as various functions have different priorities. There were issues on prioritization based on reliability and safety. MIQA needs a lot more involvement, coordination and intervention of Top management than any other element.. Here the subcommittee approach involving multidisciplinary functions became much useful and resolved the issues at a faster pace.

2.3.6. Auditing

Auditing is another area along with incident investigation which was on priority of RIL management. Establishing the auditing process was one of the major milestones in this transformation journey and whole auditing process comprising of First party audits, second party audits and third party audits were developed with the support of external experts. RIL Management expectations were clear in that a strong auditing process, nothing less than the world class is needed. A number of standards on PSM and other safety aspects were developed and implemented and the standards needed highly competent people to audit and ensure the compliance. There were two major challenges. RIL's previous experiences on auditing was that the audit findings varied with different auditors and it was required to ensure that auditors had not missed on any key aspect especially in PSM when the consequences of a process safety incidents are so high. A standardized audit checklist was developed along with audit protocols which defined how the audit will be conducted. This reduced the variation in audit observations and ensured that all critical aspects got covered..

Second big challenge for auditing was that highly competent persons from the line management trained on auditing process were needed with significant time commitment. These persons had to be Head of the department or Head of section level persons as auditing also required sufficiently senior person with good experience. These persons were critical to operations also due to their positions and the demand of time commitment was huge in the initial years as the resources were limited.. A pool of auditors from this group were trained and validated through classroom training and participation in actual audits along with the external expert auditors on process safety. An ongoing system of training auditors was also established to ensure the ongoing training and validation of auditors. There are more than 100 plants to be audited over a three year cycle , requiring a team of 5 persons conducting PSM audit of the plant for 5 days and the management ensured there were no deviations from audit plans demonstrating its commitment to process safety.

2.3.7. MOC and PSSR

Management of Change process was in practice for long in the company and though all major changes underwent this process, the issue was with the subtle changes and MOC personnel. Previously the focus was on technology changes only but while ushering in the world class standards, it was important to create the awareness on subtle changes. It was important to create an understanding of what comprises a change through extensive training at all levels of organization. The two other areas to be addressed for streamlining the MOC process were that every change should undergo the management of change process and review of the old subtle changes. A specific campaign was taken up as part of "project health check" which was launched to correct one time anomalies in the critical safety systems and all old MOCs were reviewed and updated. The process of reviewing work permits for ascertaining need to follow MOC process started at the plants to ensure strict compliance to MOC process..

At RIL, Pre startup safety reviews were already in practice before commissioning of big projects.. For example, before Jamnagar refinery was commissioned, stringent PSSR check as per global standards was carried out. However many minor projects/ changes had been commissioned without proper documentation of PSSR check. The PSSR standard was developed adopting the stringent requirements for PSSR. It was made mandatory to undergo the PSSR for all MOCs and a process was put in place that ensured implementation of category A recommendation before commissioning the change and rest of the recommendations be completed within specified time lines. Operational controls in the form of action tracking system and review process were IT enabled to ensure compliance.

2.3.8. Contractor Safety

Contractor Safety Management was one big concern area in the PSM implementation. It was designated as a separate work stream in the transformation project to bring in the required focus. There were some teething problems with respect to contractors especially in the Indian environment. On the one hand, RIL's world scale operations required the skills and competency levels of the contractors to be brought to world-class levels. On the other hand the availability of such contract workforce was a major issue. Training of contract personnel was another area of attention, as contract workers keep floating. If the contractors don't have the same values for safety, it can become a big deterrent for the process safety culture. Processes had to established to take care of such gaps right from resourcing of contractors to the actual field verifications. A six step process for contractor safety management was formulated for the group. It focused on the contracting process to ensure that only the right contractors shall be eligible for bidding and the terms and conditions related to safety and the expectations should be made clear before the bidding process itself. Simultaneously a process of contractor's field audits was initiated at the sites to ensure compliance to RIL standards of performance. The whole contractor safety management called for greater co-ordination among the stakeholders, i.e. engineering services, Operations, Contract cell and the contractors The Contractor field Audit scores are used to review the performance of contractors and this has made a significant change in the contractor's attitude towards safety.

2.3.9. Emergency Planning and Response

Emergency planning and response element has strong inter-linkage between Fire department and Operations. It is the PSM element which calls for training and periodic refresher training of employees to make sure that at the time of emergencies, people should respond as per the predefined plans. The whole activity at the sites was spearheaded by Emergency Response and planning subcommittee. The first step was to review the existing emergency response plans and to identify additional emergency scenarios and subsequently develop plans for the new scenarios. Plant level teams identified emergency scenarios for off site and Onsite emergencies and developed plant level, site level emergency response plans. The next step was to establish the process of training and retraining the employees on these emergency scenarios though a system of emergency drills and tabletop exercises. . A novel concept of discussing the PSM failures after every emergency drill exercise was started. The whole team sat together after the emergency drill and discussed on the PSM failures observed during the drill. This supported the strengthening of the PSM culture and ensured that the key gaps are understood and resolved.

Emergency Response plans are periodically reviewed and modified based on the inputs from the PHA studies undertaken by the plants.

Another focus area was the fire safety management. Though RIL always went beyond the regulatory compliance, in Fire Safety management the intention is to be world class. Again with the support of global experts on Fire safety, number of standards and procedures were implemented to strengthen the system. Fire Safety heads of all sites established a network to utilize inter-site resources and share best practices. They worked in tandem with Emergency response subcommittees at sites and established the auditing process for fire safety management to ensure the adequacy of the emergency response systems.

Special programmes and competitions are conducted to promote employee awareness on emergency response. Further this awareness at shop floor was improved through PSM focused safety observations round which provides the opportunity to supervisors to interact personally with down the line employees.

2.4. Governance and Control:

Systems for Governance and Controls enable the leadership to monitor the sustenance of the Health, Safety and Environment system, and used as tools for course correction. An important step in successfully managing Health, Safety and Environment risks is the activity and process of gaining operational control over significant Health, Safety and Environment aspects. The four tools utilized for governance and control are:

2.4.1. Operational Excellence through Operational Discipline:

Developing a high level of Operational Discipline is the foundation that supports successful management system. RIL's Operational Discipline comprises of eleven elements and is classified under three categories as:

a. **Leadership Characteristics**
   Leadership by Example
   Common shared values
   Strong Teamwork
   Pride in the organization
b. **Well Defined Managing Processes**
   Sufficient and Capable Resources
   Employee involvement
   Active Line of communication
   Up to Date documentation
c. **Maintaining High standards**
   Practice consistent with procedures
   Absence of Shortcuts
   Excellent House keeping

Operational Discipline was focussed on during the later part of the journey. A standard on Operational Discipline was developed and workshops were conducted and assessment was done to identify the areas of improvement.

2.4.2. Monitoring and Operational Controls

The next critical element in Governance and Control was the **Operational Controls**. Operational control such as Operating Procedures, Permit Process, engineering specifications, preventive maintenance programs, work instructions, Authority limits, etc were reviewed and strengthened. Apart from reviewing all the SOPs and SMPs, all technical authority limits were revised upwards to higher levels for a limited time to strengthen the controls. IT systems are utilised to strengthen the operational controls at the plant level.

2.4.3. Management review

Another element in Governance and control is the **Management Review**. This aspect is a critical success factor in RIL journey towards process safety excellence along with the planning process. A strong review system from the HSE Committee of Directors to the plant level was put in place and diligently worked to ensure the activities got completed on time. Not only the leadership team at all levels committed time and resources, but also participated enthusiastically to ensure that the Excellence Journey should accomplish its desired goals. The management of process safety is governed through subcommittees and networks formed for each of the PSM element. The implementation is further supported by Corporate PSM leadership team at the corporate level and individual PSM coordinators supporting at the respective sites. Each plant is empowered with specific plant level teams which are the working groups. A metrics of key performance indicators was established to monitor the performance in process safety. These KPI get reviewed at the plant level by respective plant HODs who are accountable for implementation at the plant level and respective subcommittees and Apex committee accountable at the Site level. The key trends get reviewed by the HSE Council in its monthly meeting. The network meetings ensure that the benefits of competencies across the company are available for smooth implementation at each of the sites.

2.4.4. Progressive Motivation:

The last of the tool for governance and control is the Progressive motivation. The most important task and often the most difficult is Motivation. The key positive motivators used are the involvement of employees. The supervisors and leaders listened to the safety concerns at every level and addressed them. The safety Observations rounds, Sub committee meeting and suggestion schemes are utilized to involve and motivate the employees. Through their actions, Leaders demonstrated to down the line employees that safety is a value to them and all are part of the same team.

Special campaigns and competitions are conducted in process safety. Special skits and films on the line of popular epics like Mahabharata and quiz shows are developed on the lines of popular Game shows and employees are rewarded for safety performance on the spot.

On the other hand, for repeated safety violations a progressive discipline process is rolled out. The primary purpose of discipline is to educate and not to punish and disciplinary process is applied to ensure employees adhere to acceptable standard of conduct. Six levels are provided in the disciplinary process from Verbal Correction Contract to Dismissal of employee. Red rules are in force to ensure compliance to basic safety rules and violation of red rules calls for severe disciplinary actions leading to dismissal

3. Conclusion

PSM serves mankind as it helps in preventing catastrophes and human suffering that can result as a consequence of industrial incidents. RIL Safety management framework helps RIL to progress in its journey to PSM excellence and helps in preventing such incidents. There have been many noteworthy successes and at the same time few failures. With every success the bar of process safety excellence gets raised further as it is a continuous Journey. The failures made RIL to learn and enrich itself by strengthening the systems to avoid future mistakes.

The key driver which led to the successes in RIL's journey is the Management commitment and the leadership's visible demonstration of safety values through their actions. The RIL management provided direction to the company through policy and principles and by setting challenging goals to stretch the performance. Leaders and the line managers took responsibility for Process Safety Management implementation.

The whole process was enabled by the World class standards for performance and a subcommittee structure that ensured that the cross functional issues got resolved quickly. The micro level planning with adequate resource allocation and two way communication between top and lower level of management ensures that the momentum is never lost. The key drivers and enablers created the foundation for implementation of process safety management operational elements and created a culture that can sustain the process safety excellence journey. The key drivers and enablers also ensured that the hurdles be identified upfront and get resolved collectively. The governance and controls played a critical role in guiding this culture. Operational controls and along with the diligent review process at all levels of management with defined performance indicators ensures that the process safety issues and concerns are identified and addressed upfront. Focus on Operational discipline along with the progressive motivation process established a corrective mechanism and a dependable culture for sustenance of process safety journey.

Journey towards PSM excellence at RIL has provided opportunity to engage all stakeholders in process safety with empowerment at all levels.

Hazards of Unplanned Power Outages: Implementing Appropriate Safeguards

Russell A. Ogle
Exponent, Inc.
Lisle, Illinois, USA
rogle@exponent.com

Delmar "Trey" Morrison, III and Tara L. Henriksen
Exponent, Inc.
Lisle, Illinois, USA
tmorrison@exponent.com
thenriksen@exponent.com

Prepared for Presentation at
American Institute of Chemical Engineers
2010 Spring Meeting
6th Global Congress on Process Safety
San Antonio, Texas
March 22-24, 2010

UNPUBLISHED

Hazards of Unplanned Power Outages: Implementing Appropriate Safeguards

Russell A. Ogle, Delmar "Trey" Morrison, III, and Tara L. Henriksen
Exponent, Inc.
Lisle, Illinois, USA
rogle@exponent.com
tmorrison@exponent.com
thenriksen@exponent.com

Keywords: process hazard analysis, layer of protection analysis, power outage, emergency power systems

Abstract

Accident root cause investigations often identify unplanned power outages as the cause of a catastrophic accident. Unplanned power outages can be an initiating event that leads to an emergency shutdown of a process unit. If the emergency shutdown is managed incorrectly, it can result in an uncontrolled release of hazardous chemicals.

To prevent a catastrophic accident, it is necessary to employ one or more safeguards to interrupt the chain of accident events. A process hazards analysis is the most effective tool for identifying these safeguards and evaluating their effectiveness prior to an incident. Examples of such safeguards are emergency shutdown procedures with operator training, emergency shutdown systems (including safety instrumented systems), and backup power supplies.

This paper explores the ways in which hazard analysis and emergency planning can be utilized to ensure a safe transition to a secondary power source during an interruption of primary power. A case study of an accidental release of hydrochloric acid vapor is presented in which a safeguard failed to function during a power outage. An interesting feature of this incident was the excessive testing of the emergency power system that led to its premature and undetected failure. Thus, when commanded to operate during a power outage, the backup system failed. We illustrate how a layer of protection analysis, coupled with the development of best practices from industrial safety standards, would have suggested a better inspection and test protocol for the emergency power system.

1. Introduction

A typical chemical plant relies heavily on electricity to power mechanical equipment, drive electrochemical processes, and implement process controls. In the absence of any safeguards, a power outage will cause an uncontrolled shutdown of the process operations. If the process operation involves the management of large quantities of hazardous chemicals, an uncontrolled shutdown can release hazardous chemicals with catastrophic consequences. One source of information about chemical releases caused by power outages is the National Response Center database. For example, in the five-year period from 2004 to 2008, an average of 110 hazardous chemical releases was attributed to power outages annually [1].

A fundamental objective of process safety is to maintain containment and control of hazardous chemical inventories [2]. This objective can be achieved through a combination of process design, operational controls, and safety protective systems [3]. A loss of electrical power can lead to a loss of control of the equipment. For example, unless safeguards are in place, a power outage means the loss of process measurement data and the inability to issue control commands to final control elements. More seriously, a loss of electrical power could result in a loss of cooling water pumps or agitation for a chemical reactor. Thus, a power outage could have disastrous consequences and must be considered as an initiating event during a process hazard analysis.

In this paper we consider some of the process hazards created by power outages and identify safeguards that can maintain containment and control of the process hazards. Next we discuss a case study in which an emergency power system failed to operate. The emergency power system was a safeguard intended to maintain operation of an acid gas scrubber during a power outage. The failure of emergency power resulted in a release of hydrochloric acid vapor. We illustrate how a layer of protection analysis, coupled with the adoption of best practices from industrial safety standards, would have suggested a better maintenance strategy for the emergency power system. Finally, we offer recommendations for developing a strategy for power outage safeguards.

2. Hazards of Power Outages

A power outage is a major deviation from the design intent for a typical chemical process plant. At an equipment level, it causes a loss of electric motors, electric actuators, transmitters, and electronic programmable systems. On a more prosaic level, a power outage will result in a loss of artificial lighting, comfort heating and cooling, computers and office equipment. To assess the impact of these equipment outages on process hazards, the equipment must be considered in terms of its functional role.

A process hazard has the potential to cause harm to people, property or the environment [2]. Process hazards can be identified by considering the effect of a loss of functioning equipment on the process. At a functional level, the loss of an electric motor on a centrifugal pump during a power outage may result in the loss of cooling water for a chemical reactor. If the chemical reaction is exothermic, the loss of cooling water may promote the occurrence of a thermal runaway reaction. Similarly, the loss of the electric motor on the agitator of the same reactor may contribute to a thermal runaway. A thermal runaway reaction may result in a loss of containment of the hazardous chemical.

The most common process hazards created by a loss of power are due to deviations from the normal or anticipated operating states. These hazards may include sudden changes in operating temperature, pressure, flow rate, level, and inadvertent mixing of incompatible chemicals.

Process hazards are unique to each chemical process facility, but can be identified by the consequences they have the potential to cause. These may include the overpressurization of piping or vessels, toxic exposure, thermal exposure, mechanical force, radiation, and electrical shock [2].

It is also important to consider accident scenarios in the context of the complete life cycle of a power outage: outage, shutdown sequence, recovery, startup sequence. For example, during the course of managing a power outage, a safe state for a piece of equipment during the outage may be an unsafe state for the startup sequence. Thus, a hierarchy for planning for power outages may take the form of the following:

1. Ensuring containment of hazards
2. Providing for process state/conditions monitoring
3. Establishing a priority for systems restoration

Scenario-based hazard evaluation techniques like HAZOP studies may be effective in identifying process hazards initiated by a power outage and in identifying strengths and deficiencies in systems according to this hierarchy.

Considerations during a PHA

This section will discuss some specific examples of concepts and considerations in a PHA, such as a HAZOP, to provide more guidance on appropriately planning for power outages. In many PHAs, a deviation such as "power outage" will be applied to this reactor, but such a broad approach may lead to missing potential hazardous states. There are at least four different power outage modes to consider during a hazard analysis:

1. Loss of incoming power to plant
2. Loss of onsite distribution systems
3. Loss of onsite power generation

Additionally, the transient effects of a power outage should be considered. What if there is a power surge (e.g., lightning strike) that accompanies the power outage? What is the anticipated timeline for restoring incoming utilities? Is there an ability to isolate internal distribution issues (e.g., redundant switch houses)? If the onsite power house goes offline, will the loss of electricity generation be immediately accompanied by the loss of steam or compressed air? Can the loss of a single phase of multi-phase power distribution be detected?

3. Safeguards for Power Outages

Before developing safeguards for a particular accident scenario, one should first establish the process safety objectives. The process safety objectives for managing large inventories of hazardous chemicals during a power outage typically fall into one of two strategies: containment or controlled release to a safe destination. Separately, one must determine which of three operational strategies will be selected: continued operation, selective operation, or shutdown. The process safety and operational strategies must be consistent with each other.
The process safety objectives are achieved through the application of the layers of protection [4]. The concept of layers of protection can be tailored specifically for power outages. Two key issues to consider are whether any given layer of protection requires an emergency power source, and if so, whether the demand can be handled by a stored electrical energy source or by an engine-driven generator power system. Stored electrical energy systems, such as batteries and uninterruptible power supplies, are appropriate for operating a distributed control system (DCS), providing emergency illumination, or powering a fire alarm system. More substantial power requirements, like driving electric motors for centrifugal pumps, will probably require an emergency generator set. Thus, if the operational strategy is continued operation or selective operation, an emergency generator set will probably be required. Recommended performance requirements for emergency and standby power systems are given in Standard 110 of the National Fire Protection Association (NFPA) [5].

Regardless of the process safety strategy, containment or controlled release, careful consideration must be given to the specification of valve states (open or closed) on loss of power. The following case study illustrates these considerations.

4. Exothermic Reactor Example

A generic reactor section from a P&ID is depicted in Figure 1. This reactor is used to process a thermally unstable liquid phase mixture that undergoes an exothermic reaction. The liquid phase reactants are supplied via centrifugal pump P-123A to the reactor and mixed by agitator AG-123A. Reactor temperature is controlled by circulating the reactant mixture through an external heat exchanger via centrifugal pump P-123B. Cooling water is supplied to heat exchanger HX-123B via pump PC-123B from a cooling tower. The process is monitored by an operator in a central control room via a computerized interface. The reactor is automatically temperature controlled via measurements from the temperature controller TIC and throughput is controlled automatically via the level controller LIC.

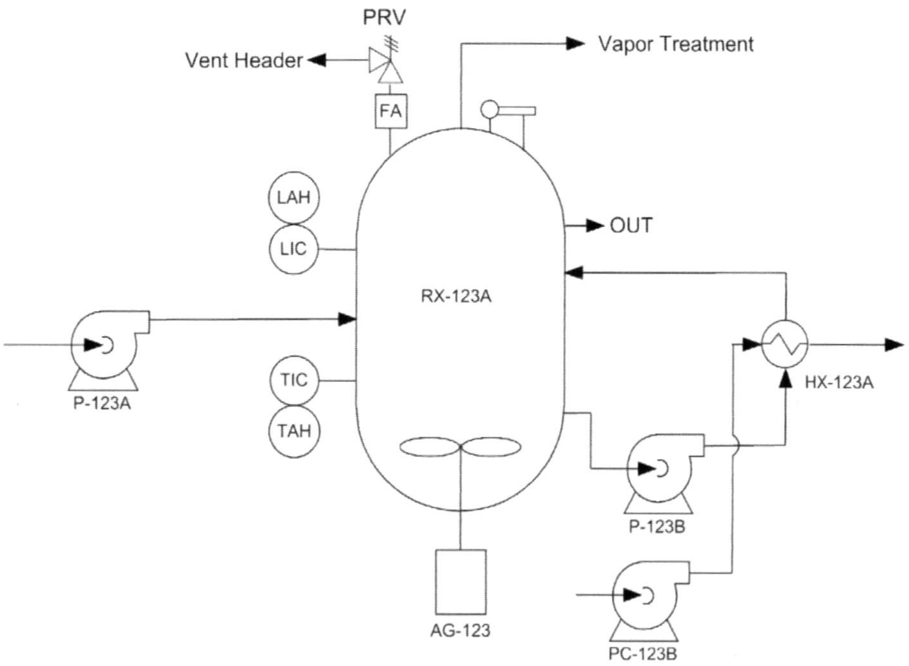

Figure 1. Example P&ID of a reactor.

Some considerations for addressing the hazards of a power outage on this reactor are described in the What If table below.

Process: Reactor RX-123A **Topic:** Power Outage				
What If?	**Hazard**	**Consequence**	**Safeguards**	**Recommendations**
1. Total loss of power	H1.1. All pumps stop	C1.1. Loss of cooling in reactor leads to thermal runaway	S1.1. Emergency dump via gravity drain upon high temperature alarm (TAH) S1.2. Emergency cooling water supply gravity flow from water tower	R1.1. Consider backup power generator for AG-123 and P-123B to complement use of emergency cooling water.
2. Loss of power to PC-123B	H2.1. Loss of cooling flow	C2.1. Same as C1.1.	S2.1. Same as S1.1. S2.2. Same as S1.2.	
3. Loss of single phase	H3.1. Damage to P-123B	C3.1 Damage not identified to pump P-123B leads to uncontrolled loss of cooling. Same as C1.1.		R3.1. Consider adding pump low amps alarm

5. Case Study: Emergency Power System Failure

This case study involves a power outage at a chlor-alkali facility. During normal operation, a gaseous waste stream containing hydrochloric acid vapor was treated in a caustic scrubber prior to discharge to the atmosphere. An electrically-powered centrifugal pump supplied caustic solution to the top of the column. The pump was connected to an emergency power system that was intended to maintain operation of the scrubber tower in the event of a power outage. The continued operation of the caustic scrubber tower was considered a necessary layer of protection during a power outage. A prior hazards analysis had determined that it may take several minutes to blow down the pressurized hydrochloric acid source. The facility operator thus considered the emergency operation of the caustic scrubber to be of critical importance; therefore, the control room operator manually tested the emergency generator set every day. This is an example of selective operation of specific equipment during a power outage with the process safety objective of providing for a safe, controlled release of a hazardous chemical.

On the day of the incident, a power outage occurred. The operator responded by activating the emergency power system and noted the positive indication on the control board that the pump was running. Several minutes later, a supervisor noticed that although the generator was running, the scrubber pump was not operating. This resulted in a reportable release of hydrochloric acid vapor.

The subsequent incident investigation found that the starter on the pump's electric motor had failed. Thus, although the emergency generator successfully operated, the motor did not start. The failure of the starter device was at first a perplexing finding. Further investigation revealed that the test frequency recommended in NFPA 110 for an emergency generator in this type of service was once monthly. By testing the emergency power system approximately 30 times more frequently, time to failure of the component was accelerated. Investigators also discovered that the red indicator lamp in the control room was a positive indication that the emergency generator was running; it was not an indication that the pump was running. As an outcome of this investigation, the facility operator adopted a monthly test frequency for the emergency power system in accordance with the NFPA guidelines. They also reconfigured the red indicator lamp so that it illuminated only when the pump was running.

6. Recommendations for developing a Power Outage Safeguard Strategy

Drawing from our hazard analysis and accident investigation experience, we offer the following recommendations for developing a power outage safeguard strategy.

6.1 Determine the "safe state" for the process unit during a power outage.

In a power outage of prolonged duration, if you select containment as your process safety objective, the process unit will eventually attain a state of stable thermodynamic equilibrium with respect to its ambient surroundings.

Is it safe to allow the process unit to approach that stable state in an uncontrolled fashion? If not, what are the hazards of an uncontrolled process unit: mechanical, thermal, or chemical? Should the operator have the ability to monitor the process unit's progress towards a stable state? What instrumentation should remain operative during the outage?

6.2 Specify how the layers of protection will operate during the emergency shutdown to bring the process unit to a safe state.

Each layer of protection should contribute towards achieving your process safety objective. By definition, each layer of protection must remain independent of other layers, but must also work in harmony with them. For example, if your goal is to achieve a controlled release to a safe destination, you may choose to dump the contents of a chemical reactor into a quench tank. Installing a pressure relief valve (PRV) that discharges into the atmosphere is not consistent with that process safety goal. The PRV should discharge into a header that leads to the quench tank.

6.3 Ensure the reliability of the layers of protection.

Clearly specify the functional requirements for each layer of protection. Consult industry standards and guidelines to identify, document, and implement generally recognized and accepted good engineering practices. Formulate a sound basis for inspection, testing, and preventative maintenance on critical components of each safeguard. Audit the implementation of these efforts and improve your mechanical integrity program as you gain new information.

6.4 Plan for a safe restart of operations.

Before the power can be restored and operations restarted, it is essential to verify that the process unit can be restarted safely. A pre-startup safety review should evaluate the equipment, chemicals, people, and procedures that will be required to resume operation of the process unit.

6.5 Audit the success of emergency operations.

Accident investigation is an important source of feedback for identifying opportunities for improvement in the process safety management system. The recommendation to audit a successful emergency shutdown may seem unusual. But a power outage in a chemical plant is a highly hazardous situation. The successful management of the event is an important validation of your layers of protection and should be documented. If for no other reason, imagine how much easier it will be to revalidate the process hazards analysis for that unit when you have documentation that the layers of protection functioned as intended.

7. Conclusion

Emergency operations safeguards and layers of protection should be evaluated regularly. The original emergency operations procedure may not be adequate when substantial changes to procedure or equipment occur. When not specifically stated in the reference literature [5], the appropriate frequency of evaluation is dependent upon industry standards and good engineering practice.

A HAZOP and a layer of protection analysis, coupled with the adoption of best practices from industrial safety standards, can aid in designing a strategy for the emergency power system. The recommendations given in this report can aid in the selection and proper implementation of emergency operations safeguards during power outages.

8. References

[1] National Response Center, Download NRC Data, calendar years 2004-2008, http://www.nrc.uscg.mil/download.html (accessed 12/31/2009; searching on the term, "power outage").

[2] Center for Chemical Process Safety, *Guidelines for Hazard Evaluation Procedures, Third Edition*, American Institute of Chemical Engineers, New York, pp. 23-28, 2008.

[3] Center for Chemical Process Safety, *Guidelines for Safe Process Operations and Maintenance*, American Institute of Chemical Engineers, New York, pp. 46-47, 1995.

[4] Center for Chemical Process Safety, *Layer of Protection Analysis: Simplified Process Risk Assessment*, American Institute of Chemical Engineers, New York, pp. 12 and 75-88, 2001.

[5] *NFPA 110 Standard for Emergency and Standby Power Systems*, National Fire Protection Association, Quincy, MA, 2005.

Near-Miss Management to develop Dynamic Leading Indicators to Predict Incidents

Ankur Pariyani (Presenter)
Department of Chemical and Biomolecular Engineering, University of Pennsylvania, Philadelphia, PA 19104 USA (Tel: 215-898-3536; e-mail: pariyani@seas.upenn.edu)

Ulku G. Oktem
Risk Management and Decision Processes Center, Wharton School, Philadelphia, PA 19104 USA (Tel: 215-573-7704; e-mail: oktem@wharton.upenn.edu)

Warren D. Seider
Department of Chemical and Biomolecular Engineering, University of Pennsylvania, Philadelphia, PA 19104 USA (Tel: 215-898-7953; e-mail: seider@seas.upenn.edu)

Prepared for Presentation at
American Institute of Chemical Engineers
2010 Spring Meeting
6th Global Congress on Process Safety
San Antonio, Texas
March 22-24, 2010

UNPUBLISHED

Near-Miss Management to develop Dynamic Leading Indicators to Predict Incidents

Ankur Pariyani (Presenter)

Department of Chemical and Biomolecular Engineering, University of Pennsylvania, Philadelphia, PA 19104 USA (Tel: 215-898-3536; e-mail: pariyani@seas.upenn.edu)

Ulku G. Oktem

Risk Management and Decision Processes Center, Wharton School, Philadelphia, PA 19104 USA (Tel: 215-573-7704; e-mail: oktem@wharton.upenn.edu)

Warren D. Seider

Department of Chemical and Biomolecular Engineering, University of Pennsylvania, Philadelphia, PA 19104 USA (Tel: 215-898-7953; e-mail: seider@seas.upenn.edu)

Keywords: Near-Misses, Abnormal Events, Leading Indicators, Warning System, Incidents, Process Safety, Petroleum Refinery

Abstract

A new methodology involving near-miss utilization and management for chemical plants is introduced to help identify an escalation in the probability of the occurrence of incidents – permitting the system to alert the operator(s) of a potential major problem likely to occur in the near future. Also, the system can detect the onset and/or presence of inherent faults, or special-causes, likely to lead eventually to incidents. Using this new methodology, the operators can be alerted up to several hours before potential undesirable events are likely to occur. Thereafter, as the special-causes progress, with the potential for trips and possibly accidents increasing, the frequency of alerts increases. To demonstrate this technique, graphs of the dynamic frequency of abnormal events for the constituent (high-, medium-, and low-priority) variables associated with a fluid-catalytic-cracking unit (FCCU) at a major petroleum refinery, are presented. Then, indicators are identified that would have projected trips that occurred during the selected time period.

1. Introduction

Every low-probability, high-consequence incident at any chemical plant is preceded by many high-probability, low-consequence events, which are recognized as *near-misses* (Pariyani *et al.*, 2009). An ideal warning system for a chemical plant will account for all these near-misses and develop leading indicators to notify the operators of undesirable incidents, likely to happen. In particular, such knowledge becomes highly desirable for unmanned plants. This paper discusses initial techniques using distributed control system (DCS) and emergency shutdown (ESD) system databases to generate alert signals to prevent incidents. Results are presented from the analysis of the DCS and ESD databases of a fluid-catalytic-cracking unit (FCCU) at a major petroleum refinery, which processes more than 250,000 barrels of oil daily.

2. Methodology

Pariyani *et al.*, 2009 introduce *operating-belt zones* to define *abnormal events* experienced by process and quality variables of a plant. Figure 1 shows a control chart for a typical variable and its high/low, high-high/low-low, and ESD alarms. The chart is divided into four belt zones, beginning with its green-belt zone (normal operation), during which the variable lies within acceptable limits. When the variable moves beyond its limits, into its yellow-belt zones, high/low alarms are triggered. When it moves beyond the limits of its yellow-belt zones, into its orange-belt zones, high-high/low-low alarms are triggered. The border between its orange- and red-belt zones is the threshold limit that triggers its ESD system. *Abnormal events* begin when process or quality variables move from their green-belt zones to their yellow-, orange- or red-belt zones – with these colors chosen to gauge the severity of the abnormal event. Clearly, these departures are precursors to near-miss consequences or accidents, when the regulatory and protection systems (that is, the DCS, human operators, and the ESD system) fail to maintain normal operation. Consequently, herein, abnormal events, for variables that return to their green-belt zones, are recognized as *near-misses*, which could have propagated into incidents. As a result, vast amounts of near-miss information becomes available for tracking and dynamic risk assessment.

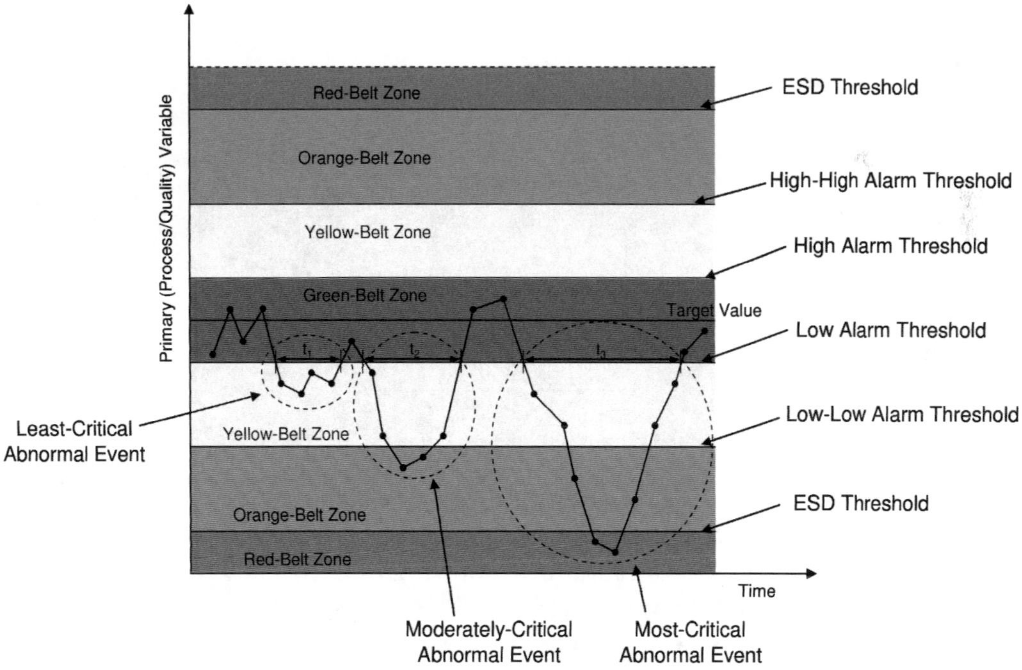

Figure 1. Control chart for a process/quality variable showing various operating belt zones and alarm thresholds

Depending upon their criticality, abnormal events are classified in three categories, as shown in Figure 1:

i. *Least-critical abnormal events*: Abnormal events that cross the high/low alarm thresholds, but do not cross the high-high/low-low alarm thresholds.

ii. *Moderately-critical abnormal events*: Abnormal events that cross the high-high/low-low alarm thresholds, but do not cross the ESD thresholds.

iii. *Most-critical abnormal events*: Abnormal events that cross the ESD thresholds.

To assist the operators during upset situations, alarms are prioritized in accordance with their actions. Typically, 15-25% of the alarms are designated as high-priority alarms, 20-30% as medium-priority alarms, and 50-65% as low-priority alarms. Accordingly, the variables associated with the low-, medium-, and high-priority alarms are referred to as low-, medium-, and high-priority variables.

Figure 2 shows a convenient representation of the dynamic frequency of abnormal events in time. Each abnormal event is represented by a unit pulse of width equal to its duration (or recovery-time). Then, as shown, the frequency of abnormal events at any instant is the sum of the unit pulses associated with the abnormal events.

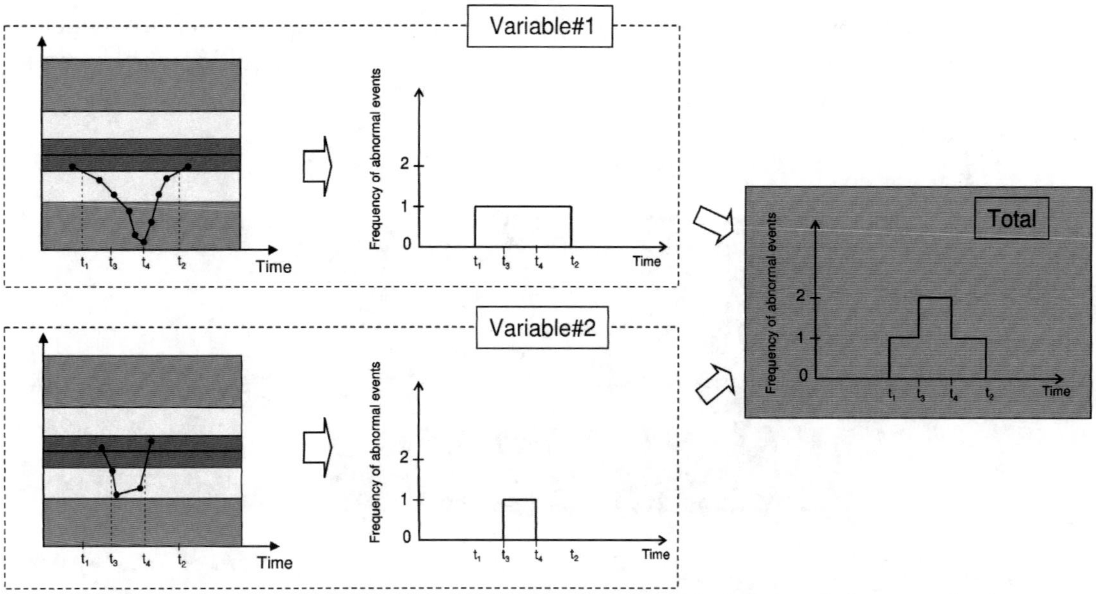

Figure 2. Representation of the dynamic frequency of abnormal events in time

Turning to the FCCU data, Figures 3a,b present the dynamic frequency of abnormal events for a selected study period (14 hours), with Figure 3b showing the expanded region near the trip. Note that the dynamic frequency of abnormal events is the number of variables out of their green-belt zones at any instant. During normal operations, the frequency of abnormal events remains low and more or less constant – for example, as shown in Figure 3b, before the trip, about 13 (total) variables were out of bounds, but just a few minutes later, a high frequency of variables (~40 in total) were out of bounds, creating a flood of alarms for the operators. According to the DCS and ESD logs, two most-critical abnormal events occurred during this time period – one of which resulted in a trip (shown as a vertical line).

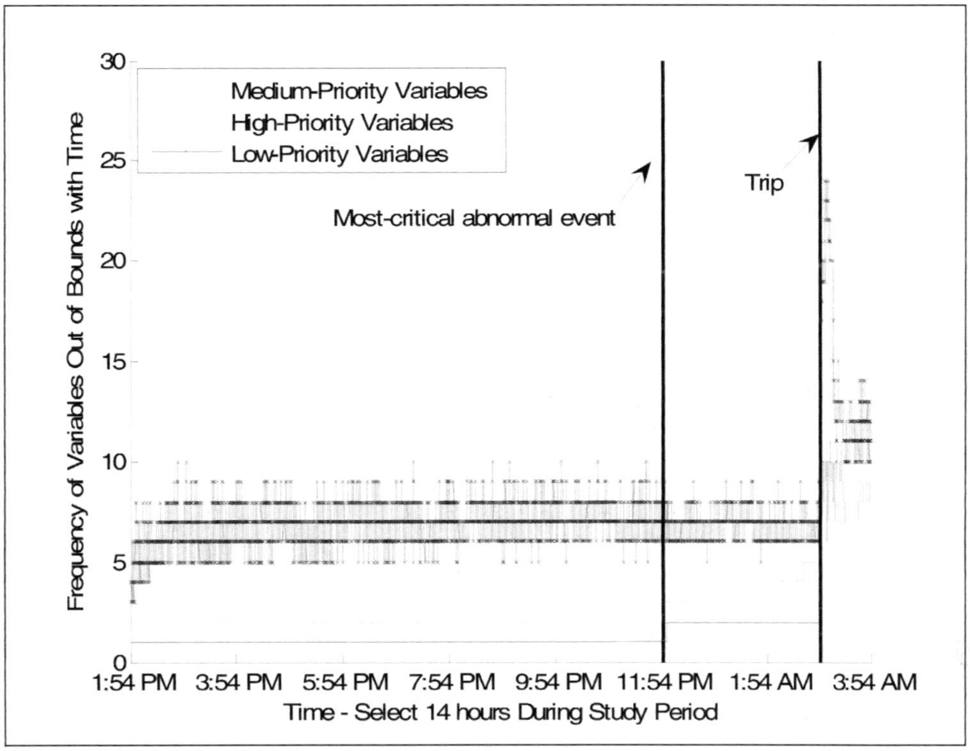

Figure 3a. Dynamic frequency of abnormal events for a select period of time

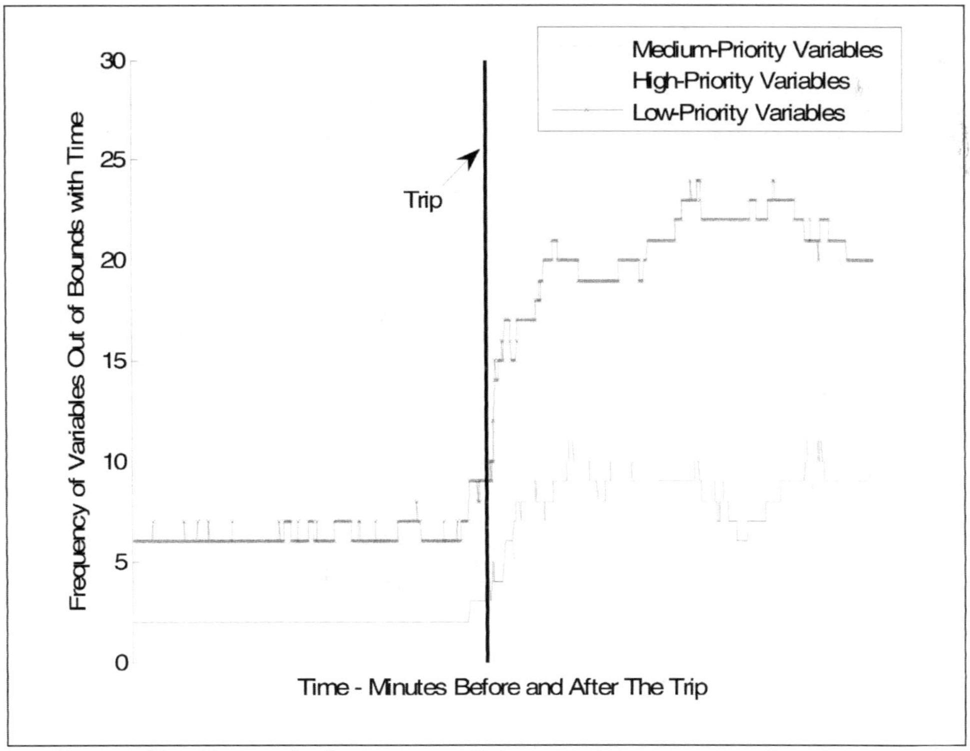

Figure 3b. Dynamic frequency of abnormal events near the trip event

The methodology introduced herein has two main purposes. First, it seeks to detect the presence of faults (or special-causes). This information reported at regular times can notify the operators of problems at their onsets, permitting the operators to take precautionary actions before most-critical abnormal events or trips occur.

The second purpose is to send *alert signals* to the operators about most-critical abnormal events or trips. The lead-times (that is, the time between the alert-notification and the occurrence of the undesirable event) for alerts can vary from hours to seconds depending upon the severity and progression of a fault (or special-cause). In these studies, the lead times range from several hours to just 7 seconds. These lead times can: (a) help the operators prevent the occurrences of undesirable events by taking appropriate actions, and (b) better prepare the operators for tackling the consequences, should an undesirable event occur.

Herein, to flag alerts, the patterns followed by the process variables in the presence of special-causes are used to identify leading indicators for the most-critical abnormal events and trips – with the frequency of abnormal events for high-, medium-, and low-priority variables tracked automatically. Alerts are flagged when a criterion (specific to a process), a function of frequency of abnormal events associated with the variables, is satisfied. Several criteria were tested for the study period data – for example, alerts are flagged when there is an *absolute increase* of at least four or five abnormal events associated with all variables synchronized with at least one abnormal event associated with each of the high-, medium-, low-priority variables. As shown next, for the data examined, the results are presented for one of the criteria that gave very few false-negative alerts and an optimal number of alerts (~5-10) before any incident.

3. Results

The premise of this paper is that had the leading indicators methodology presented herein been implemented during the study period, multiple alerts would have been proposed before the occurrence of any of the most-critical abnormal events and trips. Figure 4 presents several proposed alerts (shown with *blue* vertical lines, labeled alphabetically, A-E) for the select period, using the criterion. For the first most-critical abnormal event, four alerts, A, B, C, and D, would have been dispatched 7.7 hr, 3.9 hr, 51 sec, and 8 sec before its occurrence. Collectively, these alerts would have suggested faults/special-causes to help the operators better understand the extent of the abnormal events. Figure 5 shows the last two of these alerts, C and D.

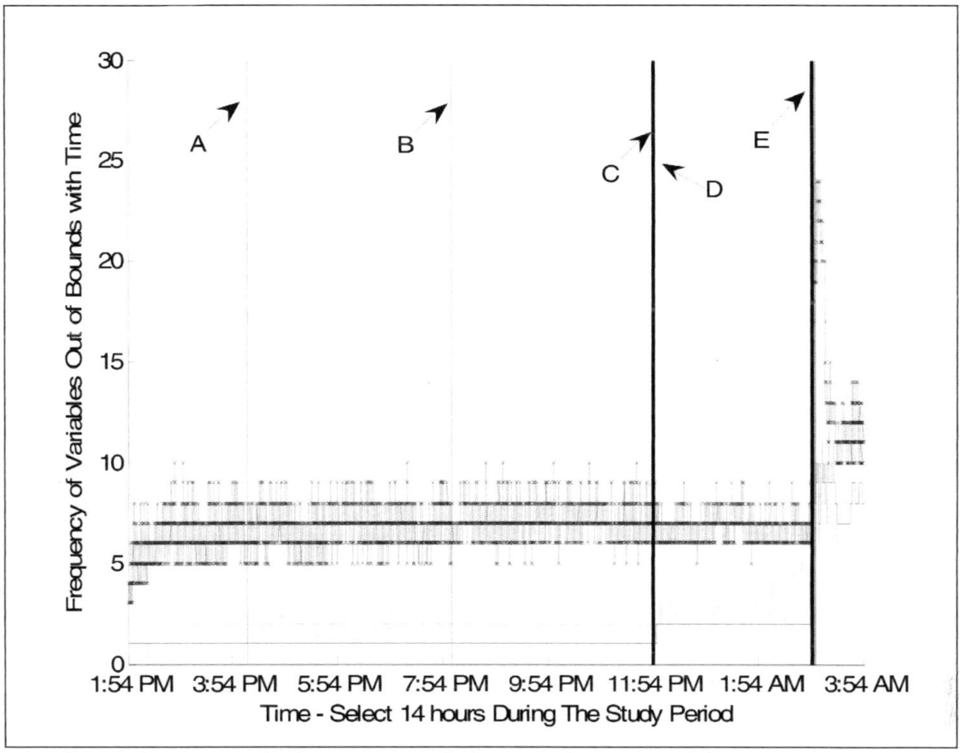

Figure 4. Dynamic frequency of abnormal events with proposed alerts

The second most-critical abnormal event occurred nearly 2 hours later, resulting in a trip. An alert, E, would have been dispatched 8 seconds before that abnormal event (and 38 seconds before the trip event). Additional alerts would have been dispatched after the trip event.

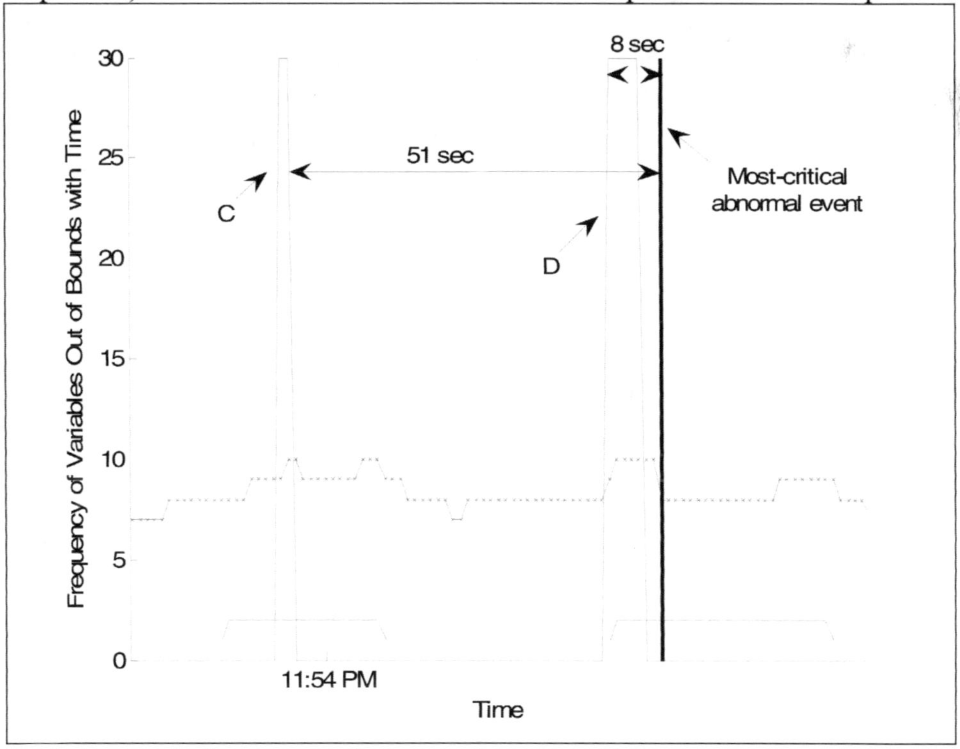

Figure 5. Dynamic frequency of abnormal events with proposed alerts and lead times

Thus, had these alerts been in place, the operators would have had opportunities to make corrections and prevent most-critical abnormal events and trips. Also, the successes of these alerts can reduce the extent of alarm flooding – with fewer alarms for low-priority variables needed, accompanied by a reduction in the frequency of alarms. This should reduce the extent of confusion, enabling operators to prioritize their actions, providing better responses to upsets. Next, a longer representative period of time is selected during which 23 most-critical abnormal events were experienced, one of which resulted in a trip. For this period, a total of 125 alerts would have been proposed using the criterion above, with 28 alerts having lead times less than 3 hr, 10 between 3-6 hr, 18 between 6-12 hr, 19 between 12-24 hr, 30 between 24- 48 hr, and 20 greater than 48 hr. The alerts having 3-hr lead times or less would have been dispatched for all but one most-critical abnormal event, for which only one alert with lead time from 3-6 hr would have been dispatched. Usually, multiple alerts are proposed over these time-gaps after the early detection of special-cause(s), eventually leading to most-critical abnormal events. Clearly, all of the proposed alerts would have deserved careful consideration by the operators as they sought to prevent most-critical abnormal events and trips.

Other criteria for flagging alerts are possible – some of which reduce the number of false-negatives but increase the number of alerts. Note that there are no false-positives associated with this methodology as all of the alerts are important and signify the presence of special-causes likely to lead eventually to most-critical abnormal events and trips.

4. Concluding Remarks

In summary, the proposed methodology has the potential to detect nearly all of the most-critical abnormal events and trips, forewarning operators about the onset and progress of special-causes by utilizing the near-miss data to flag alerts having lead times from several hours to a few seconds. Based on the results for the study period, these alerts have been shown to provide extra protection, helping operators detect upset situations and take appropriate actions well in advance. With this methodology, less than 3% false-negatives (most-critical abnormal events and trips which were not detected) were identified for this period.

5. References

[1] Pariyani, A., W. D. Seider, U. G. Oktem, and M. Soroush, "Incidents Investigation and Dynamic Analysis of Large Databases in Chemical Plants: An FCCU Case Study," submitted to *Ind. Eng. Chem. Res.*, 2009.

Learning the Lessons from Buncefield

Ian Travers
Head of Chemical Industries Strategy Unit
Health and Safety Executive

Introduction
The fire and explosion at the Buncefield fuel storage depot, Hemel Hempstead, UK in December 2005 was the largest peacetime explosion in Europe. Fortunately no one was killed and the number a people injured was small mainly because the incident happened in the early hours of Sunday morning. However, the incident has had major repercussions for the UK storage and petrochemical industry.

The Incident
The Buncefield depot is a large tank farm 3 miles (about 4.8 km) from the town centre of Hemel Hempstead, Hertfordshire. It was built and began operating in 1968. Fuel was transported to the complex through three pipelines taking products from three refineries and distributing them to the South East of England.

Early on Sunday 11 December 2005, a series of explosions and subsequent fire destroyed large parts of the Buncefield oil storage and transfer depot causing widespread damage to neighbouring properties.

Late on Saturday 10 December 2005 a pipeline delivery of unleaded gasoline started to arrive at Tank 912. At about 05:30 on 11 December the tank overfilled. The safety systems in place to shut off the supply of gasoline to the tank to prevent overfilling failed to operate. Gasoline cascaded down the side of the tank, collecting at first in the bund. As overfilling continued, the vapour cloud (formed by the mixture of petrol and air) flowed over the bund wall. The cloud dispersed and flowed west off site towards the nearby Industrial Estate. Up to 300 tonnes of gasoline escaped from the tank, about 10% of which turned to vapour that mixed with the cold air eventually reaching concentrations capable of supporting combustion. The area covered by the vapour cloud was estimated to be around 120,000m2 and the average height of the cloud was around 2m giving a volume of 240,000m3.

The main explosion took place at 6 am and was of massive proportions. It was followed by a large fire that engulfed 23 large fuel storage tanks covering a high proportion of the Buncefield site. The incident injured 43 people. Fortunately, no one was seriously hurt and there were no fatalities. Nevertheless, there was significant damage to both commercial and residential properties near the Buncefield site. About 2000 people had to be evacuated from their homes and sections of the M1 motorway were closed. The fire burned for five days, destroying most of the site and emitting a large plume of smoke into the atmosphere that dispersed over southern England and beyond. The fire and the efforts taken to fight it resulted in millions of litres of stored fuel, firewater and foam escaping from the damaged containment systems. The Fire service applied 68 million litres of water and 786,000 litres of foam concentrate to fight the fire. Some of this foam contained perfluorooctane sulphonate (PFOS) which is persistent, bio-accumulative and toxic. About 33 million litres of fire water run-off were recovered from the terminal and taken off-site in road tankers for treatment and disposal. The operation to remove fuel and firewater from the site and its surroundings continued for many weeks after the fire was extinguished.

The accident has been reported to the European Union because the area of contaminated groundwater exceeds 1 hectare. The nearest public supply borehole is about 3 kilometres from the site and it has been shut down since the accident. The clean-up is expensive and will take many years to complete.

An important aspect of the incident was the severity of the explosion, which would not have been anticipated in any major hazard assessment of the oil storage depot before the incident.

There is a weight of evidence to suggest that the source of ignition was in the emergency pump house. Overpressures within the area of the cloud were found to be consistently high. From a combination of damage assessment and comparative testing and analysis it was concluded that the overpressure within the cloud was generally greater than 200kPa; the maximum overpressure was probably much higher. These high levels of overpressure were seen in all areas; there was no distinction between different terrain (car parks, tank farms, open grassland and belts of trees). Overpressure diminished rapidly with distance away from the edge of the cloud; evidence suggests overpressures in the region of 5-10 kPa within ~150m.

The Investigation

The UK Health and Safety Commission established The Major Incident Investigation Board (MIIB), independently chaired by the Rt. Hon Lord Newton of Braintree. The MIIB carried out an extensive investigation of the incident and published eight reports, four setting out findings and recommendations. Those four reports are entitled:

- Design and operation of fuel storage sites;
- Emergency preparedness for, response to and recovery from incidents;
- The explosion mechanism advisory group report; and
- Land use planning and the control of societal risk around major hazard sites.

The MIIB's complete findings were set out in its final report[1] published in December 2008.

Working with Industry to Improve Standards

A joint industry and regulators group was established to translate the recommendations of the MIIB investigation into improved standards of safety and environmental protection at fuel storage sites and to drive forward high standards in process safety leadership. Initially the Buncefield Standards Task Group, BSTG, addressed the immediate lessons from the incident then subsequently, the Process Safety Leadership Group, PSLG[2], completed the work.

The UK regulatory regime for major hazards is the Control of Major Accident Hazard Regulations, 1999, COMAH. These regulations implement the EC SEVESO II Directive[3] in the UK. The UK legislative framework is goal setting, based upon risk assessment that requires duty holders to reduce risks so far as is reasonably practicable. This involves a striking a balance between the benefit of risk reduction set against the cost of implementing the measure. At its core the COMAH Regulations require operators of major hazard establishments to take all measures necessary to prevent a major accident and limit the consequences to people and the environment. In deciding on whether this requirement is being met the law takes account of current good practice within the industry. The approach of the PSLG was therefore to clarify standards of good practice for large scale fuel storage sites as this would then mean that duty holders would be expected to meet this good practice at both new and existing sites within the context of COMAH.

Establishing good practice and what measures are reasonably practicable has to be set against a new broader understanding that a large scale loss of gasoline could give rise, in circumstances that are not clearly understood, to a large explosion. Arriving at suitable solutions involved a high degree of commitment from both industry and the regulators (safety and environmental). This is to not only to identify the standards involved but also to ensure that all relevant fuel storage sites have a programme in place to comply with the PSLG's guidance within an agreed timescale and to monitor and publicly report on progress implementing any improvements needed. Industry and the regulators saw this as part of the challenge of improving leadership in the major hazard sector based on the lessons of Texas City and Buncefield. Applying improved standards to existing facilities rather than restricting them to new installations was viewed as the only way to make any real impact within a mature industry sector where few new storage facilities were likely to be built.

The PSLG Final Report: Safety and Environmental Standards for Fuel Storage Sites[4] was published in December 2009. It covers six key areas:

- Systematic assessment of safety integrity level requirements;
- Protecting against loss of primary containment using high integrity systems;
- Engineering against escalation of loss of primary containment;
- Engineering against loss of secondary and tertiary containment;
- Operating with high reliability organisations; and
- Delivering high performance through culture and leadership.

The Report calls for high standards of overfill protection, secondary and tertiary containment at fuel storage sites and appropriate mitigation measures based on the scenario of an explosion followed by multiple tank fires. As well improved physical standards the report also sets out a clear expectation for improved process safety leadership.

The focus in terms of primary containment is on improved standards for gasoline storage. However, the report also contains a methodology which can be used to determine whether other flammable liquids could equally give rise to a large scale vapour cloud in the event of a tank overfill. Twelve such substances were identified and giving rise to a similar risk. These are given in Annex 1.

Systematic assessment of Safety Integrity Levels (SIL)
The report requires that the overall systems for tank filling control should be of high integrity, with sufficient independence to ensure timely and safe shutdown to prevent tank overflow. It recommends the adoption of BS EN 61511 "Functional safety, safety instrumented systems for the process industry sector"[5]. The report also recommends common methodology to determine safety integrity level (SIL) requirements for overfill prevention systems for fuel storage tanks. The methodology takes account of:

- the existence of nearby sensitive resources or populations;
- the nature and intensity of depot operations;
- realistic reliability expectations for tank gauging systems; and
- the extent/rigour of operator monitoring.

Overfill protection systems (including instrumentation, devices, alarm annunciators, valves and components comprising the shutdown system) should be assessed using BS EN 61511, which sets a minimum performance for SILs. This includes the following considerations:

- design, installation, operation, maintenance and testing of equipment;
- management systems;
- redundancy, diversity, independence and separation;
- fail safe, proof test coverage/frequency; and
- consideration of common causes of failures.

The guidance recommends the use of Layers of Protection Analysis (LOPA) as the means for undertaking a risk assessment for atmospheric fuel storage tanks within BS EN 61511 to determine the appropriate SIL. Detailed guidance on the application of LOPA to fuel storage tanks is given in the report.

Overfill Protection for gasoline storage tanks

Given the potential of gasoline (in the event of an overfill) to give rise to an explosion the report calls for gasoline tanks filled at a rate of more than $100m^3$/hr to meet SIL 1 as a minimum (as defined in BS EN 61511-1). The overfill protection system for such tanks should preferably be automatic and should be physically and electrically separate from the tank gauging system. Automatic overfill prevention may include, but not be restricted to, measures such as automatic shutdown of the supply line or automatic diversion of the flow to another tank.

The report allows for alternative measures giving a equal level of protection where the installation of such an independent automatic overfill prevention system at an existing tank is demonstrated to give rise to other more serious safety or environmental risks elsewhere on site. Such alternative measures should include an overfill prevention system to at least BS EN 61511-1 SIL 1, combined with other measures to provide high integrity and reliability. Those that include an operator(s) as part of the overfill prevention system should demonstrate that the reliability and availability of the operator(s) can be adequately supported to undertake the necessary control actions to prevent an overfill without compromising the outcome. Operator involvement should be properly managed, monitored, audited and reviewed on an ongoing basis. It is unlikely that an operator can be included in an overfill prevention system rated above SIL 1 as defined in BS EN 61511-1.

The report contains supplementary guidance on Layer of Protection Analysis, LOPA, and automatic overfill protection systems for gasoline tanks.

Defining The Tank Capacity

The report requires tank capacities, fill levels and safety margins to be clearly determined for all fuel storage tanks. To prevent an overflow, tanks should have headspace margins that enable the filling line to be closed off in time. The set points of high level trips and alarms should allow sufficient time for the action to be taken to deal with the developing situation. A vital element is a clear definition of the maximum capacity of the vessel. This is the maximum level consistent with avoiding loss of containment (overfilling or overflow) or damage to the tank structure (eg due to collision between an internal floating roof and other structures within the tank, or for some fluids, overstressing due to hydrostatic loading).

Having established the overfill level (maximum capacity), it is then necessary to specify a level below this that will allow time for any action necessary to prevent the maximum from being reached/exceeded. This is termed the 'tank rated capacity', which will be lower than the actual physical maximum. The required separation between the maximum capacity and the tank rated capacity is a function of the time needed to detect and respond to an unintended increase in level beyond the tank rated capacity. The response in this case may require the use of alternative controls, eg manual valves, which are less accessible or otherwise require longer time to operate than the normal method of isolation.

These requirements may seem clear cut. However, in the UK prior to Buncefield specifying clear capacities and safety margins and a 'normal fill' level was not a universal practice. In a survey of 108 major hazard storage sites conducted following Buncefield it was found that the safety margin between the high-high alarm level and the and the overflow point varied between a few minutes to many days at normal fill rates with, in some cases, little correlation between these and the time needed to undertake an emergency shut down. Clearly safety margins were being squeezed to gain commercial benefit from the additional tank capacity.

Engineering against escalation of loss of primary containment

Failure of an overfill protection system places reliance on the tank to avoid the uncontrolled loss of primary containment of hazardous substances. The adoption of appropriate design standards should ensure tank integrity and suitable overflow and venting mechanisms. Throughout the life of the tank, integrity of primary containment should be maintained through a process of periodic inspection, maintenance and repair.

New tanks should be designed to BS EN 14015[6] or API 650[7] (or equivalent) as these offer up-to-date standards providing in-depth guidance on design and construction elements for vertical cylindrical atmospheric storage tanks. The report recommends that new tanks should be of single-bottom design, which can be supported by suitable inspection arrangements providing the optimum configuration for ensuring continuing integrity. This will facilitate full non-destructive examination of floor-plate welds. Consideration should be given to the overflow route from vent to bund to ensure that, within the constraints of the design code, obstacles in the overflow route are minimised.

Tanks should either be of 'frangible roof' construction, or should be equipped with an emergency vent of adequate area to prevent over-pressure under accidental relief conditions, which exclude internal explosions. EMMUA 180 'Frangible roof joints for fixed roof storage tanks: Guide for designers and users'[8]. provides a suitable standard and emergency vents should comply with an appropriate design standard (API 2000 or equivalent[9]).

EEMUA 159[10] and API 653[11] represent relevant good practice and should form the basis of minimum industry standards for tank integrity management and repair to prevent loss of primary containment. EEMUA 183 *Guide for the Prevention of Bottom Leakage*[12], should be applied particularly with regard to the maintenance and repair aspects for tanks with a double bottom configuration.

The scope of inspections, detailed in EEMUA 159 and API 653, acknowledges the typical tank failure modes including corrosion, settlement and structural integrity and provides good guidance for early detection and measurement of symptoms that could lead to failure. A written scheme of examination is required for internal/out-of-service inspections. EEMUA 159 (Appendix B2) provides an example of such a checklist. EEMUA 159 and API 653 provide guidance on inspection intervals by either fixed periodicity or by a risk-based methodology. The tables of fixed inspection intervals within this guidance can be used where there is little or uncertain tank history available. A risk-based inspection (RBI) approach allows the use of actual corrosion rates and performance data to influence the most appropriate inspection interval.

Engineering against loss of secondary and tertiary containment
While priority should be given to preventing a loss of primary containment, adequate secondary and tertiary containment remains necessary for environmental protection and safety of people in the event of a loss of primary containment of hazardous substances. The failure of secondary and tertiary containment at Buncefield contributed significantly to the failure to prevent a major accident to the environment

Tank bunds (dykes) should be impermeable and bunds should have fire resistant structural integrity, joints and pipework penetrations. This covers the preparation of the tank base and foundation plus the selection of lining systems; concrete, earth or polymeric or polymeric and mineral composites. It is important that protection from fire is included in risk assessment for selecting different types of lining systems.

Following the Buncefield incident it was recognised that the addition of steel plates to cover the inside faces of movement bund joints provided enhanced fire resistance to existing joints. Improvements to existing bunds containing gasoline tanks should be made by replacing existing sealants with fire-resistant versions and, in addition, fitting steel cover plates where physically possible to fit. This proposal of the combination of cover plate and fire-resistance sealant is recommended as a good practice retrofit solution for existing installations, where routine inspection of the sealant is carried out. New joints should be installed with metal water stops and fireproof joints and where practicable, existing joints should be upgraded to provide water stops and/or fireproof joints.

Bunds should have no pipework that penetrates through the bund floor; no pipework that penetrates through the bund walls as far as reasonably practicable. Otherwise, pipework penetrations should be provided with adequate sealing and support. Bunds should have sufficient capacity to allow for tank failure and firewater management. This will normally be a minimum capacity of either 110% of the tank rated capacity (TRC) of the largest tank or 25% of the total tank rated capacities (TRC) of all the tanks within the bund whichever is the greater. A bund should be capable of withstanding the full hydrostatic head of liquid that may arise from the loss of primary containment.

Firewater management and control measures
Well-planned and organised emergency response measures are likely to significantly reduce the potential duration and extent of fire scenarios, and so reduce firewater volumes requiring containment and management. Site-specific planning of firewater management and control measures should be undertaken with active participation of the local Fire and Rescue Service, and should include consideration of:

- bund design factors such as firewater removal pipework, aqueous layer controlled overflow to remote secondary or tertiary containment (for immiscible flammable hydrocarbons);
- recommended firewater/foam additive application rates and firewater flows and volumes at worst case credible scenarios (including severe pool fire or multiple tank / multiple bund fire); and
- controlled-burn options appraisal, and pre-planning/media implications.

Tertiary containment
Tertiary containment minimises the consequences of a failure in the primary and secondary containment systems by providing an additional barrier preventing the uncontrolled spread of hazardous liquid. Tertiary containment is achieved by means external to and independent of the primary and secondary containment systems, such as site drainage and sumps, diversion tanks, impervious liners and/or flexible booms. Tertiary containment will be utilised when there is an event that causes the loss of containment (for example bund joint failure or firewater overflowing from a bund during a prolonged tank fire), and is intended to ensure that loss of control of hazardous materials does not result from such an event.

Sites operators should assess the requirement for tertiary containment, on the basis of environmental risk, and to make site action plans for improvement. Provision of tertiary containment should also take into account safety aspects – for example the flows and accumulations of hazardous liquids on and around a site.

The measures on secondary and tertiary containment are complementary to the Competent Authority Containment Policy released April 2008 [13]

Operating with high reliability organisations
The need for high reliability organisations follows from the recommendations relating to technological improvements in hardware. Such improvements are vital in improving process safety and environmental protection, but achieving their full benefit depends on human and organisational factors such as the roles of operators, supervisors and managers.

A high reliability organisation has been defined as one that operates relatively error free over a long period of time. Two key attributes of high reliability organisations are that they:

- have a chronic sense of unease, ie they lack any sense of complacency. For example, they do not assume that because they have not had an incident for ten years, one won't happen imminently; and
- make strong responses to weak signals, ie they set their threshold for intervening very low. If something doesn't seem right, they are very likely to stop operations and investigate. This means they accept a much higher level of 'false alarms' than is common in the process industries.

The following factors should be addressed to achieve a high reliability organisation:

- Clear understanding and definition of roles and responsibilities, and assurance of competence in those roles;
- Effective control room design and ergonomics, as well as alarm systems to allow front-line staff, particularly control room operators, to reliably detect, diagnose, and respond to potential incidents;
- Appropriate staffing, shift work arrangements and working conditions to prevent, control and mitigate major accident hazards;

- Setting and implementing a standard for effective communication at shift and crew change handover; and
- Effective management of change, including organisational change as well as changes to plant and processes.

Delivering high performance through culture and leadership
Industry leaders have a critical role to play in delivering high performance in process safety management. Recent incidents at Buncefield and Texas City have shown that a culture of process safety should be actively developed, grown and championed from the top of an organisation. Industry should demonstrate a commitment to process safety leadership, and a willingness to promote the process safety agenda at all levels within an organisation, and externally with other stakeholders.

Major incidents continue to happen at an unacceptable level and many organisations seem incapable or unwilling to learn the lessons form major incidents. Why does this continue to be the case? Nothing stays still in process safety and this is even more important in today's world with increased technological know-how. Whilst process automation and management systems have brought great benefits to industry, they can also create a false sense of security at all levels. Sometimes people running the plants can start believing that everything is being controlled, the process is steady, that incidents don't happen. For much of the time that may be true. However, when unusual signals come in (alarms for example) how well are these responded to? People need to be trained to deal with the rare occurrence, and how to analyse what might be wrong.

Senior managers, CEOs and Directors of major hazard companies need to be competent in process safety and to be able to make business decisions in the full knowledge of the impact on control of major hazard risks. They need to develop an acute sense of vulnerability knowing that the systems and controls they have commissioned, paid for and put in place will deteriorate with catastrophic consequences without constant vigilance based on real-time information about process safety performance. Process safety can never be a completed task, it requires ongoing day to day attention. Managers and operators alike in major hazard companies can never have a process safety vacation – not even one day or part of a day.

To underscore the importance of culture and leadership in the delivery of high reliability organisations, the PSLG published Principles of Process Safety Leadership (Annex 2). Over the next year or so operators of major hazard companies within the UK will be expected to demonstrate that they have adopted these Principles in the day to day operation of their business.

Lessons from Buncefield (and Texas City)
In conclusion the following key lessons arise from major incidents such as Buncefield:

- Simple failures can, and often do. cause major disasters;
- Once a large loss of containment occurs it is almost impossible to control the consequence. The outcome will usually be down to chance. Therefore, high standards of primary containment of hazardous substances are essential at all times;
- One company's failure can impact on everyone in that sector. Society and the regulators representing their interests will inevitably demand a universal response across the whole sector. The major hazard sector is now global and applying the lessons from a major incident will often demand a global response;
- Major hazard businesses need to develop a sense of vulnerability –identifying and focusing on their weakest point and paying constant heed to the need to manage process safety risks;
- There are always prior warnings of impending catastrophe. A simple failure that led to a major incident would, in all probability, have occurred before but as it did not lead to any significant consequence little attention would have been paid to it. Organisations should develop real time performance monitoring for process safety risks and act on these prior warnings;

- Leadership and process safety competence lies at the heart of effective decision making. Business decisions in major hazard companies almost always will have an impact on the management of risk; and
- Culture and organisational expectations drive the behaviour of employees. Developing a positive safety culture can take many years to build but can be destroyed in an instant should managers or supervisors demonstrate a disregard for high standards of process safety.

Annex 1: Flammable liquids likely to give rise to a large vapour cloud in the event of a tank overflow

- Acetone;
- Benzene;
- Crude oils (subject to further consideration of the Reid vapour pressure);
- Raw gasoline;
- Methyl ethyl ketone;
- Naphthas;
- Reformate (worst case – light);
- Natural gas liquids (condensates);
- Methyl tert-butyl ether;
- Iso Pentane;
- Special boiling point solvent 2; and
- Toluene.

Annex 2: PSLG Principles of Process Safety Leadership

Process Safety Leadership Group (PSLG) is committed to improving process safety in the industries we represent. We believe that to achieve this, industry leaders have a critical role to play and must commit to establishing the following principles of process safety management in each business:

Principles:

- Clear and positive process safety leadership is at the core of managing a major hazard business and is vital to ensure that risks are effectively managed;
- Process safety leadership requires board level involvement and competence. For companies with boards located outside the UK then the responsibility to show this leadership rests with the most senior UK managers;
- Good process safety management does not happen by chance and requires constant active engagement;
- Board level visibility and promotion of process safety leadership is essential to set a positive safety culture throughout the organisation;
- Engagement of the workforce is needed in the promotion and achievement of good process safety management;
- Monitoring process safety performance based on both leading and lagging indicators is central to ensuring business risks are being effectively managed;
- Publication of process safety performance information provides important public assurance about the management of risks by an organisation; and
- Sharing best practice across industry sectors, and learning and implementing lessons from relevant incidents in other organisations, are important to maintain the currency of corporate knowledge and competence.

The PSLG regards these principles as fundamental to the successful management of a major hazard industry. We will work with all stakeholders to establish them as foundations to effective management of risks in our businesses via the following arrangements:

Organisation & Resources:

- Process safety accountabilities should be defined and championed at board level. Board members, senior executives and managers should be held accountable for process safety leadership and performance;

- At least one board member should be fully conversant in process safety management in order to advise the board of the status of process safety risk management within the organisation and of the process safety implications of board decisions;
- Appropriate resources should be made available to ensure a high standard of process safety management throughout the organisation and staff with process safety management responsibilities should have or develop an appropriate level of competence;
- Organisations should develop a programme for the promotion of process safety by active senior management engagement with the workforce, both direct and contract staff, to underline the importance of process safety leadership and to support the maintenance of a positive process safety culture within the organisation;
- Systems and arrangements should be in place to ensure the active involvement of the workforce in the design of process safety controls and in the review of process safety performance;
- Business risks relating to process safety should be assessed and reviewed regularly using an appropriate business risk analysis methodology;
- Leading and lagging process safety indicators should be set for the organisation and periodically reviewed to ensure they remain appropriate for the needs of the business. Information on process safety performance should be routinely reviewed at board level and performance in the management of process safety risk is published in annual reports;
- Companies should actively engage with others within their sector and elsewhere to share good practice and information on process safety incidents that may benefit others. Companies should have mechanisms and arrangements in place to incorporate learning from others within their process safety management programmes;
- Systems and arrangements should be in place to ensure the retention of corporate knowledge relating to process safety management. Such arrangements should include information on the basis of safety design concept of the plant and processes, plant and process changes, and any past incidents that impacted on process safety integrity and the improvements adopted to prevent a recurrence.

PSLG Commitment
Implementation of the above process safety leadership principles and arrangements may vary in both detail and time in different organisations. However in recognition of the essential role these principles and arrangements play in the management and sustainability of our major hazard businesses, as members of PSLG we commit to working to establish them in the industries and businesses we represent as foundations to effective process safety management and the prevention of major accidents.

Tony Traynor, Chair,
Process Safety Leadership Group

Chris Hunt, Director General,
UK Petroleum Industry Association

Steve Elliott, Chief Executive,
Chemical Industries Association

Martyn Lyons, Chairman,
Tank Storage Association

Bud Hudspith, Unite H&S Adviser
(on behalf of the Trades Union Congress)

Peter Davis, UK Onshore Pipeline Operators' Association

Martin Bigg, Head of Industry Regulation,
Environment Agency

Allan Reid, Head of National Environmental Protection and Improvement, Scottish Environment Protection Agency

Peter Baker, Head of Chemical Industries Division, Hazardous Installations Directorate,
Health and Safety Executive

References

1. The Buncefield Incident 11 December 2005, The Final Report of the Major Incident Investigation Board: 2008. HSE Books. ISBN 978 0 7176 6270 8

2. The Process Safety Leadership Group, PSLG, comprised of representatives of the United Kingdom Petroleum Industries Association (UKPIA); the Tank Storage Association (TSA); United Kingdom Onshore Pipeline Operators Association (UKOPA); Chemical Industries Association (CIA); Unite the union; Environment Agency (EA); Scottish Environment Protection Agency; and the Health and Safety Executive (HSE).

3. EC Council Directive 96/82/EC on the control of major-accident hazards involving dangerous substances.

4. Process Safety Leadership Group, PSLG Final Report – safety and environmental standards for fuel storage sites, HSE Books, 2009, ISBN 978 0 7176 6386 6.

5. Functional safety. Safety instrumented systems for the process industry sector. Framework, definitions, system, hardware and software requirements. BS EN 61511-1:2004. British Standards Institution

6. Specification for the design and manufacture of site built, vertical, cylindrical, flat-bottomed, above ground, welded, steel tanks for the storage of liquids at ambient temperature and above. BS EN 14015:2004 British Standards Institute ISBN 0 580 45461 4

7. Welded tanks for oil storage API STD 650 (Eleventh Edition) American Petroleum Institute 2008

8. Venting atmospheric and low-pressure storage tanks - Non-refrigerated and refrigerated API. STD 2000 (Fifth edition) American Petroleum Institute 1999

9. Frangible roof joints for fixed roof storage tanks: Guide for designers and users Publication 180 .Engineering Equipment Materials Users' Association 2007 0 85931 161 9

10. Users' Guide to the inspection, maintenance and repair of aboveground vertical cylindrical steel storage tanks. Publication 159 (Third edition) Volumes 1 and 2 Engineering Equipment Materials Users' Association 2003 ISBN 978 0 85931 131 1

11. Tank inspection, repair, alteration, and reconstruction. API STD 653 (Fourth edition) American Petroleum Institute 2009

12. Guide for the prevention of bottom leakage from vertical, cylindrical, steel storage tanks. Publication 183 Engineering Equipment Materials Users' Association 1999 ISBN 978 0 85931 115 1

13. COMAH Competent Authority Policy on Containment of Bulk Hazardous Liquids at COMAH Establishments. http://www.environment-agency.gov.uk/business/sectors/37107.aspx

6TH GLOBAL CONGRESS ON PROCESS SAFETY

Evergreening Your PHA

Sheri Sammons
Siemens Energy Inc.
4615 Southwest Freeway Suite 900
Houston, TX 77027
Sheri.sammons@siemens.com

Steve Higgens
Siemens Energy Inc.
4615 Southwest Freeway Suite 900
Houston, TX 77027
Steve.higgens@siemens.com

Prepared for Presentation at
American Institute of Chemical Engineers
2010 Spring Meeting
6th Global Congress on Process Safety
San Antonio, Texas
March 22-24, 2010

UNPUBLISHED

Evergreening Your PHA

Sheri Sammons
Siemens Energy Inc.
4615 Southwest Freeway Suite 900
Houston, TX 77027
Sheri.sammons@siemens.com

Steve Higgens
Siemens Energy Inc.
4615 Southwest Freeway Suite 900
Houston, TX 77027
Steve.higgens@siemens.com

Keywords: PHA, MOC, Hazard, safeguard

Abstract

The PHA, LOPA, and other process safety information have all been documented. The hazards and safeguards/independent protection layers have been identified. Some time later a device is altered. Is this device a potential safeguard? How does this affect the process and how is this documented? Is the hazard, other safeguards, or any other information affected?

Today as changes are made to operating facilities, they are documented, approved and filed away. The changes are then reviewed during the five year revalidation PHA. Critical information could be buried or inaccessible. In the future, an evergreen database of information can be used to ensure that safeguards/independent protection layers are not eliminated or altered without proper investigation, engineering, and documentation. In addition, maintenance activities need to be able to be tracked and verified if the safeguards/independent protection layers are to be relied upon. PHA safeguards and/or LOPA IPLs should be available in one easily accessible location so they can be retrieved, reviewed, and updated as changes are made to the facility. This paper will address procedures, policies and/or technologies to significantly improve access, availability, and accuracy of this information.

1. Introduction

The Process Hazard Analysis is used to identify potential process hazards in facilities that are covered under OSHA 1910.119. A team of technical and operating experts review the current state of the process to determine what potential hazards exist and what safeguards are in place to prevent a significant consequence from occurring. The OSHA standard requires a review of the PHA to be done at least every five years.

The current PHA process in many PSM covered facilities is to wait as long as possible between cyclic PHA and to start reviewing the required PSI a couple of months before the PHA is to commence. What if changes have occurred during the previous five years that affect the safeguards in a plant? Based on the current process, these inconsistencies and possible safety concerns will only be found once every five years, if they are caught at all. We propose that in the future an evergreen version of the PHA is available and updated whenever a change to any Process Safety Information occurs in a facility. This should improve process safety in the plant and lead to less burden on the operations, maintenance and engineering personnel in a plant.

2. Current State of PHAs

Issues that arise during the current PHA process:

- PHA required by OSHA every five years and a complete PHA is not required, only a revalidation
- Minimum amount of resources are typically available for revalidation
- Documentation of changes at the facility may not all be in one place – time and effort must be expended to find appropriate information and some information may get missed. Examples of this include
    - Temporary MOCs
    - Emergency situations and quick fixes
    - Replacement in kind scenarios
- Documentation of incidents and follow ups are located somewhere else
- Changes at the facility that may have affected safeguards listed in previous PHA may get missed during revalidation

Elements of PSM are often looked at as separate and disconnected. However, the safeguards, consequences and causes reviewed in a PHA are based on the effectiveness of all of the PSM elements. In addition, the Process Safety Information used during the PHA is assumed to be up-to-date and correct.

3. Evergreening of PHA Process

A more robust and realistic approach to PHAs in the future is to constantly update the PHA as changes are made in the facility. Updating the PHA as changes are made can add to the value of the PHA. It can become the go-to document for operations, maintenance and engineering for safety information throughout the plant lifecycle. Since information is added as it changes, the labor required to gather information at the cyclic PHA should be reduced. The information should be higher quality also. Below are some examples of facility changes that could affect the hazards and safeguards documented in a cyclic PHA.

3.1 MOC

The process safety information at a facility goes across a wide spectrum of functions within the plant. Engineering, operations and maintenance all will have an impact on the state of a facility over a five year period. In a PSM covered plant changes are made using Management of Change (MOC) as outlined in OSHA 1910.119. However, it is up to the MOC initiator and reviewers to confirm that there has not been an effect on process safety. A safety checklist, mini-HAZOP or

PHA may be done to review the effect the change of the MOC has on the process. However, is the existing PHA being used to confirm that changes are not affecting a different part of the process (i.e. a safeguard)? A proposed solution to this would be to have a list/database of safeguards required to be cross referenced before any MOC is approved. Also, if the MOC hazard review identifies a potential hazard, this should be cross referenced with the existing PHA to confirm that it has already been identified and if not, it should be added along with any safeguards and recommendations to an evergreen file of the PHA. This process should also be used with temporary MOCs. It may be necessary to have a 'red-lined' version of the PHA for the duration of the temporary MOC. Even in an emergency situation the PHA should be reviewed to judge any potential impacts. Otherwise, there is the potential that this information will be lost before the next PHA cycle. An untouched copy of the cyclic PHA should be kept as a reference.

3.2 Mechanical Integrity

From a maintenance and testing standpoint, Preventative Maintenance programs are referenced as safeguards for certain scenarios. If a PM program is referenced as a safeguard during a PHA, the PM database is checked to confirm that the identified equipment is included. This type of equipment may not be identified as safety critical even though it is being used as a safeguard. As updates are made to the PM process, changes may be made to the maintenance schedules of equipment used as safeguards. This could lead to a reduction in process safety. By cross referencing the identified safeguard PM database and perhaps flagging it a safeguard, this issue would be less likely.

3.3 Replacement in Kind

The criteria for 'replacement-in-kind' should be carefully used and invoked. It will be up to each operating facility/owner to determine what exactly constitutes a replacement-in-kind situation. There is potential that changes to the process could affect the PHA even if a change or modification is deemed replacement-in-kind. These changes would not be documented in the MOC system. A review of the PHA could be a determining factor. After the PHA is reviewed and it is found that the change will affect the process enough to warrant an update of the PHA documentation, it is likely that the change should be moved to the MOC system.

3.4 Controls Systems Updates

Controls system modifications that may appear to only affect the process, may actually affect a safeguard. There are some control loops that may be referenced as a safeguard for specific consequence based on how they operate when the PHA was completed. If there is a change to the control scheme, it may affect the effectiveness of the safeguard. If all safeguards were documented and cross referenced with the other information in the plant, the plant will be safer.

3.5 Relief Systems

Pressure relief system information needs to be correct, current, and available throughout the plant lifecycle. In some cases, relief valve and flare studies are completed and placed on a shelf like the PHA documentation. In other cases, the PSV documentation may be revised based on process changes but the information is not cross referenced to the existing PHA. PSVs are safeguards used throughout PHAs. Therefore, a cross check between the PSV database and the PHA should be completed whenever a change is made to the process or PSV equipment.

4. Process

At first thought, updating and maintaining this information could appear to be too time consuming and not provide any benefits since it is not a specific requirement of OSHA 1910.119. However, there are several benefits including saving time, money and improving process safety at the plant. Let's talk about improvements to the process first and then move into the benefits.

The goal of the PHA is to provide the owner/operator with quality information on the hazards and safeguards for their process. The continuing update of the PHA documentation will help with this effort. The PHA can then become the primary source for determining the effect of potential changes, a tool used to get new employees up-to-speed quickly with the process, and allow anyone at a central location to audit the facility. The PHA does not have to be a chore; it can be a useful tool. There are many areas and departments that can be affected by changes in the process. The PHA then becomes the tool to ensure that all of these areas are up to date.

The Evergreen PHA process in the future should implement the following as changes are made to the process:

- Add new hazard information to the PHA
- Confirm that existing safeguards have not been compromised
- Change MOC process to include cross checking with a PHA/Safeguard database

In addition, as revalidations to operating procedures, training, PSVs, maintenance procedures and schedules are completed, there should be confirmation that these items have not changed and not affected the validity of the PHA.

Some additional points to consider:

- The PHA and Safeguard Database must be accessible, searchable and scalable
- The data must be secure while allowing visibility to the employees and contractors who required the data
- Revision tracking should be used to maintain the integrity of the original PHA

All of these processes and systems can be put in place using today's available technologies. Using interactive databases is one way technology could be used to improve the efficiency of this process. However, an investment in the future and support from senior management will be required.

5. Benefits

The benefits of the Evergreen PHA:

- Retain technical knowledge base lost due to attrition as information has already been added to the PHA database

- Operations has quality information about the current state of the process available in one document
- The PHA process becomes less burdensome and the resulting document more valuable
- Real time hazard information can be made available to operations

6. Conclusion

Evergreening your PHA allows you to verify that changes to your process are not affecting the process safety of you facility. If changes are affecting the process safety, this information can be immediately added to the database so future changes can be checked against it. As the PHA is being updated frequently, it provides a much better representation of the hazards existing and how to prevent them from occurring. This leads to a higher quality PHA that becomes a primary resource used real time by operations. From a resource standpoint, less time should be required to gather data and perform the cyclic PHA.

In conclusion, the evergreen PHA increases overall process safety, is a valuable resource to operations, and should not increase the overall effort required to produce a high quality PHA.

7. References

[1] 29CFR 1910.119, Process Safety Management of Highly Hazardous Materials (OSHA); www.osha.gov

BART: a comprehensive Risk Management Tool for Asset Integrity

Paolo Cherubin
Tecnomare SpA,
Via Pacinotti 4, 30175 Marghera (VE), Italy
Ph. +39 041 796711
paolo.cherubin@tecnomare.it

Stefano Pellino
Sviluppo Tecnologie Industriali,
Via Mazzini 12, 56125 Pisa, Italy
Ph. +39 050 48001
stefano.pellino@stionline.it

Paolo Carnevale
Tecnomare SpA,
Via Pacinotti 4, 30175 Marghera (VE), Italy
Ph. +39 041 796711
paolo.carnevale@tecnomare.it

Riccardo Bandini
Karachaganak Petroleum Operating BV
West Kazakhstan Oblast 090300
Burlin Region, Aksai
Ph: +7 763 222 2490
BandiR@kpo.kz

Chiara Cerruti
Eni E&P Division,
Via Emilia 1, 20097 San Donato M.se (MI), Italy
Ph. +39 02 52062651
chiara.cerruti@eni.it

Prepared for Presentation at
American Institute of Chemical Engineers
2010 Spring Meeting
6th Global Congress on Process Safety
San Antonio, Texas
March 22-24, 2010

UNPUBLISHED

BART: a comprehensive Risk Management Tool for Asset Integrity

Paolo Cherubin
Tecnomare SpA,
Via Pacinotti 4, 30175 Marghera (VE), Italy
Ph. +39 041 796711
paolo.cherubin@tecnomare.it

Stefano Pellino
Sviluppo Tecnologie Industriali,
Via Mazzini 12, 56125 Pisa, Italy
Ph. +39 050 48001
stefano.pellino@stionline.it

Paolo Carnevale
Tecnomare SpA,
Via Pacinotti 4, 30175 Marghera (VE), Italy
Ph. +39 041 796711
paolo.carnevale@tecnomare.it

Riccardo Bandini
Karachaganak Petroleum Operating BV
West Kazakhstan Oblast 090300
Burlin Region, Aksai
Ph: +7 763 222 2490
BandiR@kpo.kz

Chiara Cerruti
Eni E&P Division,
Via Emilia 1, 20097 San Donato M.se (MI), Italy
Ph. +39 02 52062651
chiara.cerruti@eni.it

Keywords: risk assessment, barrier, incident, bow-tie, top event

Abstract

Large losses are typically the result of the failure of multiple safety barriers, often within complex scenarios. An innovative methodology accompanied by a dedicated software BART (BAseline Risk assessment Tool) has been set up to cover all types of onshore and offshore installations, and to allow any E&P organization to monitor all its asset status with a common and standardized approach. The BART tool combines a simplified quantitative risk assessment methodology with a bow-tie model approach to identify and assess potential hazards and associated risks, arisen from a process or activity carried out in an upstream onshore/offshore installation. The sequence of any major incident is analyzed from the identification of primary root causes up to the determination of potential effects and associated impacts. Therefore, all existent safety barriers are identified and their prevention and mitigation risk reduction contributes included in the analysis. The BART tool allows for a systematic monitoring of all major incident risks through the selection of effective control measures able to reduce the likelihood and/or to limit the potential effects of current risks. The BART methodology has been tested on some existing onshore and offshore Oil & Gas installations and the main findings have been positively compared with alternative approaches.

1 Introduction

Main goal of a hazard evaluation and risk assessment methodology is to help an Exploration & Production organization better understand the potential risks associated with process safety hazards and how to reduce the likelihood and severity of any foreseeable major incidents.

The methodology is aimed at identifying and analyzing the significance of hazardous situations arisen from a process or activity, focusing on the potential causes (equipment failures, human errors, software problems, external factors, etc…) and sequence of events that can lead to incidents (i.e. fires, explosions or unplanned release of hazardous materials) that either result in or have the potential to result in adverse impacts to people, environment or property [11].

The incident sequence must be investigated starting from the initiating cause or **threat** that has as its result a deviation from normal operating procedures or safe operating limits to an abnormal operations mode. If the abnormal situation is allowed to continue uncorrected, a **loss event** may result with possible **impacts** and an emergency situation begins.

The chain of events following an initiating cause to the loss event may be interrupted due to the presence of **preventive barriers**. The likelihood of occurrence of the loss event depends on the number of barriers in place and their effectiveness in preventing the loss event from ensuing.

In a similar way, after the loss event occurs, **recovery barriers** intervene to detect and control the release source parameters (magnitude, rate, duration, orientation, temperature, etc…) or to reduce the impacts of the released materials or energy on people, property and the environment. The severity of consequences of the loss event (expressed in terms of number of injuries and/or fatalities, extent of the environment damage, property damage, material loss, lost production and recovery costs) depends on the number of recovery barriers in place and their effectiveness in mitigating the loss event impacts.

A good way to visualize the incident sequence and understand how barriers operate is the **"bow tie diagram"** (see Figure 1). An incident sequences is represented in the bow-tie diagram with separate lines connecting each threat (potential causes/failure modes) with the top event. The propagation of each incident sequence depends on the hardware and operational barriers that are in place to prevent threat realising the top event.

The escalation incident sequences, describing what could occur in the event of the entire incident sequence being realised, are represented in the bow-tie diagram with lines linking the top event with potential consequences, considering the hardware and operational barriers that are in place to recover a degree of control and mitigate the potential effects following the occurrence of the top event.

The sequence of events showing the progression from any possible threat or cause that could lead to the Top Event (the unplanned release of a hazard) is likened to a **fault tree,** whereas the escalating incident sequence used to determine the potential outcomes or consequences after the release of a hazard is likened to an **event tree** [12].

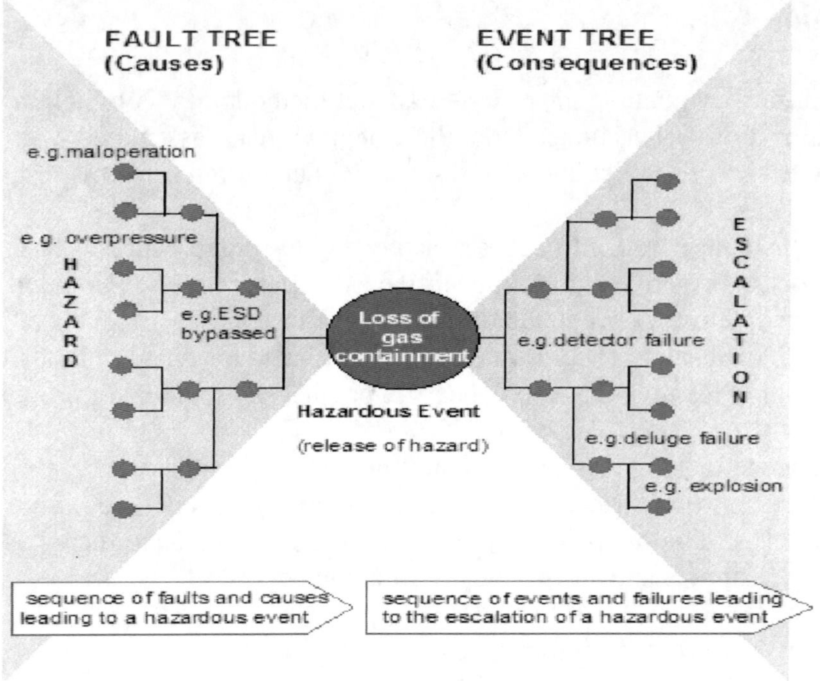

Figure 1 – Bow-Tie Model

2 Tool Methodology

The BART (**BA**seline **R**isk assessment **T**ool) methodology has been developed for the management of potential risks associated to various hazard categories, in particular process safety hazards. It is consistent with the risk management process outlined in the OGP Report N° 415 "*Asset integrity – the key to managing major incident risks*" [9]. The Tool combines a simplified quantitative risk assessment methodology with a bow-tie model approach to identify and assess any potential hazards and associated risks, arisen from a process or activity carried out in both onshore and offshore oil & gas installations.

A simplified quantitative risk assessment methodology has been developed to determine the initial risk level within each functional unit of the installation, according to the process flow, the operative conditions and the plant layout. Consistent with the unit input data, the tool calculates on the basis of specific assumptions the release frequency and the consequences for each foreseeable top event that could happen within each functional unit as if no preventive and recovery barriers are in place. Starting from these initial risks, the bow-tie approach is then adopted to evaluate the current risk levels decreasing the likelihood of occurrence and severity of consequences of each top event on the basis of the number and effectiveness of all preventive and recovery barriers in place within each functional unit.

To reduce the exposure to current risks associated to an incident scenario, the tool provides some utilities to simulate future scenarios through the implementation of corrective actions aimed at improving the reliability of the existent preventive or recovery barriers or at adding new barriers (preventive and/or recovery). The tool includes specific utilities to generate an action plan and monitor the achievement of the defined corrective actions.

2.1 *The Simplified Quantitative Risk Assessment Methodol*

The simplified quantitative risk assessment methodology has been developed and implemented in the tool to identify any potentially harmful top event inherent to a hydrocarbon installation and determine the initial risk level associated to each of them, based on the combination of two factors:

- the **frequency** of the top event;
- the **severity** of the top event.

The Plant (upstream onshore/offshore installation) is divided into functional units, according to the process flow, the operative conditions and the plant layout. Depending on their substances, the functional units can be considered as hazardous (i.e. able to potentially generate a major accident event which follows a hydrocarbon loss of containment) and thus further analysed.
The potential major accident events which may be considered in the analysis are:

- **Jet Fire** (JF) - for pressurized releases, both for gaseous and liquid phases (not considered for units at atmospheric pressure);
- **Pool Fire** (PF) - only for liquid releases (not considered for gas releases);
- **Flash Fire** (FF) - both for gaseous and liquid phases (not considered for liquid Units at atmospheric pressure);
- **Explosion** (EX) - is considered only if the unit (or an area inside the unit) has been defined as "congested",
- **Toxic Dispersion** (TX) **- only** for toxic releases (toxic liquid releases are treated as gaseous, calculating the same consequences).

2.1.1 Frequency Evaluation

Each hazardous unit is considered as a potential release source and a global release frequency is assigned. When more than one substance is present in a unit, one release source is defined for each stream and the total leak frequency is divided by the number of substances. The main equipments in the unit are counted and a specific release frequency is assigned to each (λ). For those units where no vessels are present (i.e. manifold, production headers) a piping length can be assigned instead of the number of equipments. To take into account also the contribution of the pipework F_{pw} (valves, flanges, instrument and piping) on the total leak frequency F, that is predominant in comparison with the main equipments contribution F_{eq}, it has been assumed to fix the pipework contribution as a fraction of the global release frequency. The global release frequency F assigned to the unit is given by the sum of all the single leak frequencies of the equipment categories (F_{eq}) plus the pipework contribution (F_{pw}). The database used as reference for the release frequencies calculation and apportionment is the HSE HRS 2006 [1]. The original data have been elaborated to obtain the release frequencies for all main equipment implemented in the tool. The global release frequency of the unit is divided and assigned to four hole size classes, to take into account the different dimension of the potential releases. Each release class is therefore treated as an independent top event with its own leak frequency. Once defined the release frequencies for each hole size, it is necessary to define the accidental events related to each unit.

Occurrence probabilities shall be assigned to the defined events. The method that has been used to calculate these probabilities refers to the IP Model [8]. The global ignition probability depends by the release flowrate and it is calculated according to the correlation:

$$P_{ign} = 10^{(a+b\cdot\log F)} \quad \text{[Eq. 1]}$$

where F is the release rate (in kg/s) and a, b are constants which depend by the correlation applied and by the release rate range. According to [8], the correlations implemented in the tool depend on the type of installation (e.g. Onshore Large Plant Gas, Offshore Process Liquid etc.).
Total ignition probability P_{ign} is split 50-50 between immediate ignition P_{imm} and delayed ignition P_{del} . If both jet fire and pool fire have been defined, then immediate ignition probability is also split 50-50 between jet occurrence P_{jf} and pool occurrence P_{pf}. In Figure 2 is represented the event tree implemented in the tool.

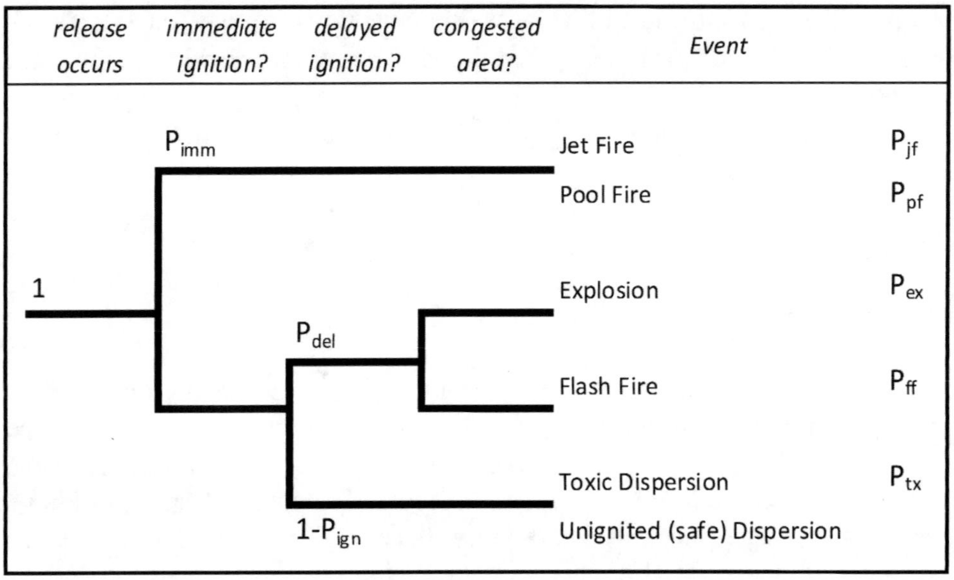

Figure 2 – BART Event Tree

From the global ignition probability P_{ign} calculated according to the IP Model correlation, the occurrences of each scenario are calculated as follows:

- Jet Fire: $P_{jf} = P_{imm} = 0.5 \cdot P_{ign}$ ($0.5 \cdot P_{imm}$ if also pool fire is considered)
- Pool Fire (if considered): $P_{pf} = P_{jf} = 0.5 \cdot P_{imm}$
- Toxic Dispersion (if considered): $P_{tx} = 1 - P_{ign}$
- Explosion: $P_{ex} = P_{del} \cdot P_{ex,del}$
- Flash Fire: $P_{ff} = P_{del} \cdot (1 - P_{ex,del})$

$P_{ex,del}$ is the explosion probability given delayed ignition occurs. The value of this parameter depends on the congestion level and the release hole size.

The likelihood of the major accident event is given by multiplying the release frequency and scenario occurrences, calculated for each top event. The likelihood is then classified on the basis of the risk matrix levels to assign a qualitative frequency level according to the following ranking criteria:

Table 1 Ranking Criteria for Qualitative Frequency

Likelihood of major accident	Qualitative Frequency
$f < 10^{-6}$ ev/yr	0
$f \geq 10^{-6}$ ev/yr and $f < 10^{-4}$ ev/yr	A
$f \geq 10^{-4}$ ev/yr and $f < 10^{-3}$ ev/yr	B
$f \geq 10^{-3}$ ev/yr and $f < 10^{-1}$ ev/yr	C
$f \geq 10^{-1}$ ev/yr and $f < 1$ ev/yr	D
$f \geq 1$ ev/yr	E

2.1.2 Severity Evaluation

The severity evaluation starts from the calculation of the physical consequences which follow an accidental release. This calculation has been performed using the DNV software 6.54 [2] and it is based on some assumptions made to allow the application of the tool on a wide range of operative conditions.

There are a lot of variables which influence the consequence of a release (e.g. fluid composition, operative pressure and temperature; release size, direction and height; discharge inventory and release time; wind velocity, atmospheric turbulence, ambient temperature, surface roughness etc.). Since these variables depend by the particular case which is being analysed and since the tool has been thought to be generally applied to oil & gas installations for coarse evaluations, some of these variables have been set up with fixed values, as it is not possible to cover all their ranges of values. All the possible combinations of the other variables, such as fluid compositions, discharge pressures and hydrogen sulphide concentrations, have been simulated to calculate the consequences of the accident events (explosion, jet fire, pool fire, flash fire and toxic dispersion). The results have been collected in a set of tables, which contain the physical effect distances for the defined damage thresholds for each fluid, event, and release class and weather category.

The last set of tables utilized by the tool for the calculation of the ignition probability are those that report the relationship between the defined operative pressures and the release rates, for each standard fluid and hole size class. All the tables built after the physical effects modelling are utilized by the tool to interpolate the consequences relevant to the particular unit conditions, which not necessarily coincide with the simulated values.

To calculate the human or asset vulnerability due to the physical effects exposure, a set of thresholds have been defined and implemented in the tool for each event according to technical literature [3],[6], [7]. Once calculated, the problem then becomes translating the information into a qualitative ranking. The effects of a top event (thermal radiation, overpressure, toxic or flammable concentration) are transformed into a distance, which is compared with a set of defined reference lengths, in order to obtain a ranking compatible with the five levels of the risk screening matrix.

To determine this geometrical variable, the physical effects of each event have been approximated with a set of **n** concentric ellipses, one for each defined damage threshold. Moreover, for conservative reason, the release has been assumed horizontal, so to maximize the event extension. As the "effect" of an event is determined both by its extension and by its vulnerability, for each identified accidental event the following parameters have been considered:

- The event damage surface ***S (a, b)***, which is the elliptical footprint of the event, within which the physical effect has a vulnerability greater than zero. This surface is assumed to be the same for each release direction (no plant layout nor geometrical constrains are foreseen) and it is characterized by its maximum downwind distance ***r*** and its maximum crosswind width ***w,*** that have been calculated according to the outlined procedure;
- The damage area ***A (d)***, which is the circular area delimited by the maximum distance ***d*** reached by potentially lethal physical effects, beyond which the fatality is therefore considered negligible.
  Since the release direction can be equally spaced on the plane, this area represents the locus of the points which may be potentially impacted by the event effects. Note that for EX, that is a non-directional event, the damage area and the event surface coincides;
- the average vulnerability $\overline{Vul_S}$ on the damage surface ***S***, defined as the probability of fatality (supposed constant) at which an individual would be exposed inside the event damage surface S:

$$\overline{Vul_S} = \frac{1}{S} \cdot \int_S vul(x, y) \cdot dx \cdot dy \quad \text{[Eq. 2]}$$

- the average vulnerability $\overline{Vul_A}$ on the damage area ***A***, defined as the probability of fatality (supposed constant) at which an individual would be exposed inside the whole potentially impacted area around the release source A:

$$\overline{Vul_A} = \overline{Vul_S} \cdot S / A \quad \text{[Eq. 3]}$$

Once all parameters are calculated, the tool combines the average vulnerability $\overline{Vul_A}$ with the damage area ***A*** to obtain a parameter which takes into account both the potential harmfulness of the event and its damage extension. This variable, called vulnerability radius (r_{vul}), has the dimensions of length and can be interpreted as the maximum distance from the release source within which an individual would be exposed to a vulnerability of 100%:

$$\overline{Vul_A} \cdot \pi \cdot d^2 \equiv 1 \cdot \pi \cdot r_{vul}^{\ 2} \quad \rightarrow \quad r_{vul} = d \cdot \sqrt{\overline{Vul_A}} \quad \text{[Eq. 4]}$$

A geometrical interpretation of r_{vul} is shown in Figure 3 for a generic 1-dimensional event, but the augmentation can be extended to 2-dimensional event to obtain the formula used in the tool.

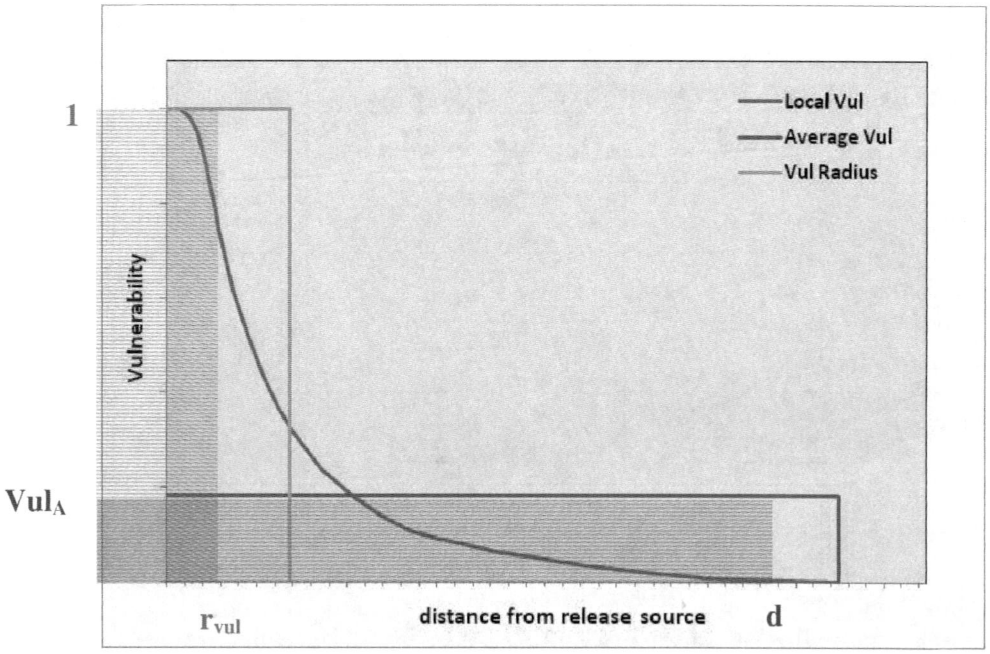

Figure 3 – geometrical interpretation of the vulnerability radius for 1-dimension

The calculated vulnerability radius is used as classification criterion to determine the severity ranking according to the risk matrix. To do this, a set of specific lengths shall be defined.
It has been chosen to define such distances taking into account the location of the analysed plant (Onshore or Offshore) and its layout, rather than assuming some constant values independently by the case study. Although these lengths are characteristic for each unit, their definition is general:

- for onshore installation:
  1. d1: half of the minimum dimension of the unit on plant layout;
  2. d2: shortest distance between the geometrical barycentre of the unit and the barycentre of the closest unit;
  3. d3: shortest distance between the geometrical barycentre of the unit and the barycentre of the closest permanently manned area.

- for offshore installation:
  1. d2: half of the minimum dimension of the deck where the unit is located;
  2. d1: half of d2;
  3. d3: height between the deck where the unit is located and the upper one; if the unit is on the Main/Upper Deck or is on another platform (e.g. adjacent satellite platform),

d3 is the distance between the release source and the Living Quarter of the main installation.

The following criteria have been implemented to classify the severity according to the Risk Matrix:

Table 2 Ranking Criteria for Qualitative Severity

Vulnerability Radius	Qualitative Severity
$r_{vul} < 1$ m	1
$r_{vul} \geq 1$ m and $r_{vul} < d1$	2
$r_{vul} \geq d1$ and $r_{vul} < d2$	3
$r_{vul} \geq d2$ and $r_{vul} < d3$	4
$r_{vul} \geq d3$	5

For offshore installations:

- if the event is an explosion and the overpressure is ≥ 0.35 barg an escalation from deck to deck with severe structural damages is assumed → G = "5"
- if the unit is on the upper deck or on another platform:
  - if $r_{vul} \geq d2$ and $r_{vul} < d3$ → G = "4"
  - if $r_{vul} \geq d3$ → G = "5"
- if the unit is on the other decks:
  - if $r_{vul} \geq d2$ → G = "4"

2.1.3 Initial Risk Evaluation

The Tool determines the initial risk for each top event on the basis of the combination of the frequency and the severity in accordance to the Screening Risk Matrix, which contemplates four risk levels:

1. Low (L, blue): Continuous Improvement
2. Medium (M, yellow) Risk Reduction Measures (ALARP)
3. Medium-High (MH, orange) Mandatory Risk Reduction Measures
4. High (H, red) Intolerable Risk

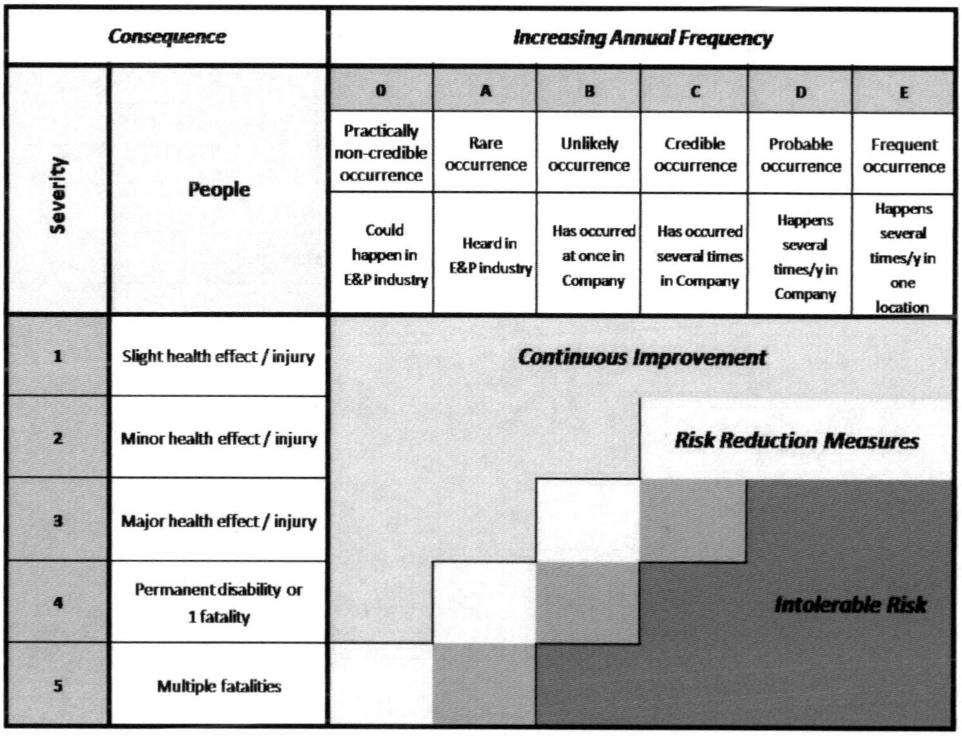

Figure 4 – BART Screening Risk Matrix

2.2 The Bow-Tie Methodology

The Bow-Tie methodology has been implemented in BART tool in order to evaluate the current risks for any top event identified by the simplified quantitative risk assessment methodology and relevant to each functional unit of the plant. The methodology requires the preliminary construction of a comprehensive Bow-Tie diagram for each top event through the identification of all independent preventive and recovery barriers in place within each functional unit, respectively to prevent the occurrence of the top event for each initiating cause (threat) and to mitigate any adverse effects whether the top event occurs. The identification process of the independent barrier is carried out through the examination of the plant available documentation, interviews with operational personnel and direct survey on locations of the onshore/offshore installations. These activities are aimed at determining which controls, both operational and hardware are in place for every foreseeable top event and at establishing how they work in order to better define the number and the composition of any independent preventive or recovery barrier. Consistent with the concept of the Independent Protective Layer (IPL) of LOPA (Layer of Protection Analysis) methodology [10], an independent barrier is so defined: "An IPL is a device, system or action which is capable of preventing a scenario from proceeding to its undesired consequence independent of the initiating event or the action of any other layer of protection associated with the scenario. The effectiveness and independence of an IPL must be auditable."

Some examples of preventive and recovery barriers are reported in the following table:
Table 3 Examples of preventive and recovery barriers

Barrier	Type	Basic Control
Equipment functionality	Preventive	Equipment Integrity
		Safety Critical Element Management
		Inspections & Testing
		Maintenance Management System
Pressure Protection System	Preventive	Pressure Protection System
Process Control System	Preventive	Process Control System
Isolation & Depressurization	Preventive / Recovery	Emergency Shutdown System
		Emergency Blowdown System
		Flaring-venting System
		Fire & Gas Detection System
Task Management	Preventive	Competence Staff/Contractor
		Contract HSE Requirements
		Process Safety Rules for operation
		Management of changes
		Operating Procedures and Manuals
		Training / awareness
Process Safety Management System	Preventive	Leadership & commitment
		Process Safety culture
		Process Safety Audit
Emergency Response System	Recovery	Contingency and Emergency Response Plan
		Communication Systems
		Emergency Response Team
		Emergency Supply System
		Escape, evacuation and rescue equipment
Fire Management	Recovery	Fire Crew
		Auxiliary fire protection systems
		Fixed fire-fighting systems
		Mobile and portable fire fighting system

Once identified, these barriers and their established association with all the functional units in which they are implemented, make it possible to build for all the functional units of the installation a comprehensive bow-tie diagram relevant to each top event considering all potential threats.

Since the evaluation of the current risk level depends on the number of the independent barriers in place and their effectiveness in preventing the loss event from ensuing or in mitigating the loss event impacts, the next step required by the methodology is the determination of the reliability of each barrier. This can be accomplished via evaluating the reliability of each control by means of dedicated checklists, which take into account all factors that can affect its performance in terms of functionality, availability, reliability and survivability (design specifications, inspections, functional testing, and maintenance requirements as well as operational procedures, operator training and competence, management of change, etc). A weight percentage is assigned to each question of the checklist depending on the importance of the factor to be analysed. By calculating the sum of all questions related to one control, it is possible to determine the maximum value of reliability associated to that particular control. The current value of reliability for the same control is constituted by the sum of all questions with affirmative answer.

The ratio between the current value and the maximum value of reliability for one control determines its level of reliability as per the following criterion:

- LOW - $R \leq 40\%$
- MEDIUM - $R > 40\%$ & $R \leq 70\%$
- HIGH - $R > 70\%$

The reliability level of each barrier is consequently determined considering the lowest reliability level of its associated controls. Once the number and the reliability of all preventive barriers in place for each threat is known, the initial risk frequency which is estimated by the simplified quantitative risk assessment methodology for any foreseeable top event, is decreased in accordance with a risk reduction criterion based on the following assumptions.

- Any HIGH reliability barrier reduces the initial frequency/consequence of the event by one order of magnitude;
- LOW or MEDIUM reliability barriers can act as a HIGH barrier depending on their number, as shown in the table below:

Table 4 BART Risk reduction criteria

		INDEPENDENT BARRIERS			
		4	3	2	1
RELIABILITY	HIGH	4H	3H	2H	1H
	MEDIUM	2H	1H+1M	1H	1M
	LOW	1H	1M+1L	1M	1L

Similarly, the risk reduction criterion is applied to decrease the initial risk severity on the basis of the number and the reliability of all recovery barriers in place to mitigate impacts on people or assets. The tolerability of current risks can be evaluated with reference to a Screening Risk Matrix.

3 Conclusion

BART methodology has been tested on different onshore and offshore installations and it has demonstrated to be useful for a systematic screening and monitoring of all major incident risks. In particular, it enables an E&P organization to identify which barriers could fail and to establish corrective actions that are able to reduce the exposure to current risks associated to any foreseeable incident scenario so far as low reasonably practicable (ALARP). The benefits relevant to the implementation of these corrective actions can be simulated within the tool through the creation of future scenarios where the current risks are re-assessed taking into account more reliable preventive and/or recovery barriers or adding new barriers with the aim to minimize the residual risk. The results of these simulations can be easily compared with each other in order to identify the most effective corrective actions to solve any identified potential risk providing to the E&P organization's management all the elements necessary for the decision making process.

4 References

[1] Health & Safety Executive, Offshore Hydrocarbon Releases Statistics, 2006

[2] Det Norske Veritas, PHAST Proprietary Consequence Analysis Software, Version 6.54

[3] Netherlands Organisation for Applied Scientific Research TNO, Guidelines for Quantitative Risk Assessment (Purple Book) CPR 18E, 2005;

[4] Frank P. Lees, Loss Prevention in the Process Industries, Second Edition, 2001;

[5] Baker W.E, Cox P.A, Westine P.S, Kulesz J.J, Strehlow R.A., Explosion Hazards and Evaluation, Elsevier Scientific Publishing Company, 1983;

[6] API RP 752, Management of Hazards Associated with Location of Process Plant Buildings – Second Edition, 2003;

[7] Health & Safety Executive, HID Semi Permanent Circular SPC/Tech/OSD/30, 2008.

[8] IP Research Report, Ignition Probability Review, Model Development and Look-up Correlations, Energy Institute, 2006

[9] OGP Report N° 415, Asset integrity – the key to managing major incident risks, 2009

[10] Center for Chemical Process Safety, Layer Of Protection Analysis: Simplified Process Risk Assessment, 2001

[11] Center for Chemical Process Safety, Guidelines for Hazard Evaluation Procedure, 3rd edition, 2008

[12] ISO 17776:2000, Petroleum and natural gas industries - Offshore production installations - Guidelines on tools and techniques for hazard identification and risk assessment

6TH GLOBAL CONGRESS ON PROCESS SAFETY

Objective Analysis of PSM Audit Data

Jerry J. Forest
Lyondell Chemical Company
a LyondellBasell company
2502 Sheldon Road, Channelview, TX
jerry.forest@lyondellbasell.com

Prepared for Presentation at
American Institute of Chemical Engineers
2010 Spring Meeting
6th Global Congress on Process Safety
San Antonio, Texas
March 22-24, 2010

UNPUBLISHED

Objective Analysis of PSM Audit Data

Jerry J. Forest
Lyondell Chemical Company
a LyondellBasell company
2502 Sheldon Road, Channelview, TX
jerry.forest@lyondellbasell.com

Keywords: Audit, database, Access

Abstract

In developing systems for Process Safety Management, compliance audits provide essential verification that practices and procedures are developed and are effective for compliance. Data collected during compliance audits is often subjective and findings are descriptive in nature. It is therefore difficult to complete objective analysis of audit data across standards/procedures to surface systemic management system gaps and to compare site performance across a company. Complete analysis of PSM audit data is analogous to incident investigation causal factor assignment, data analysis, and cause elimination.

This paper describes a database solution that:

1. Provides a data collection method for compliance audits that allows for easy database input, thereby reducing overall data handling
2. Structures PSM procedures by management system element so that these elements can be examined across various PSM procedures
3. Groups compliance audit findings by common types for Pareto analysis

The benefits of objective analysis of PSM audit data include more effective future audits and continuous process safety performance improvement at the site level.

The presentation will include a brief review of the Microsoft Access database structure.

1. Introduction

A typical PSM audit will last from one to four weeks and consist of external and internal people examining management systems looking for something wrong. In fact, an auditor hasn't done a thorough job unless system gaps are identified! Auditors typically make a list of things to accomplish, and then rely on those audited to complete each item before the next cycle.

If the audit team has done a thorough job, closure of the audit findings will result in an improvement of process safety and serve to prevent process safety events. However, there is much more utility associated with audit data if an objective analysis could be performed on the data. Audit findings tend to be subjective and therefore difficult to analyze.

This paper will present a method to structure corporate and site standards and procedures so that management system elements can be examined as a whole across the PSM system. A database solution that includes audit data collection and input to reduce data handling and data analysis is presented. Logical groupings for findings are then presented for descriptive statistical and graphical analysis. By understanding common failure modes from audit data, process safety management systems can be improved, thereby helping to prevent process safety events.

2. The Problems with Audit Data Analysis

Although auditors may take great pride in presenting objective data to demonstrate gaps in management systems, analyzing the gaps as a whole can be problematic. The data analysis problem often starts with the audit report format. That is, audit reports, including findings, are typically distributed in word processor or PDF format, making data manipulation difficult. The data must first be transformed into a media that is designed for data handling, such as Microsoft Access or Excel.[1] A second problem is simple taxonomy. How can we produce a Pareto chart from findings such as: "Six of 10 P&ID's sampled had errors when compared to a field review of the drawings"? The finding itself is objective; it states the gap in terms of the sample size and the nature of the gap. However, a gap worded in this way cannot be easily queried against similar gaps within a single audit and certainly cannot be easily compared to similar findings across different sites, divisions or countries. The first step for objective analysis of audit data is to select an electronic platform to use in order to make the analysis. The appropriate selection will also minimize error in data input.

The taxonomy problem starts with the standards and procedures being audited and extends to the categorization of audit findings into logical groups for analysis. The former is particularly relevant when doing data analysis for systemic management system gaps. For example, your company may have a training requirement associated with each standard or procedure. In order to test the health of training across various demographic segments, consistent taxonomy for training needs to be developed to easily query findings from that section of the various standards.

[1] Due to the popularity of Microsoft products, Access and Excel are mentioned exclusively from this point forward. Specific references are made to 2007 Office products.

3. Data Input Method

Due to the popularity of Microsoft products, Access can be used as an easy database solution. A prerequisite is that at least a couple of people are moderately competent in database design and use of Access. There are a couple of items to note: Access isn't necessarily the best electronic solution, but it certainly is one of the more popular and easily accessible solutions; and the data analysis suggested here can be done with built-in wizards in Access – no visual basic (VBA) programming is necessary. Database design is discussed below.

A consideration for the overall database selection is how the data is input. Although Access can easily be designed with various menus and error checking at input, many people are intimidated by its functionality. Excel seems to present a format that is more recognizable with a perception of being user-friendly. More importantly, input tables can be designed that match Access tables. The data is then easily imported from Excel to Access. Having said that, consider inputting data directly to Access if at all possible. Due to the diversity of population inputting our data, we have chosen the intermediate Excel input scheme.

In performing audits of PSM data, LyondellBasell has defined protocols that suggest various interviews and data collection activities in order to objectively measure management system performance in the field against a standard or procedure. The system in place is noted by the auditor. That is, the auditor defines the actual practice and/or procedure that the site does in order to be compliant with a particular requirement. If the actual practice and/or procedure ~~differs~~differ from the requirement, a gap is noted as objectively as possible. Based on the auditor's experience, a recommendation is made to close the gap. We also have a risk-based approach to prioritize the finding, thereby dictating length of time for closure. Included are a few unique fields to "score" the audit, which include the importance of the requirement and an overall weighting of the item compared to others in the standard or procedure. Each of these inputs becomes a column in Excel that corresponds to a field in the Access database. Appendix 1 is an example of the Excel input spreadsheet for PHA audits and each field is discussed in table 1.

Table 1 – Example Excel Audit Data Input Fields

Field Name	Data Type	Comment
Section	Pre-defined	This is a field used to query results by line item requirement of a standard or procedure
Description	Pre-defined	This is a field used to query groups of similar line item requirements within a standard or procedure
Requirement	Pre-defined	Requirement is a copy of the line item requirement from the standard or procedure.
System in Place	Memo	Auditor describes the policy, procedure or practice in place in order to comply with the requirement.
Gap Identified	Memo	In as objective language as possible, the auditor describes the gap from practice in field to system requirement.
Recommendation	Memo	Based on the knowledge of the system requirement and experience of the auditor, a suggested solution to close the gap is recommended.
Priority	Pull-down number	An Excel pull-down menu selection of this number field is used for sorting/ queries in the database.
Importance	Calculated text	This field is redundant field to priority, but is used to calculate

		a text string corresponding to the priority. E.g., Priority 2 and 3 calculate a text "Very Important." This is another field that can be used to query data.
Type	Pull-down text	Based on analysis of thousands of records from hundreds of audits, we've characterized gap types into about ten different types of gaps. The gap type field provides the consistent taxonomy to query and group gaps, provide simple descriptive statistics and graphical analysis.
Score	Calculated number	Calculated number based on the LyondellBasell scoring system. The requirement score is eventually used to score the standard, and can be rolled up to an overall audit score. E.g., our scores are expressed as "Needs Improvement", "Good", etc.

Of course, the final Excel input sheet is developed only after the database is designed so that the column headings *exactly* match the Access field names, thereby minimizing errors in import. Furthermore, depending upon the data you wish to analyze, the database designer may have more or fewer input fields. The data presented here is only an example. A few things to note:

- The fields shown above as "pre-defined" are not imported into the Access database as they are already defined in a table. Inputting this data would be redundant.
- Since people are familiar with Excel it is inevitable that they will change formatting, rename fields, add columns, rows and cells, thereby making import of this data to Access difficult. Therefore, the recommendation is that the protection features of Excel should be turned on, allowing free access to input fields only. An instruction tab can be included with the protection password that explains the reasons for protection and provides general instructions to NOT change field names, pull-downs, and formats.
- The Excel to Access data import method can be easy, but the Excel data should be reviewed by a competent Access person to verify data type and check for possible errors before import. The errors are much easier to fix in Excel than in Access. An example of an error is numerical data from the field Priority input as text. Access treats text fields differently than number fields.

The import from Access to Excel is made from the *External Data – Import – Excel* tab selection.

4. Management System Design

Before we move on to database design, the design of management systems should be addressed, since the design of the requirements affects the design of the database. By management system, I am referring to PSM requirements that your company may have defined in standards, procedures and guidelines. If these are designed well, the database can take advantage of taxonomy so that management system performance can be analyzed across seemingly divergent standards. Indeed, this type of analysis is an indirect measure of the process safety culture of an organization. A few thoughts about management system design follows.

Regardless of how your organization defines management systems, these systems usually have common elements: Management review, site procedures, reports and recordkeeping, training, performance-based audits, and accountability. Compliance with the hundreds of requirements associated with these elements is easily achieved if the information is placed into a single section or document. If your organization doesn't have similar groups of management systems in one place, they should at least be located in the same section of each document.

A common element of management systems is management review. These are requirements for management review of key performance indicators, audit results, and other elements of performance. Management review is designed to insure that an organization is doing what they say they are doing. Gaps identified during PSM audits should never be a surprise to plant managers because they should always know their own state of compliance. They know their state of compliance through management review.

A second important element of the management system is requirements for site procedures. Site procedures clearly define each person's role so that the details of the PSM system are completed. People at the site level know what to do regarding PSM activities because roles and responsibilities are clearly defined in the site procedures, and the procedures clearly explain what is required.

LyondellBasell's higher level reporting requirements describe over 180 different reports and recordkeeping requirements. Many are US-based regulatory requirements, but some are not. Audit results show that the company does a very good job with this management system because most reports and records are regulatory in nature. The point is, there are many reports and records required by our PSM systems that must be managed.

While our site procedures define who does what, people involved with PSM activities must have the appropriate training so that they know how to complete the activities. Much of the training is regulatory required, and this important management system is necessary in order to achieve excellent process safety performance.

Another common element of PSM management systems that we audit are performance-based field audits. For example, your company may require annual verification of IPLs identified in the PHA, or quarterly CAR-Seal audits. Compliance with your internal requirements is audited and the results of these audits communicated to management for continuous improvement.

A final common management system is accountability. These systems are in place to ensure critical compliance items are being addressed (e.g., extensions for PHA recommendations.) The accountability systems move decision making up to the appropriate level of management to ensure proper resources are assigned to problems, and appropriate risk identification and mitigation has occurred. The main elements of accountability systems should include: documented approvals, documented extensions, and documented deviations.

Again, single lists of required management system requirements would be great tools to help individual sites comply with corporate requirements. In the absence of these single lists, the overall management system can at least be defined in a consistent way so that the management systems can be compared across standards to determine the overall health of the system. Table 2 illustrates.

Table 2 – Common Management System Design

Standard/ Procedure Section	MOC	PSSR	PSI	PHA
1.0 Purpose				
2.0 Scope				
3.0 References				
4.0 Definitions				
5.0 Responsibilities				
6.0 Requirements				
7.0 Training				
8.0 KPIs				
9.0 Reports				
10.0 Revisions				
11.0 Forms				

An example of how management system design integrates with database design for data analysis is seen with section 7.0 training. Table 2 implies that each of the PSM standards for MOC, PSSR, PSI, and PHA have training requirements defined in section 7.0. Take that a step further and assign section 7.1 as new hire training requirements, ongoing training in section 7.2, and certification requirements are 7.3. Since common elements of training are associated with the same section name across each PSM element, the database can easily query out desired sections for analysis. As gap data is populated from audits on section 7.1, new hire training, we can perform queries that pull out all new hire training performance across all PSM standards and all demographic groups to determine common gap failure modes or management system issues related to performance.

Well-designed management system documents lead to a database design that allows data analysis of common elements across PSM standards. This type of analysis helps determine systemic management system issues.

5. Audit Data Grouping

As pointed out in the data input section above, audit findings are generally memo fields that have text descriptions of the findings. This type of text offers no consistent taxonomy that facilitates data analysis. In the management system section above, a consistent taxonomy across standards is defined through a consistent application of section numbers. Audit findings can be categorized so that a consistent taxonomy can be used for data analysis.

In order to create the gap taxonomy, analyze each gap individually in order to find logical groups of data. This analysis should surface common failure modes of the gaps. The categorization includes both management system failure and specific requirement failure. Table 3 is a simplified example of possible gap categories.

Table 3 – Gap Characterization

Procedure Inadequate
No Procedure
Procedure not Followed

Training
Documentation
Communication
Other

This example is by no means inclusive. Rather, it represents a sample of higher-level grouping for gaps. Similarly, each category might have sub-categories depending upon your data analysis needs. It is recommended that each company conduct its own historical analysis of gaps to determine input categories and subcategories. If the Excel data input method shown above is used, these categories can be used as pull-down input cells so that the consistent taxonomy for gap analysis is used in the database.

6. Database Design

A good database design should be thoroughly documented. A common failure of Access databases is failure to document. Appendix 2 is a description of the tables and fields from a sample database, and Figure 1 is a summary of the relationships in the database.

Figure 1 – Database Relationships

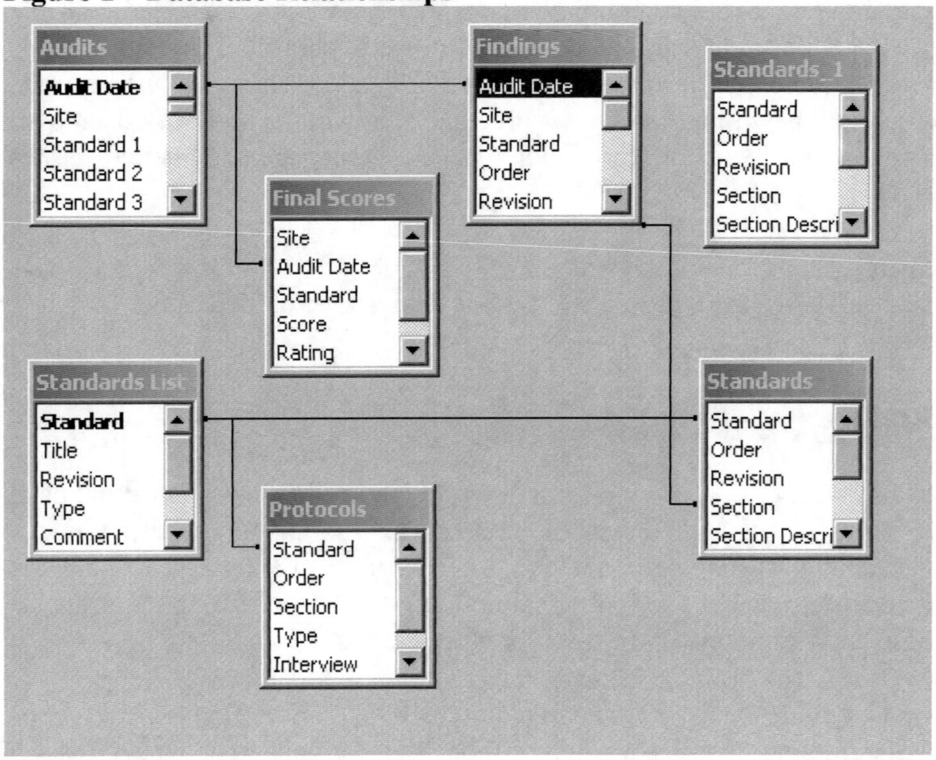

Two tables of interest are Standards and Findings. The Standards table is a copy of the standard/procedure requirements written into the database and is primarily used for the various reports that the database produces. As the name suggests, the Findings table is the final destination for either data directly input from the audit or imported data from Excel, as defined previously. The two tables are linked by the section field which is common to both. For example, if a gap is input into the Findings table from section 7.1 initial training, we can describe the requirement from the standards table in the various queries and reports that can be produced with this data.

As with any document control system, there are bound to be updates to the documents — either minor or major. Revisions to documents could represent differences in standard requirements and therefore queries of audit data may not represent a direct comparison of requirements. Several database solutions can be programmed including a separate data mapping table to "map" one revision to the next. A simple solution for data analysis is to keep the section numbers the same among revisions and only add sections. However, this isn't always possible. Another solution is to simply input a new standard in the standards table and make it unique by the standard name field. For example, LyondellBasell's PHA standard is entitled TEC-03. To account for the revisions, the database title is changed to match the revision: TEC-03r1, TEC-03r2, etc. There are other solutions such as including a number field that describes the revisions. However, this solution has unique problems that might require VBA programming to correct. A renaming scheme is a very easy fix. In data analysis, you are then required to "force fit" comparisons between revisions.

The Audits table is a list of topics audited by site and is used in pull-down menus for ease of selection in queries and other data analysis. Similarly, the Final Scores table is used as an easy way to report out on scoring an audit. With a little time and ingenuity, an advanced Access programmer could easily calculate the final scores table from the data in the Findings table; the Final Scores table is therefore redundant. However, this is acceptable to avoid VBA programming. The Standards List table is a listing of all standards and procedures in the database and is used in reporting and queries to prevent redundant data input.

LyondellBasell has written documents that describe the audit protocol used. There are two tables in the database that summarize the protocol data in two forms: interview questions and testing protocol. The tables represent a one-to-one match between standard requirements and protocol actions. These tables were added so that custom protocols could be produced by linking the protocol, standard, and findings tables to produce a protocol based on previous findings. Thus, an auditor can understand the high priority areas to focus on in future audits based on past performance. This is an example of database functionality and flexibility.

7. Data Analysis

Once the database structure is in place and includes a consistent taxonomy among standards, procedures and appropriate categorization of audit findings, the data analysis should be easy. It isn't the intent of this paper to teach Access programming, so the definition of how to write simple queries is not presented here. Explained below are a few data queries that may be useful.

7.1 Reports

With use of the Standards table, company standards and procedures can be produced directly from the database. In addition, a gap document can be produced from this table (see Appendix 1) to aid in note taking during the audit. Also, if protocol tables are included, full or abbreviated (based on previous findings) protocol reports can be queried and produced. At the end of the audit, after the findings have been either input or imported into the database, a full or partial audit report can be produced in or exported to Word. Appendix 3 shows an example from an Access report.

7.2 Pivot Tables and Queries

As with any database, there are an infinite combination of queries that can be developed in order to analyze data and produce various descriptive statistics and graphical analysis. The fact that there are so many different combinations presents a challenge in itself for the database designer; what should be programmed to make everyone happy? As stated above, begin with the end in mind and at least query the users of the database to determine what data they might want to extract. Then build queries around that.

A pivot table/pivot chart can be quite useful in data analysis. See Appendix 3 for a hypothetical example of a pivot chart. Pivot tables and charts enable flexibility of data review in a variety of ways. The example queries all findings by year selected, and queries out particular standards (procedures) by site, and/or by priority of findings. Other levels such as a procedure section could have been added to the query. There is significant flexibility associated with the pivot table and chart. Appendix 3 also shows an example of a pivot table that displays hypothetical audit scores by year of audit broken into the topics audited.

Of the infinite number of queries that can be developed, you may wish to program a few fixed ones. Appendix 3 shows an example of a query menu that uses a form with pull downs and input fields to describe various criteria used for queries that pull findings by standard or by site.

The data analysis section of the database should be written with a little flexibility in mind. This is best achieved through pivot tables and providing the user with the ability to input (or pull down) criteria.

8. CCPS Pillars

The CCPS has published *Guidelines for Risk-based Process Safety*. In this book, various activities associated with process safety are categorized by pillars: Commit to Process Safety, Understand Hazards & Risks, Manage Risk, and Learn from Experience [1]. Each pillar has elements defined and each element has principals that should be applied to achieve process safety excellence. The book goes on to define essential elements that should be in place to ensure principals are established.

The CCPS pillars are mentioned here because they provide an excellent framework to categorize audit topics. In doing so, another level of data analysis by process safety topic is provided and adds more flexibility to the database. Table 4 is a high-level summary of the CCPS pillars.

Table 4 – High Level Summary of CCPS Pillars

CCPS Pillars & Elements	Principals
Commit to Process Safety	
Develop a Process Safety Culture	PS as a core value; strong leadership; high-= performance standards; zero tolerance for willful violations; reward PS performance
Compliance with Process Safety Standards	Standard owners
Process safety Competency	Improve knowledge; up-to-date PSI; apply learnings
Workforce Involvement	Competent personnel; work activities
Understand Hazards & Risks	
Process Knowledge Management	Catalog process knowledge; protect & update knowledge; use process knowledge
Hazard Identification & Risk Analysis	ID hazards & evaluate risks; assess risks and make risk-based decisions; complete assessment results
Manage Risk	
Operating Procedures	ID what procedures are needed; develop procedures; use procedures to improve human performance; ensure procedures are maintained
Safe Work Practices	Control non-routine work
Asset Integrity and Reliability	ID in-scope equipment; maintain knowledge, skills, procedures and tools; ensure continued fitness for purpose; address equipment failures; analyze data
Contractor Management	Conduct work activities; monitor contractor management system
Training	ID required training; provide effective training, monitor worker performance
Management of Change	ID potential change; evaluate impacts; decide whether to allow the change; complete follow-up activities
Operational Readiness	Conduct appropriate reviews; follow through on decisions, actions, and use of readiness data
Conduct of Operations	Control operations activities; control the status of systems and equipment; develop required skills & behaviors; monitor organizational performance
Emergency Management	Prepare for emergencies; periodically test the adequacy of plans and level of preparedness
Learn from Experience	
Incident Investigation	ID potential incidents for investigation; use appropriate techniques to investigate incidents
Metrics	Conduct metrics acquisition; use metrics to make corrective action decisions
Audits	Conduct element work activities; use audits to enhance RBPS effectiveness
Management Review & Improvement	Conduct review activities; monitor organizational performance

9. Conclusion

Audit findings are traditionally represented as text and audit reports submitted as a word processing file. This data type tends to be subjective and therefore difficult to analyze. Detailed analysis of audit data can provide direction in areas of improvement for systemic management system gaps, weaknesses across a company in a particular standard or procedure, and improvement areas at particular sites.

In order to convert the subjective audit data and reports into a form that can be objectively analyzed, a database platform must be selected and designed. Popular Microsoft products such as Excel and Access can be used to input and analyze data.

The first level of analysis is to create a consistent management system design and taxonomy for corporate and site standards and procedures. Consistency allows data query across any company demographic for descriptive statistical and graphical analysis. Analysis of data across standards and procedures surfaces systemic management system gaps.

The second level of analysis is to create consistent taxonomy for data input and recommendation type. Consistency here allows for descriptive statistical and graphical analysis of data from individual audits and by demographic.

Full utility of the audit database can be achieved by designing a protocol query so that future audits focus on gaps found in past audits. Furthermore, with the right taxonomy and database structure defined, data analysis is only limited by the creativity of those analyzing the data.

10. References

[1] Center for Chemical Process Safety, Guidelines for Risk Based Process Safety. American Institute of Chemical Engineers, 2007

11. Appendix 1 – Example - Excel Audit Data Input Template

Section	Description	Requirement	System in Place	Gap Identified	Recommendation	Priority	Importance	Type	Score
TEC-03r2 5.2	PHA Methodology	One of the following methodologies is used: 5.2.1 What-If 5.2.2 Checklist 5.2.3 What-If/Checklist 5.2.4 Operational Safety Standard (OSS) 5.2.5 Hazard and Operability Study (HAZOP) 5.2.6 Failure Mode and Effect Analysis (FMEA) 5.2.7 Fault Tree Analysis 5.2.8 Any other methodology accepted by a recognized institution expected to possess such expertise such as the AICHE							
TEC-03r2 5.5.2	Scope	Operating Modes: Initial Startup Normal Operations Temporary Operations Emergency Shutdowns Emergency Operations Normal Shutdown Startup Following a turnaround or emergency shutdown							

12. Appendix 2 – Table Design

Table	Field Name	Description	Data Type	Purpose
Audit				The audits table is used in pull down list in order to select unique audits for query
	Audit Date	First day of audit in DD/MM/YYYY format	Date	Primary key
	Site	Site input as standard company 3 letter site designation	Text	Field from which to query results by site
Final Scores				Quick overall look at audit scores – this table could be generated by the astute data-base designer!
	Site	Site input as standard company 3 letter site designation	Text	Field from which to query score by site
	Audit Date	First day of audit in DD/MM/YYYY format	Date	Primary key
	Standard	Standard Name	Text	
	Score	The numerical score calculated per procedure	Number	Pareto analysis of performance
Findings				Findings table is the destination of audit data imports
	Audit Date	First day of audit in DD/MM/YYYY format	Date	
	Site	Site input as standard company 3 letter site designation	Text	
	Standard	Standard Name	Text	
	Section	Standard section as it appears in document, TEC-03 5.2	Text	A unique identifier for standard element data analysis
	System in Place	Audit result of procedure or practice in place to comply with standard requirement	Memo	
	Gap Identified	Gap of practice/procedure in place with standard requirement	Memo	Imported from audit
	Recommendation	Auditors recommendation to close gap	Memo	Imported from audit
	Priority	RBPS priority	Number	Used to query/ sort findings by priority
	Importance	Most Important, Very Important, Important, or Less Important	Text	A secondary grouping field for priority
	Type	The categorization of findings	Text	A query field for categorized or grouped findings
Protocol				Table is a question bank

Table	Field Name	Description	Data Type	Purpose
Interview				for interviews used during audits
	Standard	Standard Name	Text	
	Section	Standard section as it appears in document TEC-03 5.2	Text	A unique identifier for standard element data analysis
	Interview	Position to be interviewed	Text	Helps the auditor structure interviews during the audit
	Protocol Action	The interview question	Memo	
Protocol Testing				Bank of activities for use during audits to test compliance with standards
	Standard	Standard Name	Text	
	Section	Standard section as it appears in document, TEC-03 5.2	Text	A unique identifier for standard element data analysis
	Test	Type of verification (procedure review, activity, etc)	Text	
	Protocol Action	The verification Step	Memo	
Sites				
	Region	North, South, East, etc.	Text	A unique identifier by company used to query/ group data
	Site	3 Letter identifier of site audited	Text	A unique identifier to query data
	Site Name	Full name of site	Text	For Reports
Standards				
	Standard	Standard Name	Text	
	Section	Standard section as it appears in document, TEC-03 5.2	Text	A unique identifier for standard element data analysis
	Section Description	A field used to group similar requirements in a standard or procedure	Text	To query like requirements within a standard and between standards
	Requirement	Description of the actual standard/ procedure requirement	Memo	
	Weight	1, 2, or 3	Number	Used in scoring
Standards List				
	Standard	Standard Name	Text	
	Title	Standard/ Procedure title	Text	For Reporting
	Revision	Revision	Number	To insure appropriate match up of standard and findings table
	Type	Safety, Process Safety, Environmental, etc	Text	A query/ sort field
	Comment	Comment	Memo	To note unique things about the standards

13. Appendix 3 – Example - Database Functionality

13.1 Example Access Report

Audit Report BMO 1/20/2003

Standard GEN-02r2

Section	**Requirement**	**System in Place**	**Gap Identified**	**Recommendation**	**Priority**	**Impor**	**SIP**	**Rating**
GEN-02r2 5.2	A Closure Plan is developed, approved, issued and input as the appropriate audit in IMPACT for each covered audit/benchmark/self-assessment, meeting the following requirements.	The plant has a database in place to track action items to closure. Plant procedure SP-54 describes the closure requirements.	No Gap Found				S	1
GEN-02r2 5.2.	Action items to implement the recommendations and remediate the findings are clearly stated.	The database tracks all action items including OE, PSM, PHA, Local Audits, Loss Prevention, and Total Incident Reporting.	No Gap Found				S	1
GEN-02r2 5.2.	Responsibilities for each action item are assigned to specific individuals.	Plant staff assigns responsibilities to specific individuals	No Gap Found				S	1
GEN-02r2 5.2.	For findings that require engineering solutions or capital projects to resolve, closure dates should be established consistent with the level of risk of the finding and closure requirements of TEC-03-P1.	SP-54 describes the requirements and is consistent with Gen-02 for OE and PSM action items.	1. SP-54 does not require risk ranking of Loss Prevention Action Items. (HSE GEN-02 1.1)2. A review of closed action items from the 2001 OE, PSM audit, and Loss Prevention showed a. 18 of 40 OE items were closed after the target date. (5.2.3)b. 8 of 17 PSM action items were closed after the target date. (5.2.3)c. 1 of 4 Loss prevention action items were closed after the target date (5.2.3)d. 1 of 8 randomly selected OE action items was listed as closed based upon the elimination of OPR-16. OPR-16 has not been eliminated. (5.2.3)	1. Revise SP-54 to include the requirement to risk rank Loss Prevention Action Items2. a.) b.) c.) Implement a process to bring more attention to action items that are coming due, with a goal of zero past-due action items. d. Re-open all action items that we based upon the elimination of OPR-16.	6	Less Importa nt	PS	2

Tuesday, January 12, 2010 Page 1 of 23

13.2 Example - Pivot Chart

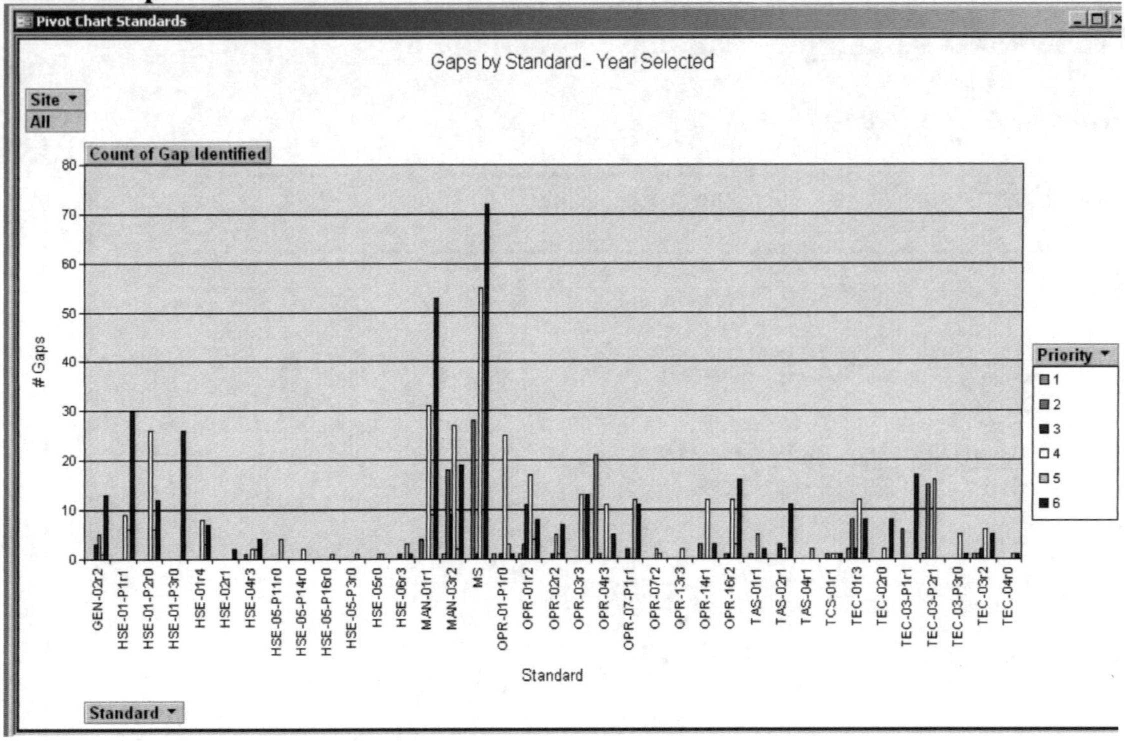

13.3 Example - Scorecard (Pivot Table)

Pivot Table Scores

Audit Date By Month: All

Type	Standard	Region: East								Region: North				
	Site	BLO		BMO				CVO		CLO		FPO	MIO	
	Audit Date	11/3/2003	2/13/2006	1/20/2003	11/1/2004	9/17/2007	4/23/2008	9/21/2004	6/18/2007	8/2/2004	7/9/2007	7/18/2005	7/7/2003	12/4/2006
		Rat	Rat	Rat	Rat	Rat	Rat	Rat	Rat	Rat	Rat	Rat	Rat	Rat
Audit Follow-up & Closure	GEN-02r2											Adequ		
	GEN-03r0													
HSE	HSE-01-P1r1													
	HSE-01-P2r0													
	HSE-01r3												Adequ	
	HSE-01r4													
	HSE-02r1													
	HSE-04r3													
	HSE-05-P11r0													
	HSE-05-P14r0													
	HSE-05-P16r0													
	HSE-05-P3r0													
	HSE-05-P4r0													
	HSE-05-P9r0													
	HSE-05r0	Adequ												
	HSE-06r3													
	HSE-07r0													
Logistics	TAS-01r1													
	TAS-02r1					Adequ								Adequ
	TAS-04-P1r0													
	TAS-04r1						Adequ							
Maintenance Performance	MAN-01r1									Adequ	Adequ			
	MAN-03r2		Adequ								Adequ			
	MS													
Management of Change	OPR-02r2													

13.4 Example - Query

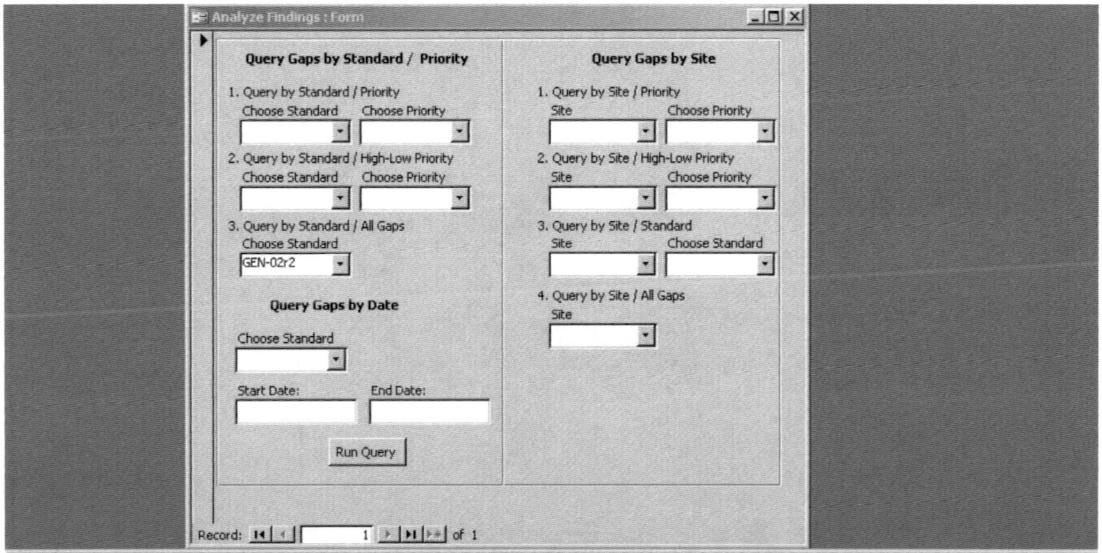

Analyze Findings2 : Select Query

Audit Date	Standard	Section	System in Place	Gap Identified	Recommendation	Prior	Importanc	SIP	Rati
8/4/2003	GEN-02r2	GEN-02r2 6.2.2	Closure documentation is retained in a hard copy file and in the Impact electronic system.	2. Closure documents could not be located for 1 of the 30 files reviewed. (6.2.2)	2. Using the list of missing documents, update the closure files.	6	Less Important	PS	2
2/17/2008	GEN-02r2	GEN-02r2 6.2.2	Action items are closed in IMPACT, but documentation supporting this closure, it's location and the extend of	3. Procedure 701.4 "REPORTS AND RECORDKEEPING" requires closure	3. Update procedures 701.4 to define types and examples of support	6	Less Important	PS	2
11/3/2008	GEN-02r2	GEN-02r2 6.2.2	The closure documentation, i.e., details of closure, PRP approver and Plant Staff member/action item	7. In a review of 32 IMPACT action items associated with the 17 VERY IMPORTANT	7. Update SP-089 to specify that links or reference to supporting	6	Less Important	PS	2
6/12/2005	GEN-02r2	GEN-02r2 6.2.2	All closure documentation is retained in IMPACT	2. In reviewing 30 action items of a population of 100 past	2. Update procedure to reflect practice in place	6	Less Important	PS	2

Record: 1 of 167

AUTHOR INDEX

American Institute of Chemical Engineers
3 Park Avenue
New York, New York 10016-5991

CURRAN ASSOCIATES INC
proceedings.com